W9-ATH-149

Adjustment: Models and Mechanisms

ADJUSTMENT
MODELS AND MECHANISMS

Irving F. Tucker
UNIVERSITY OF MARYLAND

ACADEMIC PRESS
New York • London

ACADEMIC PRESS, INC.
111 Fifth Avenue, New York, New York 10003

United Kingdom Edition published by
ACADEMIC PRESS, INC. (LONDON) LTD.
Berkeley Square House, London W1X 6BA

LIBRARY OF CONGRESS CATALOG CARD NUMBER 73-112294

PRINTED IN THE UNITED STATES OF AMERICA

In Memory of
Professor Kenneth W. Spence

Contents

Preface

Writing a textbook on human adjustment is a formidable task. In recent years there has been an enormous expansion of research dealing with selected areas of personality and behavior, biopsychology, motivation, human and animal learning, and behavior modification. There has been a corresponding shift from broad general theories to an interest in more circumscribed situations, variables, and mechanisms, and an increase in specific theoretical models used to interpret them. The proliferation of data from these diverse areas—all of which have a definite bearing on human adjustment—has made the search for simple, global interpretations of human behavior somewhat anachronistic. However, despite a massive accumulation of data, at the present time generalizations to natural circumstances can be made from the experimental laboratory and from theories and models of limited aspects of behavior on a provisional basis only. Nevertheless, it is possible and, I believe, desirable to integrate our present knowledge in order to obtain a cross-sectional and relatively comprehensive picture of it.

It is common for the student to find the principles and knowledge he has acquired from the core courses of the usual psychology curriculum largely abandoned in courses such as the psychology of adjustment, personality, or abnormal psychology. On the other hand, some psychology of adjustment texts propose that a small number of simple principles derived from early studies of the learning and motivation of lower animals carry most of the explanatory burden. Again, it is my belief that the relatedness of courses in general psychology, experimental psychology, learning, physiological and social psychology to the psychology of adjustment ought to be stressed, rather than minimized, and that this should be done without denying the student exposure to relevant observations of descriptive psychiatry and clinical psychology and without presenting an overly simplified or pretentious "scientific" interpretation of such complex subject matter.

This text is designed, therefore, to present a systematic and comprehensive account of present-day scientific knowledge of human personality and behavior. A wide range of research findings and topics relevant to a scientific interpretation of human behavior are introduced. Throughout the text there is a gradual and progressive buildup of the resources necessary for a scientific understanding of the mechanisms underlying human personality and adjustment.

Chapters 2 through 5 deal with the basic principles of biology and of learning. Chapter 2 presents the fundamentals of the structure and functioning of the human nervous system, a discussion of the body's endocrine system, the biological dynamics of stress, and a synopsis of laboratory and other research findings on the topic of stress. Chapter 3 continues the presentation of relevant physiological topics: mechanisms of arousal, the primary drives, our present-day neurophysiological understanding of emotion, and a discussion of psychoactive drugs.

Chapters 4 and 5 deal with the fundamentals of the psychology of learning. Chapter 4 outlines the basic principles of respondent and operant conditioning: extinction, generalization and discrimination, delay of reinforcement, varied reinforcement, and the effects of different schedules of reinforcement on behavior. Chapter 5 deals exclusively with the topics of punishment and anxiety.

Chapters 6 through 8 present the application of these fundamental principles to more complex aspects of human behavior. Chapter 6 deals with the topics of frustration and conflict. Chapter 7 is concerned with extending laboratory findings to complex aspects of human adjustment: social learning and imitation, identification, and the learning of dependency, aggression, and sexual behaviors. Chapter 8 deals with the topics of verbal conditioning and verbal behavior. These topics are related to very complex aspects of human adjustment, such as the nature of psychic driving, the implicit mediation of emotional responses, and repression.

Finally, Chapters 9 through 12 are concerned with personality structure and personality dynamics. Chapter 9 deals with the learning of the self-concept and its relation to various aspects of personality and behavior. It also deals with the experience of anxiety. Relevant research is presented. Material relevant to an understanding of personality defense is presented in Chapter 10. The defense mechanisms are discussed within the framework of the repression-sensitization dimension of personality. Relevant research is also presented. Chapter 11 integrates much of the previous material in a discussion of complex patterns of human adjustment along a continuum from optimal adjustment and maturity, through the classic patterns of neurotic disturbances, to severely disordered behavior. Finally, techniques of behavior modification and treatment are presented in Chapter 12.

References have been placed at the end of each chapter. However, each one is cited briefly in the text by using the author's name and year of publication. I am grateful to many publishers and authors for permission to mention, quote or otherwise use material from their publications. Appropriate credit is given in the list of references and in figure headings throughout the text.

Many people contributed to the production of this book. It is impossible to thank them all adequately here. Among those who deserve special mention are my professors at the University of Iowa: G. Bergman, J. S. Brown, I. E. Farber, M. E. Rosenbaum, R. W. Schulz, and the late Kenneth W. Spence. Although their influence was not direct, whatever professional competence the text reflects is due largely to the educational experience they provided. Dr. John Goudeau, Professor Richard Scott, and Professor Harry Lindgren gave me substantial support, each in his own way, during the writing of the text. Professor Brendan Maher reviewed the early drafts, chapter by chapter, with competence and understanding.

IRVING F. TUCKER

Tokyo, Japan
September, 1969

Adjustment: Models and Mechanisms

Chapter 1

The Behavioral Science Approach

HISTORICAL PERSPECTIVE

Interest in the scientific study of behavior began to awaken following the Darwinian revolution. Darwin believed that over aeons of time new species developed because genetic mutations allowed some organisms to adapt better to their environment. The fittest tended to survive and propagate. In the biological sciences evidence for such evolutionary trends was abundantly accumulated and recognized.

One of the major avenues of scientific inquiry opened by the Darwinian thesis was the study of how inherited characteristics are transmitted and modified. Recent advances in microbiology and research tools like the electron microscope have begun to make a scientific approach to these questions realizable.

Another major avenue of scientific inquiry opened by Darwin's ideas was the study of the nature of the adaptation process and the mechanisms underlying the adjustment of organisms to their environment. It was soon found that adaptive responding is characteristic of organisms very low in the evolutionary scale. Take, for example, the tiny organism called a Stentor (Fig. 1-1):

> Stentor, a tiny protozon, is attached to the substratum at the lower end of its tube. At the top of the tube are cilia or hairs that draw water containing food particles down into it. If a few drops of red ink are introduced into the water near the animal, a series of responses is initiated. First, the Stentor bends to one side, avoiding the ink. If several bend-

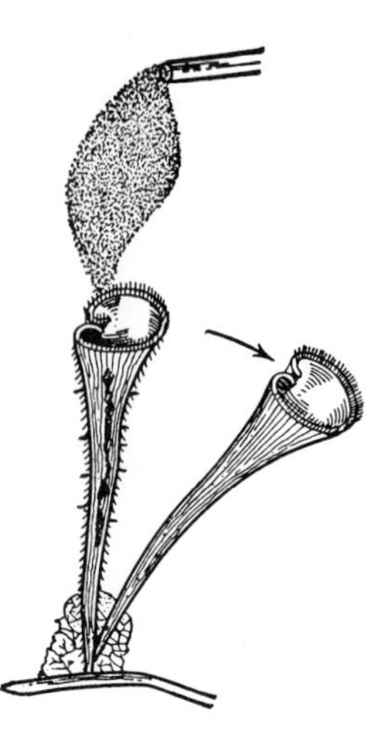

FIG. 1-1. Adjustive behavior of Stentor: introduction of a drop of ink near the mouth of the tiny organism elicits the adjustive response of turning away from the noxious stimulus. (Jennings, 1906.)

ings do not bring relief, the movement of the cilia is reversed, pushing the water away instead of drawing it in. If further adjustment is necessary a third response is made, or contracting into its tube. Finally, when all those responses prove unavailing, the Stentor releases itself from support and floats away. [Shaffer & Shoben, 1956, pp. 124–125]

Given evolutionary continuity, would it not be plausible that the adjustments of higher animals, including man, represent systematic changes in response to changing biological and environmental stimulation? In a culture that values progress, achievement, and adaptation, it is not surprising that such an hypothesis stimulated the study of the behavior of organisms and their adjustment to the environment.

Darwin's theory gave further impetus to the study of behavior and adjustment because the suggestion that man had inherited his body from animal ancestors also raised the question of whether there might be a continuity between the cognitions and emotions of lower animals and those of man. The acceptance of such a belief in continuity between the mind of lower animals and that of man fostered an interest in animal research and experimentation as a basis for the scientific study of man. Research with lower organisms had many advantages. It allowed for free experimentation; it made control over the relevant conditions possible; and it was practical.

Psychologists became interested in comparative psychology, in the question of why organisms behave as they do, and in the adjustment process. They subsequently established the school of psychology known as *functionalism*. It was brought to the attention of the public in an address delivered by James Rowland Angell in 1907 as president of the American Psychological Association (Angell, 1907). Two features of that address are of special interest: first, Angell presented the adjustment of organisms to changing environmental conditions as the subject matter for a science of psychology. Secondly, considerable optimism was expressed concerning the role of a science of psychology in eventually being able to say something about how people should live. Both of these ideas have remained essential themes throughout the development of American psychology.

One of Angell's students, John B. Watson, made an important contribution by outlining methodological techniques for the study of the adjustment of organisms. The work of the Russian psychologist, Ivan P. Pavlov, was also recognized and incorporated by American psychologists. Later some cognizance was taken of the Freudian movement, *psychoanalysis*. Generally, though, psychoanalysis has lacked an experimental basis and hence verifiability of many of its ideas is also lacking. However, occasional references to psychoanalysis will appear in this text, because its influence, although diffuse and indirect, has been profound.

Much of the vigor, optimism, and naïveté of the young science of psychology can be seen in the following amusing and homely commentary from Watson's classic *Behaviorism* (Watson, 1924):

> Without taking anyone into my counsel suppose I once trained a dog so he would walk away from nicely-ground, fresh hamburger steak, and would eat only decayed fish (true examples now at hand). I trained him (by use of electric shock) to avoid smelling the female dog in the usual canine way—he would circle around her, but would come no closer than ten feet (J. J. B. Morgan has done something very close to this in the rat). Again, by letting him play only with male puppies and dogs and punishing him when he tried to mount a female, I made a homosexual of him (F. A. Moss has done something closely akin to this in rats). Instead of licking my hands and becoming lively and playful when I go to him in the morning, he hides or cowers, whines and shows his teeth. Instead of going after rats and other small animals in the way of hunting, he runs away from them and shows the most pronounced fears. He sleeps in an ash can—he fouls in his bed, he urinates every half hour and anywhere. Instead of smelling every tree trunk he growls and fights and paws the earth, but will not come within two feet of a tree. He sleeps only two hours per day and sleeps these two hours leaning up against a wall rather than lying down with his head and rump touching. He is thin and emaciated because he will eat no fats. He salivates constantly (because I have conditioned him to salivate to hundreds of objects). This interferes with his digestion. Then I take him to the dog psychopathologist. His physiological reflexes are normal. No organic lesions are to be found anywhere. The dog, so the psychopathologist claims, is mentally sick, actually insane; his mental condition has led to various organic difficulties such as lack of digestion: it has "caused" his poor physical condition. Everything that a dog should do—as compared with what dogs of this type usually do—he

does not do. And everything that seems foreign for a dog to do he does. The psychopathologist says I must commit this dog to an institution for the care of insane dogs; that if he is not restrained he will jump from a ten-story building, or walk into a fire without hesitation.

I tell the dog psychopathologist that he doesn't know anything about my dog; that, from the standpoint of the environment in which he was brought up (the way I have trained him) he is the most normal dog in the world; that the reason he calls the dog "insane" or mentally sick is because of his own absurd system of classification. I then take the dog psychopathologist into my confidence. He becomes extremely angry. "Since you've brought this on, go cure him." I then attempt to correct my dog's difficulties, at least up to the point where he can associate with the nice dogs in the neighborhood. If he is very old, or if things have gone too far, I just keep him confined; but if he is fairly young, and he learns easily, I undertake to retrain him. I use behavioristic methods. I uncondition him and then condition him. Soon I get him eating fresh meat by getting him hungry, closing up his nose and feeding him in the dark. This gives me a good start. I have something basal to use in my further work. I keep him hungry and feed him only when I open his cage in the morning; and the whip is thrown away: he soon jumps for joy when he hears my step. In a few months time I not only have cleared out the old, but also have built in the new. The next time there is a dog show I proudly exhibit him, and his general behavior is such an asset to his sleek, perfect body that he walks off with the blue ribbon.

All this exaggeration. Almost sacrilege! Surely there is no connection between this and the poor sick souls we see in the psychopathic wards in every hospital! Yes, I admit the exaggeration, but I am after elementals here. . . . [Watson, 1924, pp. 244–246]

As the new psychology progressed interest became increasingly centered upon the scientific study of behavior. The biological basis of behavior has been investigated and both successful and unsuccessful adjustment have come to be viewed as the joint result of biological and learning processes.

The science of psychology is still in its infancy. Practically speaking it is not more than 30, and certainly not more than 60, years old. However, the concerted efforts of physiologists, psychologists, and other social scientists have, especially in recent years, brought the outlines of a scientific interpretation of human behavior within focus.

BEHAVIORAL SCIENCE AND THE LAYMAN

Scientific enterprise has generally received support in our society. There have been certain deterrents, however, to progress toward a scientific understanding of human behavior. Antiscience attitudes are not reasonably based. Nevertheless, certain sources of tension created by the behavioral sciences are understandable. For example, there appears to be a general uneasiness about "technicians." In the case of the behavioral scientist, this appears to involve anxiety produced by what might be called "social engineering." No one likes to be intentionally manipulated by others. Our beliefs concerning the democratic way of life have always included a concern with checks and balances, so that any one faction, or any one person, can be prevented from obtaining too much influence.

The application of scientific methodology has, however, proved again and again to be highly successful. Applied to the study of man, there is little doubt that it will result in man's increasing ability to apply techniques of prediction and control to himself and to others. The natural sciences have been shown to be applicable for man's benefit or destruction: surely the uses to which the behavioral sciences are put must eventually be considered. It would seem that the science of psychology is hardly advanced enough as yet to create grave concern about its uses. Some psychologists, however, have already raised the issue (see Rogers, 1956; Skinner, 1959).

Another difficulty involves the acceptance of the scientific frame of reference itself. The nonscientist frequently appears to lack an understanding of the abstract character of scientific knowledge. Specifically, conceptual difficulties center around the following aspects of the scientific frame of reference: determinism, reductionism, and operationism.

DETERMINISM

The position that neither physical events nor human choices and behavior are uncaused, but that they are the result of physical and psychological antecedent conditions, is called determinism. Science seeks to discover the conditions under which various kinds of events occur and to formulate these conditions in general terms. That is, phenomena are explained by means of general laws and by reference to specific conditions. The statements of such determining conditions constitute the explanation of the events to which they correspond.

Philosophers of science have developed very precisely formulated accounts of the logical and empirical conditions of adequacy for scientific explanations. The basic pattern of scientific explanations can be seen in a relatively simple illustration provided by Hempel and Oppenheim (1948):

> A mercury thermometer is rapidly immersed in hot water; there occurs a temporary drop in the mercury column, which is then followed by a swift rise. How is this phenomenon to be explained? The increase in temperature affects at first only the glass tube of the thermometer; it expands and thus provides a larger space for the mercury inside, whose surface therefore drops. As soon as by heat conduction the rise in temperature reaches the mercury, however, the latter expands, and as its coefficient of expansion is considerably larger than that of glass, a rise of the mercury level results. This account consists of statements of two kinds. Those of the first kind indicate certain conditions which are realized prior to, or at the same time as, the phenomena to be explained; we shall refer to them briefly as antecedent conditions. In our illustration the antecedent conditions include,

among others, the fact that the thermometer consists of a glass tube which is partly filled with mercury, and that it is immersed in hot water. The statements of the second kind express certain general laws; in our case, these include the laws of the thermic expansion of mercury and of glass, and a statement about the thermic conductivity of glass. The two sets of statements, if adequately and completely formulated, explain the phenomenon under consideration: they entail the consequence that the mercury will first drop, then rise. Thus, the event under consideration is explained by subsuming it under general laws, i.e., by showing that it occurred in accordance with those laws, by virtue of the realization of certain antecedent conditions. [Cited from Feigel & Brodbeck, 1953, p. 320]

Scientific explanation appears to have only a very tenuous relationship to the experience of everyday life. For example, the perception of a brilliant sunset may seem to have little to do with a physicist's description of it in terms of electromagnetic principles. Yet the perception can be shown to depend upon these principles. The relationships between, let us say, the results of behavior and the way in which subsequent behavior changes, are obscure to a naive observer. However, a wide range of human behavior may be shown to depend upon such relationships.

As an example of the layman's attempt at explanation, consider a white rat in a T maze. Suppose a psychologist had trained him to run down the left arm of the maze by always placing a food reward at the end of that arm of the maze, and never anywhere else. If the complete course of training had been observed by an intelligent layman, he might be able to give a fairly adequate account of the animal's behavior with reference to the food reward and the hunger. If he only observed the final performance of the animal, a naive layman would attempt an explanation in terms of "wants" or "wishes" that would themselves ultimately have to be explained. That is, he might say that the animal "wants to go down the left alley" because he "knows there is food there." If the psychologist had conditioned the rat to run at a relatively slow rate by making the reward contingent upon slow running or walking, the explanation might be that the animal really doesn't "want to run very fast" or, it might be declared that the animal is "lazy." The statements do not, of course, contain general laws; nor are they capable of test by experiment and observation.

Mechanistic Processes

How often do psychologists hear the protest that human beings are not machines, nor are they machinelike! Sometimes this represents a refusal to share the deterministic frame of reference. However, the difficulty is frequently verbal. There is a language bridge from mechanism, to machinelike, to associations involving simple, rigidly determined processes. Antecedent conditions and consequent events are then construed as a one-cause-equals-one-effect type of process. Such a process is clearly at variance with the facts of human behavior.

A mechanistic process is one in which there is an orderliness and regularity among events in nature: when certain conditions are present certain events follow. In processes that are said to be mechanistic a present state of affairs is said to be "determined" by certain antecedent conditions.

Unless the subject matter of scientific study is lawful, determined, and orderly, rather than capricious, a scientific interpretation of that subject matter is not possible. In the beginning a scientist must merely *assume* that there is regularity and orderliness underlying the complex events which he observes. Indeed, this is the basic assumption, or frame of reference, of all scientific endeavors. Needless to say, this is not a specious assumption, for in the history of science application of the scientific method to phenomena of man's experience has demonstrated the mechanistic nature of those phenomena often enough to justify considerable faith in the continuing search for relevant conditions and their effects, both within established areas of science and in other realms of man's experience where the method has not yet been employed.

Variables and Functional Relationships

Let us look more closely at the "causes" and "effects" of mechanistic processes. The terms are in quotes because they are seldom employed any more either by those people who interpret the scientific enterprise (the philosophers of science) or by scientists themselves. Instead, within science specifiable variables or conditions and the functional relationships obtaining among them are employed as conceptual devices. What is a variable? It is a highly abstract idea: a variable is anything that changes. More specifically, a variable is any characteristic of objects, events, or people, that can assume different values. Thus, mass, size, weight, and distance, are all variables. Height is a variable. Objects in the world, including people, differ in this characteristic. The level of an individual's performance in a learning situation is a variable; so is the amount of fear exhibited in a given situation.

The aforementioned variables may be quantifiable. Since they represent continua of possible change, numbers may be assigned to represent different degrees or amounts of variation. Other variables do not change in a systematic way. They are discrete. One can merely name the different characteristics exhibited by the things, events, or people possessing them. For instance, the labels "blue," "brown," or "gray" may be used to designate differences in the color of people's eyes. The categories "male" and "female" designate the descrete variable of "sex," and so on.

When the qualitative or quantitative characteristics of certain factors of given situations change there are corresponding changes in events to which they are related. The changes of these characteristics may represent either antecedent or concurrent conditions. *Antecedent conditions* occur prior to, and *concurrent conditions*, at the same time as, the events that they effect, which are referred to as *consequent events*. Not all the factors of a given situation are related to the consequent events: some are essentially irrelevant for the prediction of them.

Variables that can be shown to have a systematic relationship to subsequent or concurrent events are called *independent variables*. The characteristics of events that change with changes in the independent variable are called *dependent variables*, because they are dependent upon, or related to, specifiable independent variables. Thus, in our previous example of the mercury thermometer placed in hot water the changes in the mercury level in the glass tube depended upon changes in certain specifiable independent variables: variation in temperature and variation in the heat conductivity of glass and of mercury. The relationship between one independent variable (temperature) and the dependent variable (height of the mercury column) can be schematized as in Fig. 1-2.

The relationship between a given independent variable and the dependent variable is called a *dependent*, or *functional relationship*. We ordinarily say that an observed effect is a function of, or depends upon, certain conditions, or "causes." The alert reader may have inferred by now that a "cause" is an independent variable, an "effect" is a dependent variable. When a functional relationship between them is well established it is called a *law*. Laws are statements of dependencies or of established functional relationships. When we engage in scientific interpretation we show that an established relationship, or established rela-

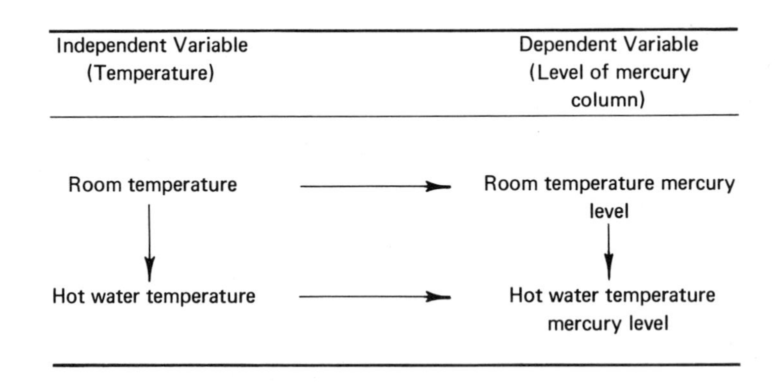

FIG. 1-2. The simple functional relationship between temperature change and change in the level of a column of mercury.

tionships, obtain between sets of variables and that they are exemplified by the particular events or occurrences that we are observing.

Consider some of the many thousands of functional relationships that might be established in the behavioral sciences. For example, an index of the individual's socioeconomic status might be found to be related to grade-point average in college. "Degree of emotionality" might be found to be a function of the amount and kind of early stimulation in infancy. The peculiar and bizzare condition called "schizophrenia" might be shown to depend upon the faulty metabolism of adrenalin, and hence an excess of the chemical substance adrenolutin in the body, instead of the normal dihydroxyindole, etc. These relationships do not happen to be laws. In fact, the last two at the present time merely have the status of hypotheses; but they are representative of the kinds of functional relationships that are established between independent and dependent variables. Relationships of this kind enter into a network of relations by means of which the behavioral scientist explains the various phenomena that he observes.

Multiple Causes and Multiple Effects

Most of the events that psychologists and other behavioral scientists are called upon to interpret are far too complex to be explained by a single cause-and-effect relationship. In fact, even events that do not require the statement of a number of laws for their interpretation seldom involve a single independent and a single dependent variable. Of course, in the laboratory, and even under natural conditions of observation, the scientist attempts to hold all other factors constant while he manipulates or observes an independent variable and measures or observes its effect on a given dependent variable. (The dependent variable for the psychologist is almost always some aspect of behavior.) However, under natural circumstances a single event may have many effects. On the other hand, many events may have a single effect.

As an illustration of the fact that a single event may have more than one effect on behavior, let us consider the effects of not rewarding behavior when the person or organism in question has continuously received reward for the same behavior in the past. The reader is undoubtedly already familiar with the effects of such an event. What happens to a spoiled child when rewards are not forthcoming? He becomes angry. He may even have a temper tantrum. Laboratory animals that have been taught a simple response to obtain reward, usually food, also show evidence of emotional behavior when the reward is for some reason delayed or discontinued.

But there is a second effect of nonreward. Behavior following nonreward tends to decrease the probability that the particular behavior will recur. In other words it weakens that particular habit or set of habits. The multiple effects of nonreward under these conditions can be schematized as in Fig. 1-3.

These two effects of not rewarding behavior which has been continuously followed by reward in the past have the following result: At first the ensuing heightened emotional state contributes to an intensification of behavior. The child who has almost always been given what he asks for will, if he is refused (not rewarded), intensify his demands to the point of becoming annoying. At the same time, assuming that others do not concede to his demands, his teasing behavior is undergoing continual weakening. A burst of teasing and emotionality may be followed by a pause, which may be followed by another outbreak of teasing, etc. However, each outbreak of teasing and emotional behavior is less intense than the previous one. Within a relatively short period of time, assuming that refusal continues, the child stops making demands entirely.

This behavioral example might be compared with the physical example of a thermometer placed in hot bathtub water which we discussed earlier. In both instances a single event leads to a relatively complicated observation: fluctuation of the mercury column in the physical example; fluctuations in the intensity of a given behavior, coupled with a progressive overall weakening of the behavior, in the psychological example. (We will see later that the progressive weakening of a given behavior following nonreward is technically referred to as extinction.)

We might also note that the explanation of these mechanistic processes involves a statement of the laws, or functional relationships, and

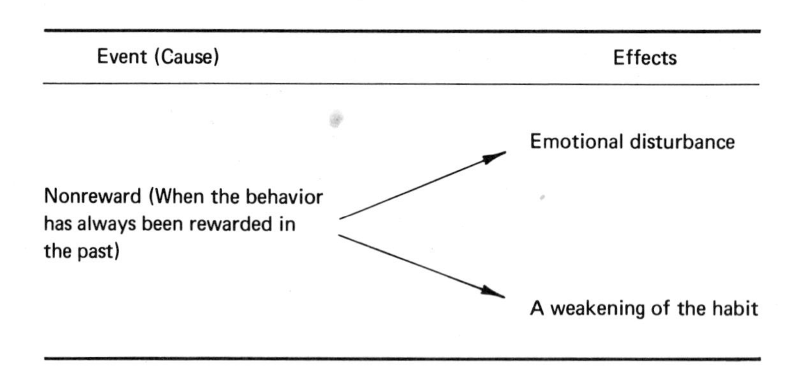

FIG. 1-3. The single event of not rewarding a previously rewarded habit has the dual effect of creating emotional disturbance and weakening the habit.

of the antecedent conditions. The relationship between nonreward and emotionality, and between nonreward and a weakening of response strength have previously been measured and established by scientific research, just as the coefficients of expansion of mercury and of glass have.

Both of these examples further allow us to appreciate the complexity of these observed events. Thus, perhaps it is important to stress the fact that these laws only hold *under certain conditions*. Different antecedent conditions might have very different effects. For instance, we will see in our discussions of learning in Chapter 4 that when a given behavior has a history of partial or intermittent reward, rather than continuous reward, almost the opposite effects occur: nonreward is followed by little heightened emotionality and there is considerable resistance to extinction, or weakening.

Multiple causes also commonly combine to produce a single effect. Let us take an everyday sort of example. In spite of the wide publicity given to horrible aircraft accidents, the fact that passengers on jet airliners are not assigned parachutes, and the travel at incredible speeds, most people find jet travel convenient and comfortable. They actually feel quite pleasant and safe. The passenger does not particularly think about the many steps airline companies take to create such an atmosphere. To mention only a few: the interior decoration of the jetliner is in good taste. It is designed to give the passenger feelings of warmth and stability. The colors may be salmon, pink, or other warm, pleasing colors with just a touch of red in them. Certain shades of blue may be used because they are very restful. While the plane is taxiing after the landing, pleasant dinner music is piped in to the passengers through the loudspeaker system. The airline companies go to a great deal of expense to find and train personable, attractive young women as stewardesses. The food served during the flight is not only of high quality, but it is also attractively arranged. Passengers are given magazines, pillows, beverages. Every whim is attended to. All of these factors contribute to a single effect, namely, the reduction of anxiety. More technically appropriate: they all produce responses on the part of the passengers that are incompatible with anxiety. The relationships involved in this example can be briefly summarized as in Fig. 1-4. Certainly the reader can think of many other examples of behavior that are overdetermined in the sense of being the result of many contributing factors.

REDUCTIONISM

Within science it is frequently necessary to fragment the phenomenon under investigation, to abstract variables from the natural context and

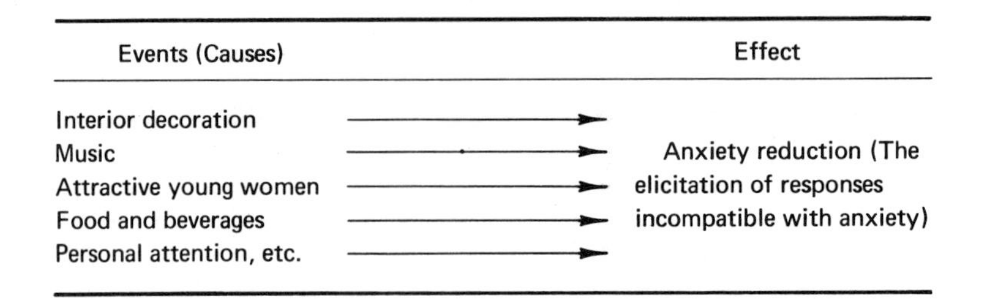

FIG. 1-4. The contribution of a number of different events to a single effect.

study a small number of them at a time in the laboratory where they can be controlled and systematically manipulated. Galileo's contemporaries undoubtedly thought that his rolling small lead balls down inclined planes on a tabletop was singularly indicative of mental derangement. Yet the laws so discovered had such scope as to be applicable to the path of a projectile, the flight of birds, rolling stones, and planetary motion. Some laws depend upon highly specific conditions being satisfied, and as such, they are relatively trivial. On the other hand, others turn out to be very wide in scope. They hold over a very wide range of phenomena. It is these highly generalizable relationships that have allowed the remarkable achievements of experimental science.

The idea of reduction and the great generalizability of certain relationships are inseparable. The scientist may reduce or fractionate phenomena and study them in artificial laboratory situations where he can have a high degree of control over the conditions surrounding them. Having done that, his findings may be generalizable to natural phenomena of much more complexity. Psychology has not as yet developed a set of basic principles comparable in scope to those established in the natural sciences. On the other hand, it does appear that certain of the basic principles that have been established primarily in the animal laboratory will prove to have very wide applicability to complex human behavior. For example, few psychologists would be surprised if it were eventually to be found that such diverse areas as perception, motivation, personality, and social psychology were able to be integrated and explained by principles derived from our study of learning.

The reader will develop a much better appreciation of how certain basic principles that have been established under highly artificial and controlled conditions may account for a wide range of human behavior as he progresses from one chapter to the next throughout this book. For the present let us take but a single example.

In Fig. 1-5 the performance of a simple escape response by two groups of experimental animals is shown. In this very simple experimental situation all of the relevant factors were controlled so that the only difference between these two groups was the intensity of shock, from which they learned to escape by running. The speed at which the animals ran is plotted on the ordinate (vertical axis) of the graph. Blocks of trials are plotted on the abscissa (horizontal axis). Note that with increasing number of trials the running speed for the group receiving the low shock intensity increased slightly, whereas the running speed for the group receiving more intense shock increased drastically. Notice that over successive trials the group receiving more intense shock ran faster and faster, whereas the group that received the less intense shock ran at about the same speed after the second block of trials.

In a wide variety of simple learning situations with animals similar results have been obtained using low vs. high intensities of motivational stimuli such as shock, hunger, or water deprivation. These conditions apparently have the effect of increasing the animals' general level of excitement, or emotionality. This increased excitement is sometimes referred to as *drive*. It greatly intensifies the animals' performance. Emotional arousal appears to energize behavior.

How general is this principle? Would it hold true for humans in similar situations? Psychologists have attempted to answer these questions. If a

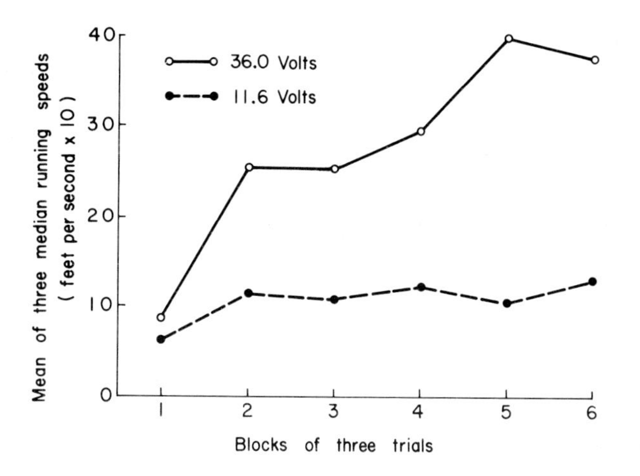

FIG. 1-5. Learning curves of a running-to-escape-shock response under two different intensities of shock. The animals receiving the more intense shock ran faster and faster in successive trials. (Ketchel, 1955.)

measure of general emotionality could be found for human subjects, then the performance of high vs. low emotional individuals could be studied under controlled conditions in order to establish the generality of these findings. Taylor (1951) derived a paper-and-pencil test of emotionality which she called the Manifest Anxiety Scale. It agreed very well with the ratings of manifest anxiety by experienced clinicians. Figure 1-6 shows the performance of anxious and nonanxious college students in the learning of an involuntary eyelid response. Here, just as in the data from lower animals, the learning is greatly intensified for the emotional, or anxious, subjects. Notice that the curves of the two groups continue to diverge up through the 51st–60th blocks of learning trials. Thus, drive does appear to intensify, or energize, behavior.

Other experiments, too numerous to mention here, have indicated that whatever behavior has been learned in the past history of the organism is intensified when drive is increased. In more normal life situations many of these behaviors may be incompatible with the most desirable performance in any given situation. Everyone is familiar with the fact

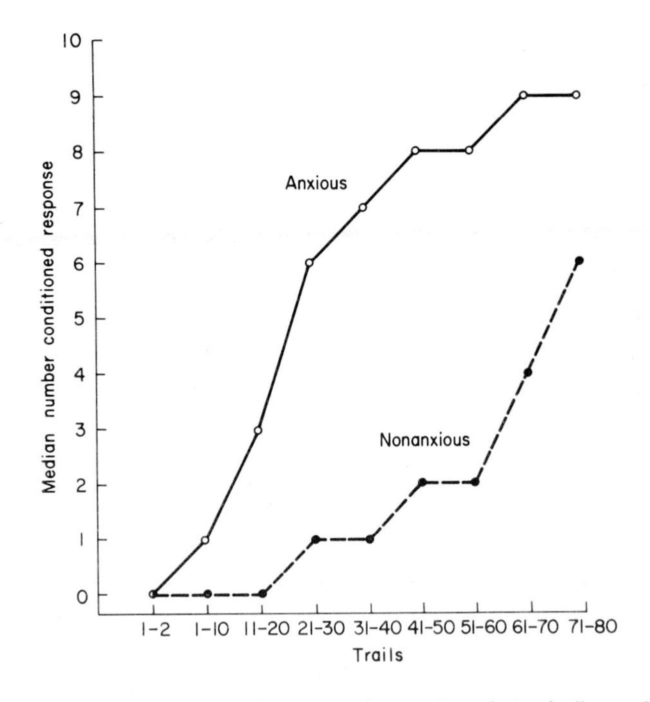

FIG. 1-6. Learning curves showing the median number of classically conditioned eyelid responses of anxious and nonanxious subjects. (Taylor, 1951.)

that under stress behavior generally becomes energized. Under the motivation of strong fear people can perform enormous feats of running, lifting, climbing, etc. However, this increase in emotionality does not always intensify problem-solving, goal-directed behavior. It may intensify quite the opposite kind of behavior, if such task-irrelevant habits happen to be the ones which are the strongest for the person because of his previous learning history.

This was shown very early in an experiment by Hamilton (1916). He, and later Patrick (1934), both found that when they placed human subjects in a room with several exit doors and then frightened them in various ways so that escape was the only possible solution, many subjects would try to escape through the wrong (locked) door. Instead of trying other doors, one of which the experimenter had arbitrarily selected as the correct (unlocked) door, they showed extremely stereotyped behavior. They continued to struggle frantically attempting to escape by way of the first door they tried, even though the response was clearly unadaptive.

Psychologists have further explored the ideas of Taylor's experiment with anxious and nonanxious subjects. Individuals were selected who scored high and low on the Manifest Anxiety Scale and systematically the strength of various habits which were "correct" in that they facilitated learning, and habits which were "incorrect" in that they interfered with learning in a given situation was manipulated. Generally it was found that when the habits with which the individual enters the learning situation happen to coincide with appropriate responding, then anxious individuals perform very well—because those habits which are intensified are also the ones that happen to be desirable in that particular situation. On the other hand, when the strongest habits with which the person enters a situation happen to be incompatible with "correct" responding, then the subject's emotional level energizes these "wrong" responses.

These basic principles of the effects of arousal that have been established by reducing phenomena and studying them under artificial laboratory conditions may have important implications for certain other human behavioral processes, namely, neurotic and psychotic behavior—the breakdown of personality and behavioral organization and functioning. We can say but little here before the student has been exposed to many of the chapters that follow. We will see, however, that there is a very high relationship between clinical diagnoses of neurotic behavior and drive, or emotionality, as measured by instruments such as Taylor's Manifest Anxiety Scale. Similarly, the findings of a very large body of research in psychosis, particularly in schizophrenia, indicate that a basic process underlying these complex behavioral patterns may involve a breakdown of mechanisms that would ordinarily inhibit excitement, resulting in a consequent intense degree of emotional arousal.

OPERATIONISM

Operationism refers to the process of defining concepts by a set of operations or measurements performed by the scientist. Such operations must always be reducible to things that can be pointed at—scales, rulers, clocks, and the like—and therefore are open to public inspection and verification. Theoretically, the most abstract concept of physics, let us say, ought to be explainable to, and verifiable by, an intelligent 12-year-old.

Consider the concept of "force" in physics. If a physicist were asked the meaning of the word "force" he would probably reply that force is mass times acceleration. If pressed further, although our hypothetical physicist might display a bit of impatience with his questioner, he would explain mass in terms of scales, pointers, weights, and the like. Acceleration might be reduced to clocks and yardsticks. This is what philosophers of science have variously called the physicalistic meaning base, the "point-at-able," or the "observable-thing" language.

The case is no different for the psychologist. When questioned about intelligence, for example, he might say that IQ is equal to mental age divided by chronological age. Further pressed, he would point at a calendar and describe how the age in terms of years and months since birth was calculated. Mental age would be described in terms of the content of the test item. He might tell you: "I asked the person a series of questions and asked him to perform a series of tasks. For example, I asked him, 'In what country is the city of Copenhagen?,' 'Who wrote Moby Dick?', or questions such as, 'Why is a license required to drive an automobile?', 'Which direction is North?' If the person answered these correctly I gave him a score of 1, if not, I put down 0. I also asked him to repeat a series of digits after me, and then to repeat them backwards; to put together a series of multicolored blocks to form the same pattern that I showed to him on a card; to tell me what was missing in a series of pictures. I gave him a whole series of such tests, entering the appropriate score for each one. After a large number of such items, I added up all the numbers and divided by the person's chronological age. That is what I mean by IQ. . . . "

Operationism actually involves two kinds of meaning (Bergmann, 1954). The first refers to the set of operations performed. A somewhat absurd example of this is given by the K coefficient, which is equal to the cube root of a man's blood pressure divided by the density of white corpuscles in his body, times his age. That is a clear definition. It is open to public inspection; it is repeatable and verifiable. However, such a concept has little significance because it is not related to anything else.

The second type of meaning involves establishing the significance of the concept. It is given by the relationships and laws into which the concept enters. Thus, if it is known that an individual has an IQ of above 140 on the Stanford-Binet intelligence test there is a high likelihood that he will go further and be more successful in school than the average person. Compared to the average person it can also be predicted, with varying degrees of accuracy, that he will read more, write more, have a higher earned income as an adult, be better adjusted, physically healthier, etc. (Terman, 1925).

Levels of Conceptualization

An operation, or set of operations, can give rise to various levels of conceptualization. Let us take a particular set of operations as an illustration: It will be recalled that nonreward tends to weaken behavior (see chap. I, pp. 9–10). The reader may also recall that the technical term for this is *extinction*.

Ayllon and Michael (1959) reported the use of extinction to eliminate the psychotic talk that had persisted over a period of 3 years in a female patient. The hospital staff had served as sympathetic listeners every now and then. Hence, unwittingly they had been rewarding the patient's delusional talking on an intermittent schedule, a schedule of reward (reinforcement) which is known to give rise to persistent behavior. The hospital staff was instructed not to talk to the patient nor to give the patient any kind of attention. Attention and rapport were only to be given when the patient talked sensibly.

In Fig. 1-7 it can be seen that in 8 weeks under these conditions the relative frequency of the patient's psychotic talk was approximately one-fourth of what it had been before treatment. The abrupt rise in the relative frequency of psychotic talk after the 10th week was apparently due to the fact that a social worker inadvertently paid attention to the patient. This resulted in the patient subsequently talking psychotically to everyone in her surroundings for the few weeks that followed.

Extinction procedures can be conceptualized on a number of levels, each being further removed from the actual operations performed. On the first level, extinction in the example given above, would refer to the operation of stopping giving attention to psychotic talk. The analog of this in the animal laboratory would be the operation of flicking a switch, and thus cutting off, by breaking the electrical circuit, the working of a food dispensing mechanism that delivers pellets of food to an animal

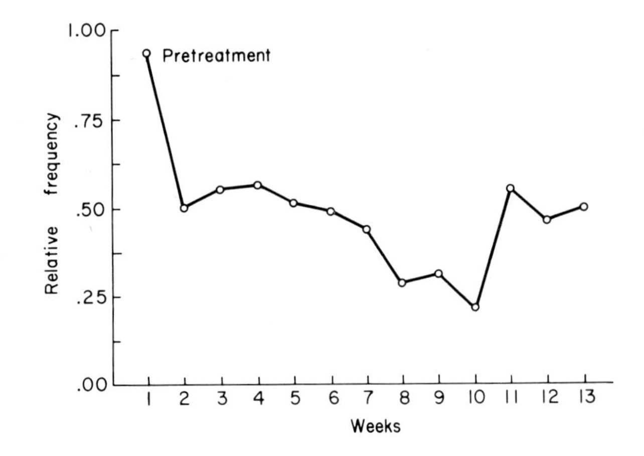

FIG. 1-7. The gradual weakening (extinction) of psychotic talk with continued nonreinforcement over a period of weeks. (Ayllon & Michael, 1959.)

when it makes a particular response, such as pressing a bar. That is to say, the meaning of the term extinction is given by the operations performed. Notice that even if no change in behavior followed this procedure the term would still be appropriately defined in the first sense of operational meaning outlined above. However, meaning in the second sense—in terms of the significance of the concept—would be absent.

A second level of conceptualization identifies the phenomenon in operational terms. Extinction, on this level, refers to the process of doing something *and measuring the changes that take place.* For our example the term extinction would refer to the diminishing frequency of psychotic talk.

At higher levels of conceptualization explanatory states or processes may be introduced which are not, in fact, observable. Thus, under the conditions in our example, namely, those of nonreinforcement, it might be postulated that a gradual buildup of inhibition occurs, yielding a gradual decrease in observable response. Physiological mechanisms may also be postulated to underlie such a process. At even higher levels of abstraction, concepts may be introduced which summarize the interaction of other postulated processes in an explanatory system. In our discussion of reduction the concept called drive was representative of such an abstraction. These elaborate concepts, must, however, always be reducible to, or specifiable by, operations that are observable, such as applying shock to an animal or administering a test to college students. It is this operational base that allows the relationships defining abstract concepts to be repeatable, verifiable, and open to public inspection.

THE EXPERIMENTAL APPROACH

Observation and hypothesis testing in some sciences are primarily nonexperimental, because the type of phenomenon observed excludes the careful and systematic manipulation of variables. Such phenomena cannot be made to occur in the laboratory. The scientist must wait for events to happen under natural circumstances and then make observations. Perhaps there are no sciences that are purely nonexperimental. This becomes true when the powerful reductive aspect of the scientific frame of reference is taken into consideration. Astronomers may have to wait for astronomical events to happen, but the basic laws of planetary motion were, as we have previously noted, integrated and interpreted by the systematic variation of the mass, velocity, etc., of lead balls on a tabletop. After the laws of such miniature, artificial systems have been worked out in the laboratory, further extensions, called "composition laws," have often enabled the interpretation of natural occurrences.

The experimentalist turns his back on phenomenalistic observation and attempts to formulate and discover the basic underlying laws by means of the systematic manipulation of variables under highly controlled conditions. The experimental approach has proven fruitful whether the practical problems confronting the scientist have been the conquering of space, disease, limitation of food supply, or mental health. The basic reason for the great prestige which the experimental method enjoys can be stated succinctly by a colloquialism: "Results talk."

In psychology, people may be employed as experimental subjects, but frequently lower organisms are. Researchers can control the experiences of animals in ways that would be impossible with human beings. Moreover, significant experiments may of necessity involve procedures that are harmful to the subjects.

The question must arise (just as it does in experimental medicine, genetics, etc.); Are the results obtained from animal experimentation applicable to humans? Of course, results of animal experiments are rarely directly generalizable without taking other factors into consideration. After having noted that, however, we can give a generally meaningful reply: That is an *empirical* question. That is to say, the answer can be approached scientifically, the generalizability of the results can be tested, and the applicability of the results to humans can be proved or disproved.

THE CLINICAL APPROACH

The clinical psychologist deals with personality appraisal and with the observation and treatment of emotional disturbances. Pressing demands for the solution of human adjustment problems have been placed on the psychologist long before he has adequate tools with which to cope with the demands scientifically. Nevertheless, the clinical psychologist could hardly be expected to wait for the development of a comprehensive science of psychology before attempting to do anything about important human problems. Although a tendency toward the formulation of highly speculative and often untestable theories of human behavior has been evidenced at times, there is a much more positive and important aspect of the clinical approach: the scientific contribution in the form of observation and study of the lives of people and in the form of clinical research.

In recent years increasing scientific investigation and the development of ingenious techniques to quantify, manipulate, and control variables related to complex personality functioning have resulted in an extensive body of research in the areas of personality and social psychology. At the same time attempts have been made to reconcile differences in various conceptual preferences and frames of reference between experimental, academic psychology, and clinical-personality approaches to human behavior. This reconciliation has frequently involved the translation of terms and concepts into the language of the learning laboratory, and the reformulation of the hypotheses in order to make scientific tests of them possible. Similarly, clinical psychologists themselves have furthered the rapprochement in two directions: first, they have appreciated the implications of certain laboratory findings; and second, they have made efforts to undertake research and gather data to test hypotheses and techniques.

The possibility must not be excluded that responsible clinical speculation, although basically outside of science, may have some heuristic value for research. Applied psychology does have the distinct advantage of making observations under relatively naturalistic circumstances. Furthermore, not enough is known at the present time either about productive thinking, or about the scientific enterprise itself, to be able to reject creative theorizing as totally divorced from scientific development.

SUMMARY

We have considered briefly the historical background of the modern scientific approach to the interpretation of personality and behavior. Certain aspects of the scientific frame of reference are discussed, along with some examples of the application of these concepts to the study of behavior.

The cultural impact of Darwinism created an atmosphere for the development of an interest in the scientific study of the adjustment of organisms to their environment. The school of functionalistic psychology, its associated movement, behaviorism, and certain influences from other theoretical approaches have since provided a framework for the modern scientific approach to human behavior.

By its very existence a behavioral science may create some misunderstanding and suspicion on the part of laymen. However, the major obstacle to an appreciation of scientific approaches to personality and behavior has been a lack of understanding of certain aspects of the behavioral scientist's frame of reference. In general, there is a failure to appreciate the abstract character of scientific knowledge, as opposed to particularistic, literary, or humanistic knowledge.

The scientist's observations are explained by reference to certain specific conditions, and by means of general laws relating those conditions to the observations. Conditions relevant to the prediction of events are called independent variables. The observed conditions that are dependent upon them are called dependent variables. The dependent, or causal relationships between two sets of variables are called functional relationships. Well-established functional relationships are called laws.

Processes that are lawful are said to be mechanistic. The simple equation of the words "mechanistic" and "machinelike" is often inappropriate and confusing. In the prediction of behavior there are multiple causes and multiple effects. A single event may have multiple effects, and a single effect may be the result of multiple causes. Furthermore, complex relationships may be modified by the conditions under which they occur.

The work of arriving at scientific explanations frequently involves studying processes under controlled, and highly artificial conditions. The scientist "makes it happen" in the laboratory. It is frequently the case that principles established under these artificial conditions may be abstracted from them and applied for the prediction of events as they occur in natural and much more complex settings. Whether or not such relationships are applicable to the more complex natural events is an *empirical* question; that is, it is a question which can be answered by making the appropriate observations to test the predictions based on them. Sometimes principles established in the laboratory using lower animals as subjects are generalizable to human behavior.

In order that the concepts and findings of the scientist can be meaningful they must be formulated in such a way that they are open to public inspection and the experiments must be repeatable by anyone who wants to verify the results. There are two kinds of meaning in a scientific sense. When we ask the scientist what is the meaning of a given concept, we are asking, "What set of operations does the concept refer to?" There is a second kind of meaning also. Once we know what the scientist did in order to arrive at a particular concept, we may ask what significance the concept has. That is, into what relationships and laws does the concept enter? The definitions and concepts of the scientist may be highly abstract, but they must always be reducible to public, observable operations and events.

Finally in recent years there has been a rapprochement between those whose chief interest has been the investigation of selected aspects of behavior in the laboratory and service workers who have had to deal with and interpret human adjustment problems without adequate scientific tools with which to cope with the problems. The former have shown more interest in extending laboratory findings to complex human behavior and in studying it directly, whereas the latter have shown more interest in being able to verify their concepts and observations, and they have recognized the need for research.

REFERENCES

Angell, J. R. The province of functional psychology. *Psychol. Rev.*, 1907, **14**, No. 2, 1–19.

Ayllon, G., & Michael, J. The psychiatric nurse as a behavioral engineer. *J. exp. anal. Behav.*, 1959, **2**, 323–334.

Bergmann, G. *The metaphysics of logical positivism.* New York: Longmans, Green, 1954.

Hamilton, G. V. A study of perseverance reactions in primate and rodent. *Behav. Monogr.*, 1916, No. 13.

Hempel, C. A., & Oppenheim, P. Studies in the logic of explanation. *Phil. Sci.*, 1948, **15**, 135–178; also in H. Feigel & M. Brodbeck (Eds.) *Readings in the philosophy of science.* New York: Appleton, 1953. p. 320.

Jennings, H. S., *Behavior of the lower organisms.* New York: Columbia Univer. Press, 1906.

Ketchel, Rhoda. Performance in instrumental learning as a function of shock intensity. Unpublished masters dissertation, State Univer. of Iowa, 1955.

Patrick, J. B. Studies in rational behavior and emotional excitement. I. The effects of emotional excitement on rational behavior in human subjects. *J. comp. Psychol.*, 1934, **18**, 153–195.

Rogers, C. R. Implications of recent advances in prediction and control of behavior. *Teach. Coll. Record*, 1956, **57**, 316.

Shaffer, L. F., & Shoben, E. J., Jr. *The psychology of adjustment.* Boston: Houghton, 1956.

Skinner, B. F. *Cumulative record.* New York: Appleton, 1959.

Taylor, Janet A. The relationship of anxiety to the conditioned eyelid response. *J. exp. Psychol.*, 1951, **41**, 81–92.

Terman, L. M. (Ed.) *Genetic studies of genius.* Vol. 1. Stanford, Calif.: Stanford Univer. Press, 1925.

Watson, J. B. *Behaviorism.* New York: Norton, 1924.

Chapter 2

Biology and Stress

INTRODUCTION

A knowledge of the basic biological mechanisms of the body is indispensable for a thorough interpretation of personality and behavior. There is a continuous interplay of effects between environmental events and bodily reactions, and between bodily conditions and our responses to the environment. The biology of the individual enters into the determination of the kinds of experiences he will have and the ways he will react to them. It provides the very mechanisms that mediate his perception and his interactions with the environment, and it sets certain limits to his experiential and behavioral possibilities.

In this chapter we will review some of the fundamentals of basic biology, and of the structure and functioning of the human nervous system. Then we will consider the body's endocrine system in relation to certain aspects of personality and behavior. Finally, we will discuss stress, individual differences in response to stress, and relevant research findings.

BASIC CELL STRUCTURE AND RESPONSE MECHANISMS

The fundamental unit of all living organisms is the cell. In general, all cells are made up of a nucleus, cytoplasm, and a surrounding membrane.

The Nucleus

The central focus of organization within the cell is the nucleus. It contains the chromosomes and genes. Their composition involves a complex substance of special significance: desoxyribonucleic acid (DNA).

The structure of the DNA molecule carries the genetic code. The DNA molecule has an exceedingly complex structure, involving sequences of bases joining sugar-phosphate strands. Varying combinations of these elements and varying relations among them serve to encode different sets of genetic information. The development and biochemical functions of the cell are governed by the DNA molecule.

Cytoplasm

The main mass of the cell is made up of cytoplasm. It surrounds the nucleus and is bounded by the cell membrane. There is a very great range of variation of structure and chemical composition of cytoplastic material among different cells of the body. Material may be manufactured by chemical reactions that occur within the cytoplasm and the end products may be transmitted through the semipermeable cell membrane. Waste material may also be passed out of the cell through this membrane.

The conversion of food molecules into energy and into chemical forms for storage also takes place in the cytoplasm. This is called *metabolism.* In the cytoplasm the manufacture of many enzymes and proteins occurs within molecules of ribonucleic acid (RNA). Certain aspects of the structure of the RNA molecules duplicate the structure of the DNA molecules found in the cell nucleus. Cytoplasm as a whole is a seat of continuous activity. Molecules within the cytoplasm change size and shape and move about so as to change the shape of the entire cell. Thus, the cell is said to possess the property of *contractility.*

The Cell Membrane

The cell membrane, which makes up the outer boundary of the cell, is semipermeable. Only certain substances are allowed to pass through it. At the membrane there is a relatively static balance, or equilibrium. Although there is an exchange of molecules across the membrane, there is a balance of pressure and movement.

The membrane may show a property that is of particular interest to us: It may be in a state of polarization. Positive ions of electrolytes collect on one side and negative ions on the other. When this happens any change in the number or in the charge of the ions on one side of the membrane disturbs the equilibrium. When a state of disequilibrium occurs across the membrane, it causes a rearrangement of adjacent molecules. If the rearrangement of molecules in the immediate vicinity of the membrane is not sufficient to restore equilibrium, the reaction is not confined, but spreads to neighboring regions along the membrane. The property of reactivity to the disturbance is called *irritability.* The spread of the reaction along the membrane is referred to as *conduction*. We will return to this topic shortly, when we consider the nature of excitation and conduction in neurons.

Receptors, Effectors, and Neurons

Throughout evolutionary development cells become increasingly specialized. Very primitive multicellular life possesses specialized cells. Some serve the special function of irritability. These become the highly specialized *receptor* cells of higher forms of mammalian life. Receptors all respond to electrical stimulation, and they may respond to other kinds of intense stimulation by stimuli for which they are not specifically adapted. However, on the whole, both in structure and function the receptor cells of higher animals are specifically developed for each of several different kinds of stimulation: light, temperature, chemical, and mechanical stimulus events. They also can be artificially stimulated to respond by the direct application of an electrical stimulus. That is, their more primitive irritability is retained. In man, types of receptor cells vary from unspecialized neurons to specialized epithelial cells and specialized neurons embedded within receptor structures.

Effector cells have similarly evolved from lower forms of cells specializing in contractility. There are two major kinds of effector cells in the body: muscular and glandular. They are the means by which the body responds to its environment. There are three types of muscles: smooth muscles, striated muscles, and cardiac muscles. Striated muscles are more elaborately structured than the simpler, spindle-shaped smooth muscles. They have special contraction fibrillae within an elastic membrane which give them a striped appearance. Cardiac muscle is similar to striated muscle.

Glands respond to changes in the internal environment by secreting chemical substances which are important for the body's equilibrium. Some glands empty their secretions into special cavities of the body. They are called duct glands. Ductless glands, on the other hand, secrete their chemical products directly into the blood. These chemical products circulating in the blood have very widespread effects on the body, especially on nervous activity.

A third special kind of cell of particular interest to us is the *neuron*. Some neurons are part of receptor structures, but the majority of them form connections between receptor and effector cells and organs. The structure of a typical neuron or nerve cell, shown in Fig. 2-1, consists of dendrites, a cell body, and an axon. The dendrite is the receiving end of the neuron. It receives impulses from other neurons. They are transmitted from the cell by way of the axon. After the axon leaves the cell body

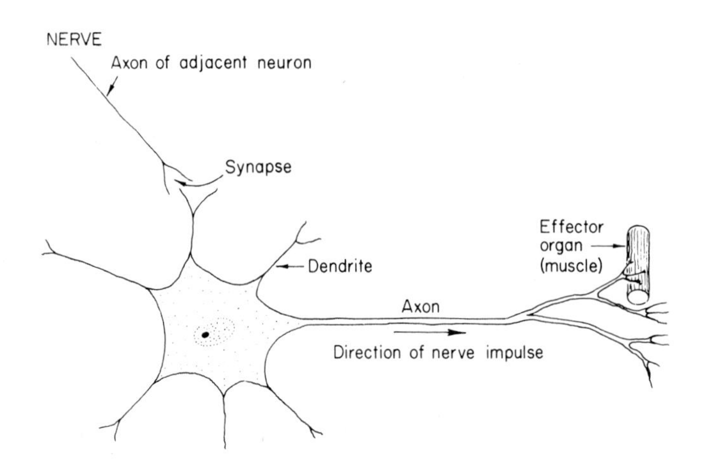

FIG. 2-1. The structure of a typical neuron. (From Rogers, 1961.)

(sometimes called the soma) it may send off collaterals (not shown in Fig. 2-1).

Axons larger than 1 micron in diameter have a thin covering called the *neurillemma* and a fatty sheath between it and the fiber called the *myelin sheath.* Within the central nervous system special kinds of supporting cells generally intertwine among neurons. They are *neuroglia cells,* which occur especially where the neurilemma is not found. They are known to be a common source of tumors, but they may also be important in the excitation and conduction of nerve impulses.

Transmission

In order to understand the transmission of nerve impulses it is necessary to turn our attention to the cell membrane. Polarization occurs across the membrane of neurons. Most physiologists believe that this is because the membrane in its resting state is relatively impermeable to sodium ions (Na). Hence, there are many more sodium ions outside the membrane than inside. The sodium ions are positively charged. Thus, a positive electrochemical field surrounds the membrane. Similarly, the presence of negatively charged potassium ions (K) inside the membrane creates a negative electrochemical field. When there are more ions of a particular kind on one side of the membrane than on the other side the membrane is said to be *polarized.* The measurable potential difference across the membrane in its resting state is called the *resting potential.*

When the neuron is stimulated, either by electrochemical impulses from adjacent neurons or by receptor activity, the membrane at the point of stimulation becomes more permeable. Positively charged sodium ions pass from the outside through the membrane into the neuron. This disruption of the resting potential—the mixing of the positive and negative electrical charges within the neuron—continues in the area immediately adjacent to the point of stimulation, the area immediately adjacent to that, and so on, until the disturbance has traversed the length of the neuron (see Fig. 2-2). This has been called a *propagated disturbance* or an *action current.* Electrical recordings of stimulated neurons indicate that a corresponding wave of negative electrical potential travels the length of the neuron. This is called an *action potential* or, more commonly, a *nerve impulse.* Soon after the action potential passes, the balance of ions across the membrane is reinstated and the membrane returns to its resting state.

When the action potential occurs, it occurs in full strength. That is, it obeys an all-or-none law: either the neuron fires, or it does not. Once

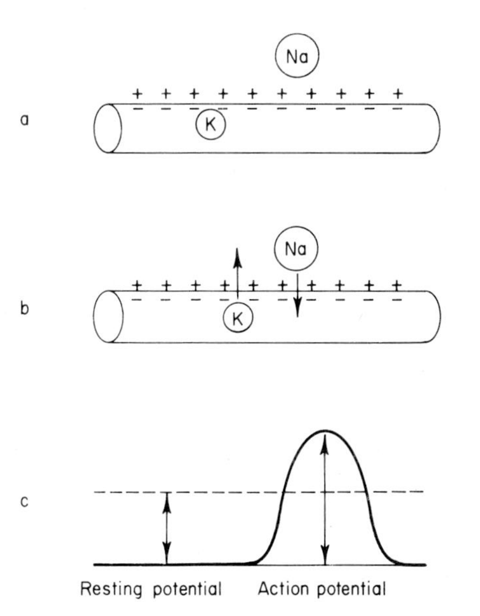

FIG. 2-2. The resting neuron has a balance of positive (Na) and negative (K) ions across the membrane (a). Stimulation of the neuron produces a disruption which traverses the membrane (b). Electrical recordings from the resting neuron indicate a potential across the membrane, called a *resting potential.* Recordings of the stimulated neuron indicate a much larger potential which traverses the membrane. This is called an *action potential.* The resting potential and the action potential are schematized in (c). (Adapted from Katz, 1952.)

the action potential occurs, it propagates along the whole length of the neuron.

When the neuron has thus been activated, there is a brief period during which no stimulus, no matter how intense, can arouse a response in the fiber. This is known as the *absolute refractory period.* Following this the neuron begins its return to normal sensitivity. This return period is known as the *relative refractory period.* During the relative refractory period the neuron may be fired again by a relatively intense stimulus. The intensity required becomes less and less until the final return to the neuron's resting state is accomplished. In some neurons it is possible to record further swings to supernormal sensitivity and then back to subnormal sensitivity before the neuron returns to its original resting state.

In order for the action potential to occur at all the stimulus must be of a certain minimal intensity. It is possible, however, for the effects of two or more subliminal stimuli to combine and cause the action potential if they occur within a period of approximately .5 milliseconds. This is called *temporal summation.* There is another kind of summation—*spatial summation.* In this case several different sources of stimulation, each one too weak to create an action potential, converge on the same point, spatially, and combine to create an above-threshold action potential. Within the nervous system both of these types of summation occur at once.

Finally, it should be noted that receptor cells are frequently exceptions to the all-or-none principle. That is, they are capable of transmitting graded potentials approximately proportional to the size of the stimulus provoking them. Certain neuronal membranes, such as those of the soma and dendrites of specialized receptors, are not electrically stimulable, but respond only to specific excitants, which may generate an action potential in the electrically stimulatable membrane of the axon, or of other fibers.

Synaptic Transmission

Neurons do not directly connect with one another. Typically, one of the axon terminals delivers excitations to the dendrites or the soma of an adjacent neuron across a very small gap (of the order of 100 angstroms). This gap is called a *synapse.* The synaptic crossing of an impulse requires a specific chemical excitant, generally called a *chemical transmitter.* Depending upon the chemical conditions at the synapse, impulses from adjacent neurons may be inhibited or facilitated. The chemical transmitters appear to be synthesized and stored in the terminals of axons.

Two chemical transmitters have been identified: acetylcholine and norepinephrine. These chemicals appear to cause excitatory potentials in adjacent neurons in some instances, and inhibitory potentials in others. The precise nature of their action is at the present time only poorly understood. It appears that transmission at the synapses is determined by a complex of conditions, including an interaction between chemicals of the axon terminals and those of the postsynaptic membrane. In any event, it is now established that synaptic transmission is chemical in nature; that the action potential arriving at the presynaptic terminals affects this chemical secretion; and that these chemical transmitters diffuse across the synaptic gap and build up an excitatory or inhibitory postsynaptic potential by interacting chemically with the postsynaptic membrane.

THE CENTRAL NERVOUS SYSTEM

The reader may be familiar with the structure and functioning of the human nervous system on the basis of previous courses in psychology or biology. Nevertheless, because of their relevance for certain important aspects of personality and behavior, some of the basic features of the central and autonomic nervous systems will be reviewed here.

Traditionally the brain and the spinal cord have been said to compose the *central nervous system.* Neural connections between these structures and the rest of the body are referred to collectively as the *peripheral nervous system.* Figure 2-3 shows in a highly schematic manner a simple reflex circuit involving a sensory neuron, a connector neuron, and a motor neuron. A single sensory neuron is shown entering the dorsal horn of the spinal cord, where it synapses with a connector neuron. The connector neuron synapses, in turn, with a motor neuron. Some reflexes are even simpler: the sensory neuron synapses directly with the motor neuron. In other reflex connections, after the impulse enters the dorsal horn it is carried to the other side of the spinal cord by a short connector fiber, only to synapse there with a motor fiber that leaves through the ventral horn on that side.

Most impulses do not leave the spinal cord immediately, either on the same or the opposite side. Instead, incoming impulses may ascend the cord, cross, and leave at a higher level; or they may travel up the cord to the great thalamic relay station of the brain—and thence to the cortical projection areas where they are interpreted as a specific kind of experience. Similarly, motor fibers originating in the brain are relayed down the cord, cross at some level, and leave through the ventral horn to in-

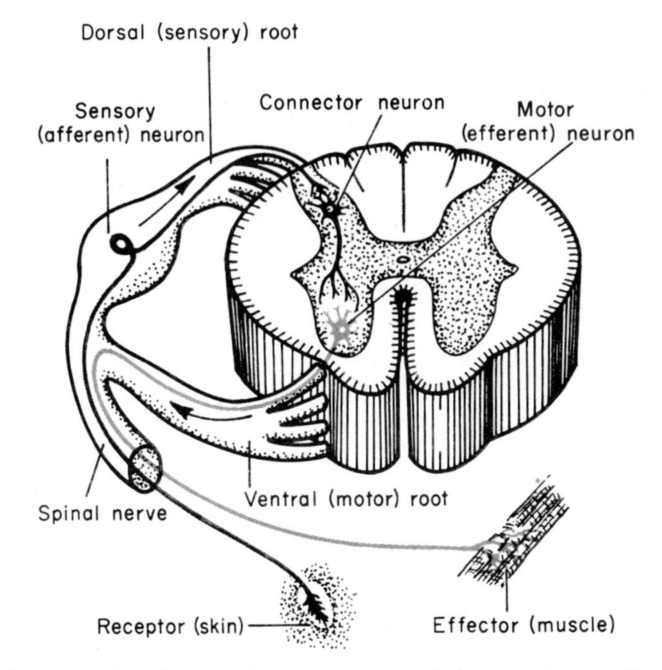

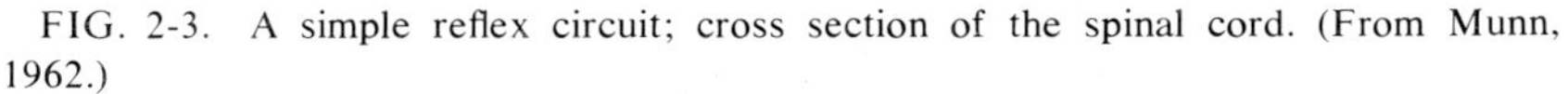

FIG. 2-3. A simple reflex circuit; cross section of the spinal cord. (From Munn, 1962.)

nervate specific effector systems. These transmission pathways are illustrated in Fig. 2-4.

The neurons combine in the cord to form massive ascending and descending columns, or tracts. Each tract is a functional unit. For example, the *lateral spinal thalamic tract* is composed of thousands of neurons that mediate sensory experiences such as touch and pressure. The *ventral cerebrospinal tract* is made up of neurons that course down the cord from the brain, mediating muscle movements. Other tracts mediate pain and temperature, muscle tone and coordination, etc.

Certain projections from the brain, particularly those of the brainstem reticular formation serve to integrate and modify many of the simple reflex circuits lower down in the cord that do not themselves involve fibers going to the brain. Thus, if the cord is severed at any level, reflexes below the cut do not function for hours or even days. This phenomena is known as "spinal shock." It is as if the connector neurons involved in the simple reflexes were long fibers going to the brain and back instead of following the shorter and more direct pathways. This is typical of the functioning of the central nervous system. The activity of one segment of the nervous system is rarely completely isolated from other, particularly from the more central aspects of the system.

The Somatosensory Projection Areas

For gross descriptive purposes the brain is divided into four lobes. These are shown in Fig. 2-5. They are the frontal, the parietal, the temporal, and occipital lobes. Incoming sensory impulses, relayed to the cortex from the thalamic relay station, are translated into a sensory experience according to the somatosensory projection area that they reach. Thus, just behind the central fissure there is an orderly arrangement of areas in which the sensory impulses are received. These are shown schematically in Fig. 2-6. When impulses arrive over pathways leading to, let us say, the area labeled "arm" in Fig. 2-6, sensory experiences are perceived as localized in the arm. If the terminal point of cortical stimula-

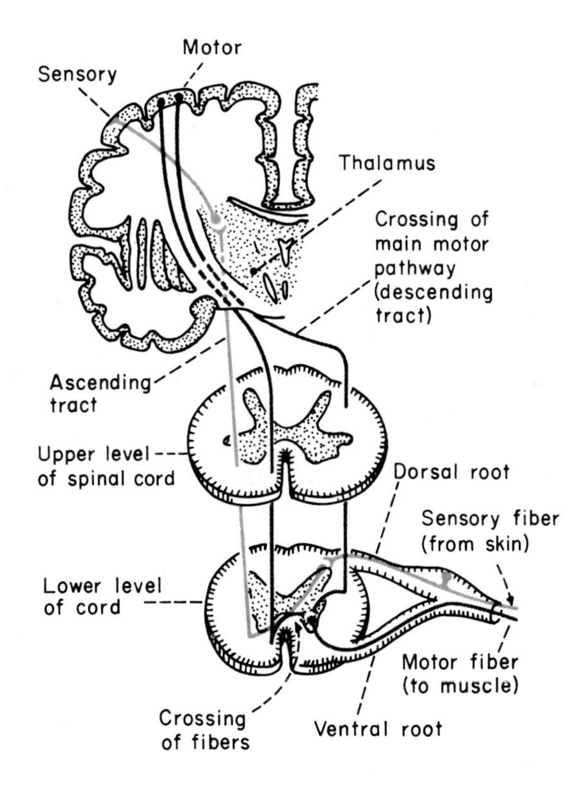

FIG. 2-4. Sensory and motor pathways within the spinal cord. (From Munn, 1962.)

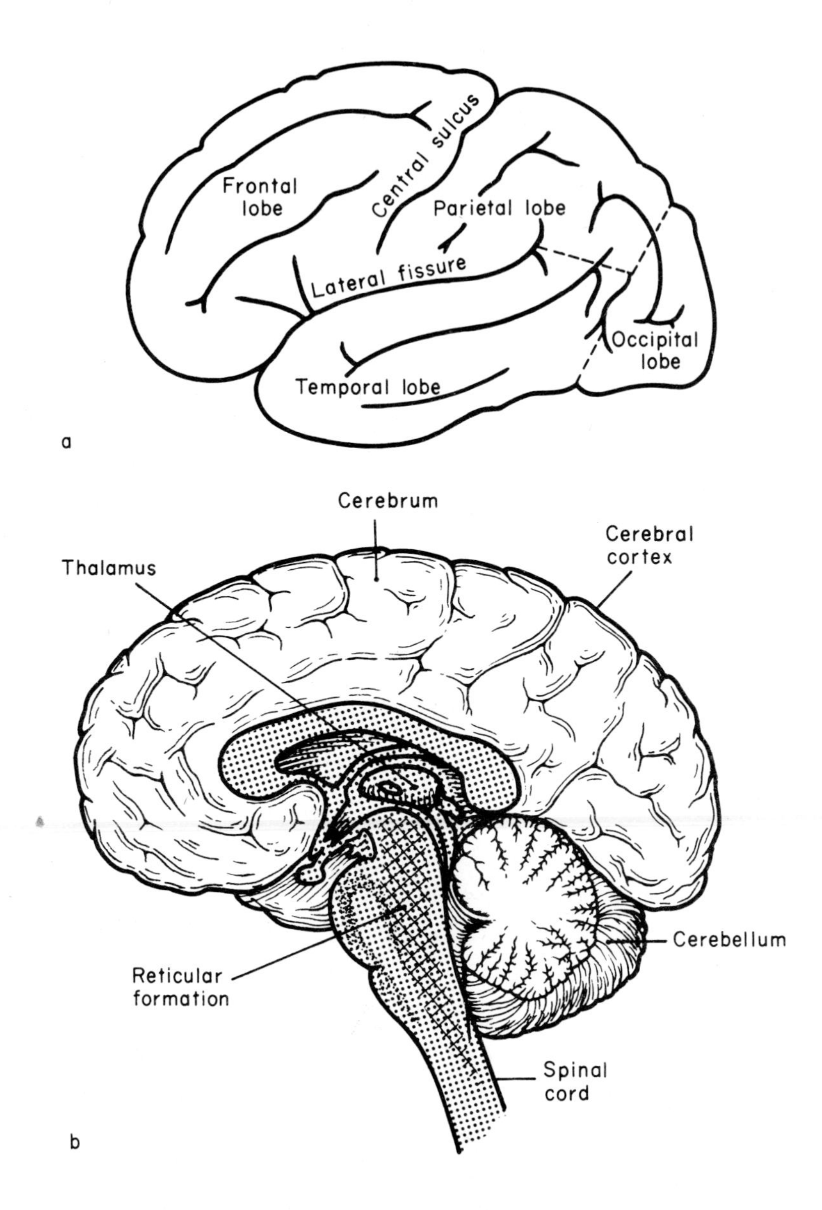

FIG. 2-5. (a) A lateral view of the human brain showing the location of the lobes, principal fissures, and sulci. (b) The relation of the brain stem to the cerebral cortex which encapsulates it, and to the cerebellum, the major motor coordinating region of the brain. [Part (b) From Munn (1962).]

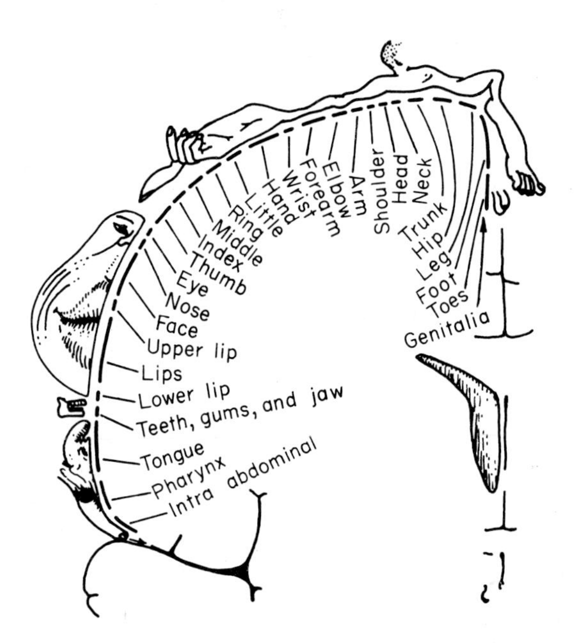

FIG. 2-6. Sensory homunculus showing the area of the cerebral cortex where sensory fibers terminate, giving rise to sensations from the areas of the body indicated. The sizes of the parts of the body are drawn in proportion to the sizes of their sensory projection areas. Note that the face has more cortical representation than areas such as the trunk or leg. [After Penfield (1958).]

tion is slightly below that area, sensations are perceived as coming from the elbow; below that, the forearm; and so forth. The person's elbow need not be stimulated in order for him to experience a sensation or a series of sensations as coming from that region. Electrical stimulation of the brain in the area indicated will have the same effect.

The enormous complexity of the neuronal structure of the cortex, with its estimated nine billion cells, unfortunately does not allow for localization more precise than shown in the sensory homunculus in Fig. 2-6. Cortical localization is, therefore, far from perfect. Notice in Fig. 2-6 that certain areas, such as the face, lips, and hands, have more cortical representation than other, larger, areas such as the legs and trunk.

Impulses arriving in the occipital lobe [see Fig. 2-5(a)] are interpreted as visual experience in the sensory projection areas located there. Auditory projection areas are located in the temporal lobe. The parietal lobe receives somaesthetic (body-feeling) projections. During brain surgery a patient stimulated in these respective areas of the exposed brain will report sensory experiences corresponding to the functional significance of these areas. Such reported experiences are usually relatively crude, e.g., whirling colors, flashing lights, or warm limbs. If stimulated in the association areas adjacent to these projection loci, the patient may report a corresponding sensation, a complex of sense data, or a memory.

Senses other than vision, hearing, and somaesthesis do not have such clearly established cortical localization. Smell and taste appear to be mediated at subcortical levels. Pain, a somaesthetic sense, has no cortical representation. At least one variety of pain experience, deep pain, appears to be mediated by thalamic nuclei.

Motor Systems

Motor neurons originating in the cortex form two systems: the great pyramidal tracts and the extrapyramidal pathways. The great pyramidal tracts cross in the midbrain in a pattern which resembles a pyramid. The pyramidal tracts have few, if any, synaptic connections before reaching the motor neurons of the lower brain and spinal chord. All motor pathways originating in the brain which are not part of the pyramidal system are collectively called extrapyramidal. This system forms part of important subcortical structures, like the basal ganglia and reticular formation.

The origin of extrapyramidal pathways is diffuse. They descend from all lobes of the cortex and sometimes overlap pyramidal areas. The origin of pyramidal tracts, on the other hand, is a little more localized. In man, the area lying immediately in front of the central sulcus [see Fig. 2-5(a)] makes the largest contribution to the pyramidal system. Fibers originating in that area, forward of that area (premotor), and in the postcentral somatosensory area make up about 60 percent of the pyramidal system (Morgan, 1965).

The motor areas as well as the sensory areas have a high degree of organization. Systematic electrical stimulation of the area just forward of the central sulcus results in gross movements of the bodily effector systems indicated in Fig. 2-7. The cortical areas controlling parts such as the abdomen and chest are diminutive in size compared to those regions controlling bodily areas exhibiting a high degree of flexibility and coordination, such as the tongue, face, and hands.

Here again, localization is not precise. A second stimulation of the same point may result in a greater, lesser, or somewhat different pattern of movements. Highly integrated movements also depend upon the cooperation of motor neurons from other areas of the brain. Extrapyramidal pathways function to inhibit effector activity as well as to facilitate

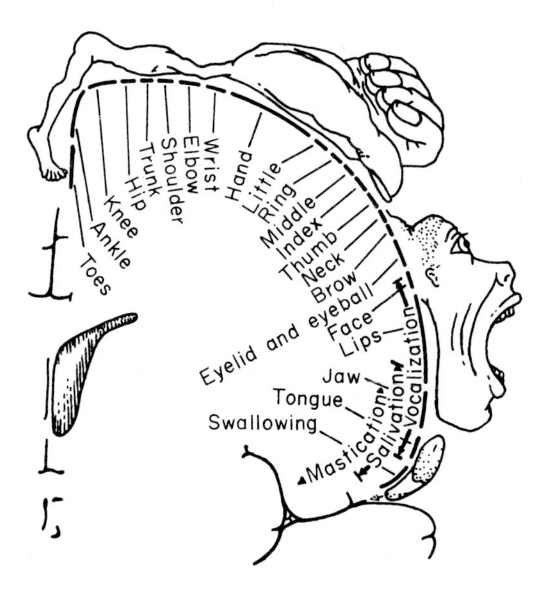

FIG. 2-7. Motor homunculus, similar to the sensory homunculus of Fig. 2-6. As in the case of the sensory homunculus, the face, hands, and feet have greater cortical representation than other parts of the body. [After Penfield (1958).]

it. Some degree of effector inhibition is always necessary for the smooth execution of complex movements.

Regulation and coordination of sensory-motor functions is accomplished by the cerebellum. It receives collaterals from all the senses and has reciprocal connections with the sensory and motor areas of the cortex, as well as the extrapyramidal system. It also influences the spinal cord directly. In some cases the cerebellum is the end station of spinal cord tracts.

THE AUTONOMIC NERVOUS SYSTEM

The autonomic nervous system functions to control the activity of glands, smooth muscle, and cardiac muscle. The label *autonomic* indicates its relative independence of voluntary control. Visceral afferents from the sympathetic horn of the spinal cord terminate in the autonomic chain ganglia outside the cord. These are illustrated in Fig. 2-8. Some postganglionic fibers leave the autonomic ganglia and return to the spinal cord. Others form connections with other sympathetic ganglia in the chain. Still others go to collateral ganglia near the muscles and glands innervated by the system.

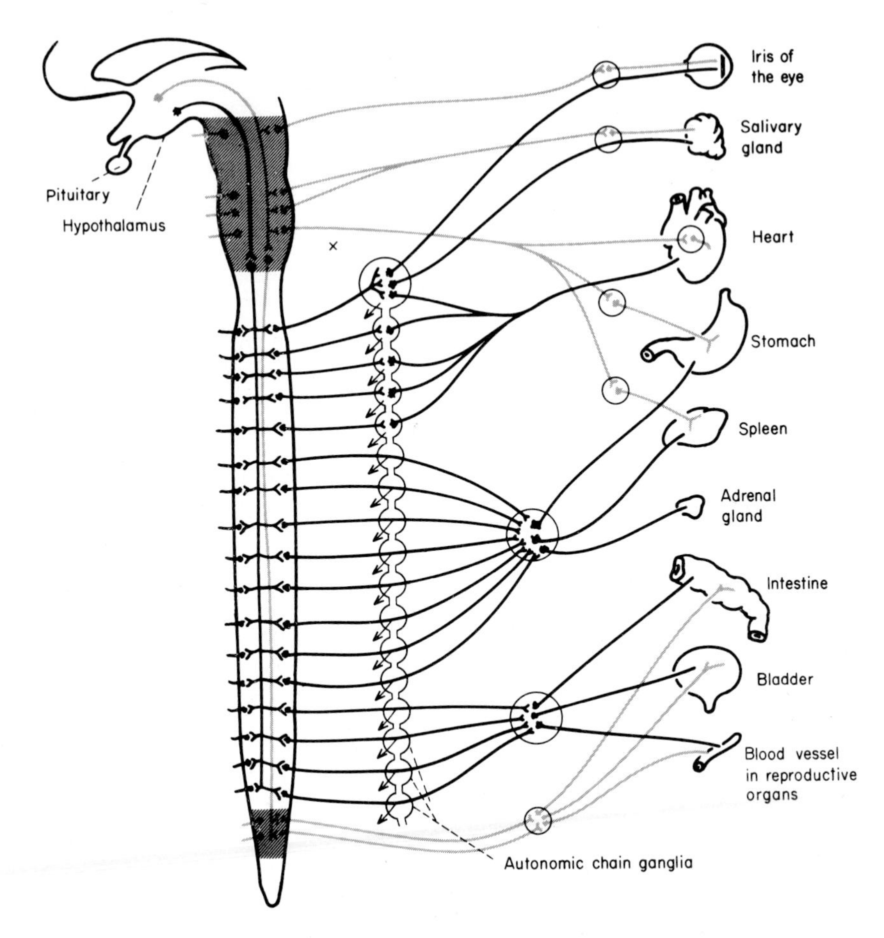

FIG. 2-8. The visceral afferents terminating in the autonomic chain ganglia outside the spinal cord and the organs innervated by the sympathetic and parasympathetic systems. Circles represent autonomic ganglia. Arrows show the course of autonomic fibers to sweat glands, blood vessels, and smooth muscles at roots of hairs. Connections with the sympathetic system are shown in black, those with the parasympathetic system in gray. (Munn, 1962.)

There are two distinct passages of autonomic transmission: the *sympathetic* and *parasympathetic* systems. The sympathetic system is also termed the thoracicolumbar system because its course of innervation is through the thoracic and lumbar areas of the spinal cord. The parasympathetic system could appropriately be called the craniosacral system, because it has its source in the cranial and in the sacral sections of the spinal cord. The parasympathetic system does not include ganglia in the sympathetic chain. Rather, the ganglia of the parasympathetic system are located near the organs innervated by this system. Fig. 2-8 indicates the organs innervated by these two systems.

The sympathetic and parasympathetic systems always act together. Their effects are generally antagonistic, although in many instances they complement each other or work in sequence. The balance of autonomic activity may swing from dominance by one system to dominance by the other. Hence, the two systems interrelate complexly. For example, under conditions producing intense fear, sympathetic innervation inhibits gastrointestinal functions, thereby restricting the normal functioning of the bladder and colon; and causing constipation. However, a temporary surge of compensatory parasympathetic stimulation results in defecation and urination. Similarly, a rapid rise in blood pressure caused by sympathetic stimulation may result in parasympathetic reflexes due to the stimulation of pressure receptors in the carotid sinus (Morgan, 1965).

All sympathetic innervations occur simultaneously. From individual to individual, however, there is considerable variability in the degree of arousal of particular organs. The sympathetic system is often thought of as functioning to prepare the organism for emergency reactions, for "flight" or "fight," whereas the parasympathetic system is thought of as acting to conserve bodily resources.

Because laboratory observations of autonomic activity have been mostly made during intense emotional states, students have a tendency to think of the autonomic system as being functional only under such extreme conditions. Actually, of course, there is a continual background of autonomic activity which changes according to the demands of the environment. Two people cannot meet each other for the first time without widespread changes in autonomic activity occurring. Even slight environmental changes which merely cause the organism to orient its receptors for perceptual clarification may elicit some degree of autonomic responsiveness.

There are individual differences in the degree of general autonomic responsiveness, and in the relative dominance of subsystems of the autonomic system under varying stimulus conditions. The autonomic responses of a given individual may also become habitual and chronic in nature.

The functions of the autonomic system do, however, appear in clearest relief under extreme conditions. Under conditions of extreme stress the sympathetic system effects widespread changes in the gastrointestinal and urinary tracts; the circulatory and respiratory systems; smooth, cardiac, and striated muscles; and in the functioning of certain organs and glands (primarily the salivary glands, sweat glands, and discharges from the adrenal medula and the liver). Sympathetic innervation causes the heart to beat faster and more strongly, and vasoconstriction to occur in the stomach and intestines. Consequently, there is a rise in blood pressure. This results in an increase in perspiration and inhibition of salivary secretion, gastric secretion, and peristaltic movement of the intestine. The muscles at the base of hair follicles are contracted; pyramidal and extrapyramidal pathways are affected; respiratory changes occur; the pupils of the eyes dilate; electrical resistance of the skin is altered; and blood sugar is released from the liver. Among the glandular changes the most important involves discharges from the adrenal medulla.

Generally, parasympathetic innervation functions to oppose the effects of the sympathetic system. It slows down heartbeat, reduces blood pressure, and produces visceral responses typical of rest and relaxation. The parasympathetic system does not have autonomic chain ganglia connections. Therefore the innervation of one organ can be independent of the innervation of another. Presumably, when parasympathetic innervation is dominant the organism is relaxed. Its functioning appears to be related to pleasant emotional states. Unfortunately, the more precise mechanisms of autonomic functioning during pleasurable emotional states and their extreme manifestations, such as those popularly referred to as delight and love, have not been as readily approachable as those involved in unpleasant emotional states like anger and fear. The problem of the physiological differentiation of psychologically different emotional states has, itself, been a very recalcitrant one. It will undoubtedly continue to be a difficult research area, and the yield will be gradual.

THE ENDOCRINE GLANDS

The glands of internal secretion maintain the hormonal balance of the body. The ductless, or endocrine, glands discharge these chemical agents directly into the bloodstream. The term hormone is derived from an ancient Greek word meaning "to arouse," or "to set in motion." Hormones initiate chains of events serving to regulate specific metabolic activities. A schematic diagram showing the location of the various endocrine glands which will be discussed in this section is given in Fig. 2–9.

Body Metabolism

General body metabolism is regulated by a multiplicity of biochemical agents which themselves have more specific effects than have heretofore been realized. It will be recalled that body metabolism involves the breakdown of carbohydrates, fats, and proteins into simpler compounds

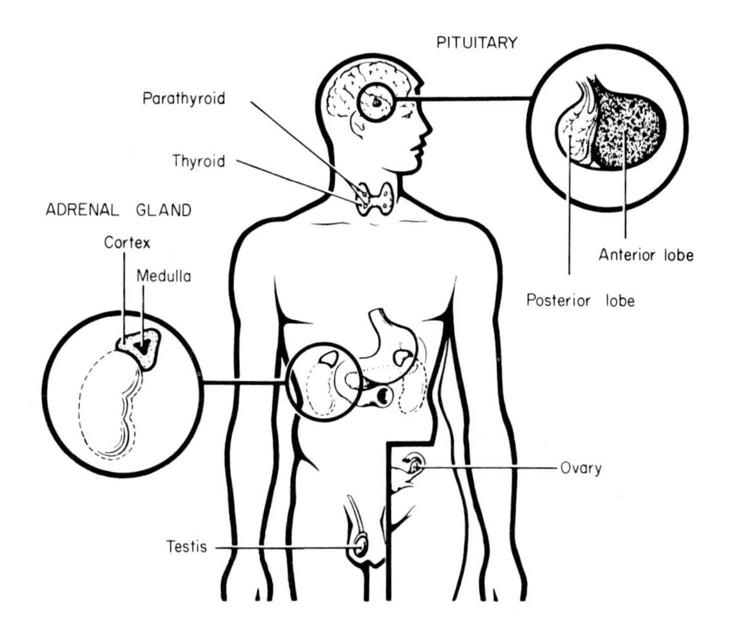

FIG. 2-9. The endocrine glands important for personality and their location. (Munn, 1962, p. 240.)

which are utilized in chemical reactions as the source of bodily energy. This process involves thousands of intricate phases, mediated by the action of enzymes which function as catalyzing agents. The action of a given enzyme is affected by factors such as amount of concentration, pH medium (degree of acidity or alkalinity), temperature, and radiant energy. Morgan (1965) has pointed out that the genes frequently transmit hereditary characteristics by affecting enzyme systems. Thus, an inherited mental defect known as phenylpyruvic oligophrenia is caused by a lack of a liver enzyme which functions to metabolize phenylalanine, an amino acid. This lack is, in turn, due to a single mutant gene.

The major source of body energy is carbohydrate metabolism. The starches and complex sugars in food are broken down into glucose, a simple sugar, which circulates in the bloodstream to the various organs of the body where it is used. Sugar may also be stored in the liver or, in small quantities in muscle. Normally, the concentration of blood sugar remains relatively stable between 60 and 90 milligrams per 100 cubic centimeters of whole blood. When the blood-sugar level is depressed (hypoglycemia) cells are deprived of the needed carbohydrates. Unlike other tissues of the body, the nervous system requires a constant supply of glucose. Consequently, if the depression of blood sugar remains below the lower limit of normal concentration, the organism will go into a coma and die. An elevation of the blood sugar above the upper limit of normal concentration (hyperglycemia) gives rise to diabetes. In this disease widespread metabolic malfunctioning occurs, as well as psychological symptoms which range from mild anxiety and depression to severe states of personality disorganization.

The hormone insulin, which is secreted by the islets of Langerhans in the pancreas, regulates blood-sugar metabolism. A deficiency of insulin causes diabetes. The deficiency can be made up by insulin injections. A large dose of insulin initiates convulsions and a coma and if the dosage is large enough, death ensues. Intense shock reactions resulting from overdosages of insulin have been found to have some psychotherapeutic value. Insulin shock therapy appears to be useful in helping to reduce the symptoms of schizophrenia, at least during the early stages. However, beneficial effects of insulin shock therapy are seldom long-lasting, although they can be prolonged when other forms of psychotherapy are employed immediately afterward.

The Adrenal Glands

The adrenal glands are located just above the kidneys. Each consists of an inner layer of cells—the *adrenal medulla*—and an outer layer of cells—the *adrenal cortex.* Cells of the adrenal medulla are genetically derived from the same tissue as the ganglia and nerves of the sympathetic nervous system. Two hormones are secreted by the adrenal medulla: *epinephrine* and *norepinephrine.* These hormones are also referred to as *adrenaline* and *noradrenaline.* The secretion of both epinephrine and norepinephrine raises blood pressure. Epinephrine raises blood pressure by increasing heart rate, whereas norepinephrine increases blood pressure by causing vasoconstriction.

Both of these hormones also effect the release of glycogen from the liver, and hence their secretion raises blood-sugar level. In this latter function epinephrine causes a greater effect than norepinephrine does. The effects of these two hormones in the so-called emergency reactions, or in strong emotion, are discussed (pp. 45–46) in connection with the functioning of the autonomic nervous system. In contrast to the adrenal cortex, the adrenal medulla is not absolutely essential to the life of the organism.

Over 30 chemically related hormones are secreted by the adrenal cortex. They belong to the chemical family known as steroids. Most of these hormones are not biologically active. The three that are most active are cortisol, corticosterone, and aldosterone. Each of these hormones performs some of the functions of the other. These hormones are active, first, in maintaining the sodium–potassium balance of the body, and, second, in carbohydrate metabolism. Cortisol functions mostly in carbohydrate metabolism, whereas aldosterone functions primarily for the maintenance of the bodily sodium–potassium balance.

Conditions of stress such as extreme physical exertion, extremes of temperature, starvation, and infectious states increase the secretion of corticosterone. In many cases of certain eye and skin infections, acute rheumatic fever, and rheumatoid arthritis a treatment may be effected by the injection of corticosterone. The psychological effects of corticosterone sometimes involve hyperactivity and insomnia, but the most common symptom is an exaggerated sense of well-being. In Addison's disease, which involves the destruction of the adrenal glands, the converse symptoms of exhaustion and depression occur. Corticosterone also functions to regulate the balance of carbohydrates and, along with cortisol, the deposition of glycogen in the liver.

Aldosterone regulates the retention of sodium salts and their excretion by the kidneys. A deficiency of aldosterone gives rise to an abnormal salt-hunger. It is possible to prolong the life of an organism whose adrenal cortex has been destroyed by the administration of excessive amounts of sodium chloride.

Some adrenocortical hormones are identical chemically with the sex hormones secreted by the gonads. If the adrenal sex hormones are secreted in excessive amounts in girls before birth the sex organs show a distinct abnormality. They become more male in character. When the secretions of this hormone are excessive in childhood the onset of puberty is accelerated in boys, and characteristics of masculinity occur in girls. Virilism in adult women, involving the growth of a beard, a deepening of the voice, and other secondary male sex characteristics is also a function of the oversecretion of these hormones. As in the case of all of the endocrine glands, there are relatively minor variations from individual to individual in the secretory activity of the adrenal sex hormones. In our culture, at least, cognizance is not taken of minor individual differences in secondary sex characteristics. Of course, characteristics which are associated with the polarities of sexuality, such as voice, gestures, and other behavioral manifestations, may also be the result of learning. Little is known about the effects of minor excesses in the secretion of adrenal sex hormones.

The Thyroid Gland

The thyroid gland consists of two lobes, one on either side of the trachea in the base of the neck. It has a very liberal supply of blood vessels, and consists of many tiny sacs which contain secreting cells. They secrete thyroxin, a chemical substance containing 65 percent iodine. Thyroid deficiency may be due to a lack of iodine in the diet, and can be prevented by adding a little potassium iodide to table salt. Disorders of the thyroid gland are common in certain areas of the world where the water is deficient in iodine, such as Switzerland and the Great Lakes region of the United States. The general function of the thyroid gland is to regulate metabolism. It affects the cellular functioning in protein synthesis and carbohydrate utilization (Weiss & Sokoloff, 1963).

Oversecretion of the thyroid gland results in an increase in metabolic activity and is accompanied by restlessness and nervousness. An underactive thyroid lowers the metabolic rate, produces a sluggishness of behavior, and has a general debilitating effect. In the adult organism a deficiency of thyroid secretion results in the condition known as myxedema. In cases of myxedema, heart rate, circulation, body temperature, and general activity are below normal. The individual's skin becomes dry and bloated and generally there is a marked trend toward obesity. Severe cases may also display schizophrenic-like reactions involving withdrawal and bizarre conduct. In mild cases general neurasthenia, chronic fatigue, and a lack of vitality may result. These latter symptoms, of course, may also result from psychological factors.

When these symptoms are a function of a hypoactive thyroid condition the patient may be treated successfully by the administration of thyroxin. A fairly accurate diagnosis of whether the symptoms are functional or organic may be secured by means of a basal metabolism test.

Hyperthyroidism is associated with restlessness, apparent anxiety, excitability, and insomnia. These symptoms may also be easily misinterpreted as having a nonorganic origin. Loss of weight commonly results from excessive secretion of thyroxin because of the increase in bodily metabolism. The condition may be treated either by surgical removal of part of the gland, or by antithyroid drugs which reduce the secretion of thyroxin. In other cases, thyroid abnormality is unmistakable because of an enlargement of the thyroid gland—called a goiter. A simple colloid goiter does not result in a concomitant secretion of thyroxin. The toxic, or "exothalamic," goiter involves a protrusion of the eyeballs and an excessive thyroxin secretion.

A thyroid deficiency in the young or growing organism has more drastic effects. It produces a condition known as *cretinism.* The organism does not mature sexually, and mental retardation is usually present. The condition results in a distinctive appearance involving a protruding stomach, short arms and legs, a puffy face, dry skin, and scanty hair. When cretinism occurs in very young children it may be the result of an iodine-deficient diet of the mother during pregnancy. If it is treated early enough by the administration of thyroid extracts the condition may be corrected entirely.

The anterior part of the pituitary gland secretes a thyrotropic hormone (TTH) which controls the production of thyroxin. As the concentration of thyroxin in the blood increases, the production of thyrotropic hormone of the anterior pituitary is reduced. This is only one of the many interrelationships among the endocrine glands. The endocrine system involves extremely complex interactions. For example, thyroid deficiency may result from defective secretion of the pituitary hormone. Similarly, an overactive thyroid may compensate for undersecretion of sex hormones.

The Pituitary Gland

The pituitary (or hypophysis) is located at the base of the brain underneath the hypothalamus to which it is connected by a stalk known as the infundibulum. It is one of the most complicated and may be one of the most important of the endocrine glands. The gland may be divided into two parts: the anterior pituitary [adenohypophysis] and the posterior pituitary [neurohypophysis]. The anterior part is not connected with the brain, nor does it have innervation. The posterior part, on the other hand, is neurally connected with the hypothalamus.

The adenohypophysis secretes six known hormones. Many of these are trophic hormones which stimulate the activity of other endocrine glands. Three of these hormones are known as gonadotrophic hormones. They function to stimulate the activity of the gonads, especially the ovaries of the female. The first of these is the follicle-stimulating hormone (FSH). It brings the follicle to maturity. The second is a luteinizing hormone (LH). It stimulates the rupture of the ovarian follicle for the release of its ovum, and the production of estrogenic hormones and the corpus luteum. The third of the gonadotrophic hormones is the lactogenic hormone (prolactin) which stimulates the production of milk in the mammilary glands.

One of the other three hormones of the anterior pituitary has already been mentioned in connection with the production of thyroxin (see above). A fifth trophic hormone of the anterior pituitary is the adrenocorticotrophic hormone (ACTH), which stimulates the adrenal cortex to produce steroids. The sixth is somatotrophin. It is perhaps better known because of its dramatic effects on growth. Somatotrophin secretion is partly under the control of blood sugar (Roth, Glick, Yalow, & Berson, 1963). A deficiency of somatotrophin production in early life causes a common type of dwarfism, whereby the affected individual appears like a miniature adult and possesses normal intellectual capacities. Excessive secretion of somatotrophin causes giantism, involving excessive growth of the long bones of the body. If an excessive secretion occurs after maturity the disease of acromegaly results. In this disease the bones having lower calcium content such as those of the chin, nose, hands, and feet show abnormal growth which frequently results in the individual's having a grotesque appearance.

The posterior pituitary contains a substance called pituitrin. Pituitrin has the effect of raising blood pressure by constriction of blood vessels leading from the arteries. It also stimulates smooth muscles, especially those of the uterus. Moreover, it functions to inhibit the excretion of urine through the kidneys.

A disease known as Froehlich's disease results from a pituitary deficiency which probably also involves the hypothalamus. Froehlich's disease is common in boys. The characteristics of the disease involve a lack of normal bone development in conjunction with obesity, and a lack of sexual maturity. Victims of this disease have a general lack of endurance, behave sluggishly, may be subject to abnormal sleepiness, and have generally very passive characters. Because such boys are frequently ridiculed and persecuted by their peers the development of various personality defenses is common. Their very appearance and behavior may predispose them, as it were, to the development of emotional disturbance.

The Gonads

Hormones secreted by the ovaries in the female and the testes in the male are collectively referred to as the gonadal hormones or sex hormones. It has already been noted that some of these hormones are secreted by the adrenal cortex and that some of the pituitary hormones are gonadotrophic. During pregnancy they may also be secreted by the placenta. The gonadal hormones may be classed as progestins, androgens, and estrogens. One of the androgen class, testosterone, is a male sex hormone secreted by interstitial cells of the testes which are embedded in tissues supporting the cells that produce reproductive spermatozoa. Testicular atrophy or castration before puberty inhibits the development of secondary sex characteristics. Thus, there is little or no beard growth and the individual's voice remains high-pitched. In some cases the body tends toward a distinct type of obesity. Generally, the chest and shoulders may remain narrow, and the general characteristics of the adult masculine form do not appear.

Castration in the adult male does not have such drastic effects. The lack of drive and agressiveness, and depression associated with older men whose gonads are going through involution may be in part due to secondary psychological effects. Of course, even following castration some of the hormone is supplied by the adrenal cortex.

Two types of hormones are secreted by the female gonads. The estrogens within the ovarian follicles are absorbed continuously into the bloodstream in small amounts, and are released periodically in large amounts during ovulation. The estrogens function to maintain normal primary and secondary sexual characteristics of the female. After the ovarian follicle is discharged from the ovary periodic secretions of another hormone, progesterone, occur. When pregnancy ensues menstruation is suppressed and the bodily changes essential to pregnancy are activated.

The effects of deficiency of the gonadal hormones in women are not as apparent as in the case of men. Hypogonadal women may show the psychological tendencies of egotisticalness, self-pity, and general resentful attitudes. These may well be secondary psychological effects of the glandular disturbance. When the ovarian hormones are absent from birth a sexless organism results, although the lack of sexual characteristics is not as noticeable as it is in the male. Removal of the ovaries following maturity does not abolish sexual drive, but it may result in a diversity of symptoms as a function of the general hormonal imbalance which may be created.

The Parathyroid Glands

In the vicinity of the thyroid gland there are very small organs which are known as the parathyroid glands. In man they most often appear in two pairs. However, there is some variation in their number and position from individual to individual. The hormone secreted by these glands controls the calcium and phosphate concentration in the blood. The level of blood calcium has direct effects on the excitability of nervous tissue. When the calcium level is depressed nerve tissue becomes more irritable. Parathyroid insufficiency causes minor muscle twitch and a hypersensitivity of the body's reflexes. Convulsions, tetany, and muscle spasms may develop when the parathyroid functioning is markedly impaired.

STRESS

The Emergency Reaction

During fear-arousing situations requiring struggle or escape, profound neural and hormonal changes occur throughout the body. The most important glandular response is the release of epinephrine (adrenaline) and of norepinephrine (noradrenaline) from the medulla of the adrenal gland. Epinephrine is sympatheticomimetic—it produces those sympathetic changes associated with emotional arousal: dryness of the mouth; inhibition of digestive processes; increase in heart rate; constriction of blood vessels; an increase of sweat gland activity; contraction of muscles at the base of hair follicles; respiratory changes; and a change in the electrical resistance of the skin.

Norepinephrine increases the resistance to blood flow by causing contraction of small blood vessels. This adds further increments in blood pressure to the elevation caused by the epinephrine-produced increase in cardiac output. Both epinephrine and norepinephrine also directly affect the release of glycogen from the liver, thus raising the blood-sugar level. This is especially true of epinephrine.

It is impossible to make a neat separation between the neural and hormonal aspects of the emergency reaction. Although the discharge of epinephrine and of norepinephrine from the adrenal medulla is neurally initiated, neural transmission itself is chemical in nature, involving the selective secretion of chemical transmitters at synaptic junctures (see pp. 28–29). We have seen that epinephrine and norepinephrine are discharged directly into the bloodstream from the adrenal medulla, and that they effect complex responses involving primarily the sympathetic branch of the autonomic nervous system. The increased secretion of neural transmitters is a second event, independent of this.

Synapses employing acetylcholine are called *cholinergic*. Those employing norepinephrine are called *adrenergic*. Adrenergic transmission is characteristic of the sympathetic branch of the autonomic nervous system. The primary chemical transmitter of the parasympathetic branch of the autonomic nervous system, and of the autonomic chain ganglia, is acetylcholine.

During the emergency reaction the sympathetic branch of the autonomic nervous system is stimulated, the blood supply to the brain and skeletal muscles is increased, and the skeletal musculature is activated. This latter aspect of the emergency reaction is primarily a cholinergic reaction of the autonomic nervous system, involving an increased secretion of acetylcholine at the juncture of motor axons and striated muscles. This activation appears to be responsible for the generalized inhibition that may accompany the fear response.

Tonic immobility sometimes occurs as a natural response to extremely noxious, painful, or frightening stimulation. It may also be learned in response to stimuli that are not innately fear-producing. The human being is able to anticipate the consequences of his behavior on a symbolic level. Through learning, he may become afraid merely by having thoughts that anticipate frightening or unpleasant events.

The Local Adaptation Syndrome

Some stress-producing stimuli have quite circumscribed physiological effects. Tissue damage resulting from lacerations, burns, excessive acidity, etc., belong in this category; so does tissue damage produced by the invasion of an infectious agent.

Local tissue injury produces inflammation. More precisely, it produces an inflammatory pattern involving swelling, pain, heat, and reddening. This inflammatory pattern has come to be known as the *local adaptation syndrome.* The connective tissue cells and fibers multiply rapidly in response to irritation or infection. They form a thick protective barrier that serves to prevent irritants from entering the blood. The adjacent blood vessels become dilated, and blood cells and other chemical substances emigrate toward the irritant. This is what produces reddening and heat. Swelling is produced by the proliferation of the connective tissue and by the cells and fluids of the dilated blood vessels entering the surrounding tissue.

The General Adaptation Syndrome

Perhaps the major contributor to work in stress has been Hans Selye. On the basis of numerous studies carried out at his laboratory at the University of Montreal, and on the basis of his lifetime work on the topic of stress, Selye has outlined a three-phase sequence of responses to stresses that endure for a relatively long time, such as heat, cold, fatigue, chronic tension, worry, and frustration (Selye, 1960). It has come to be known as the *general adaptation syndrome* (GAS). Experiments supporting the occurrence of this three-phase sequence have employed stressors ranging from cold to forced muscular work and frustration.

In Fig. 2-10 the straight horizontal line represents the level of normal resistance. The downward dip of the curve represents the first phase of the GAS, the *alarm reaction* (AR). This is essentially synonymous with the emergency reaction. The difference is that the AR includes effects that occur over a longer period of time. Adrenalin circulating in the bloodstream stimulates the activity of the sympathetic branch of the

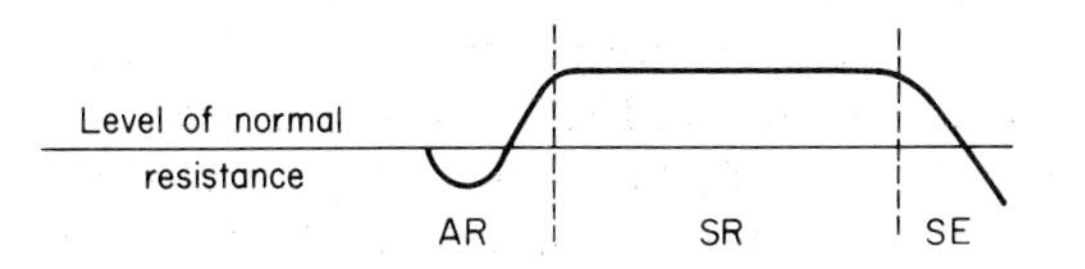

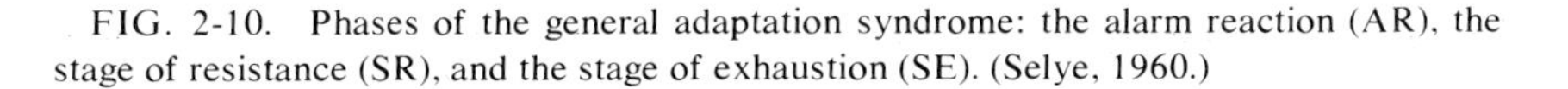
FIG. 2-10. Phases of the general adaptation syndrome: the alarm reaction (AR), the stage of resistance (SR), and the stage of exhaustion (SE). (Selye, 1960.)

autonomic nervous system, and the selective secretion of acetylcholine and norepinephrine occurs throughout the nervous system.

Stressors like heat, cold, fatigue, and frustration activate the pituitary to produce ACTH during the AR phase. This has a peculiar effect: any inflammation present in the body due to previous local stress tends to be cleared up. This is because the adrenal cortex, stimulated by ACTH, secretes excesses of both antiinflammatory and proinflammatory hormones. When both are present in the blood the antiinflammatory hormones always have the predominant effect.

Another hormonal change that occurs during the AR is the increased production of the thyrotropic hormones, which are among the most potent accelerators of chemical reactions in the body. They stimulate metabolism as a whole.

In the AR phase, the lymphatic cells tend to disintegrate in the thymus, the lymph nodes, and in the circulating blood. At the same time eosinophil cells tend to disappear from circulation.

The second phase of the general adaptation syndrome is the *stage of resistance,* labeled SR in Fig. 2-10. During this phase, in contrast to the alarm reaction phase, blood sugar increases and the adrenal cortex becomes laden with an unusually large number of fat droplets. The organism becomes adapted to the stressor. During the SR phase even sharp increases in the level of stress can be withstood.

Finally, after a long period of time, the adrenal cortex rapidly loses its store of fat droplets and blood sugar is sharply reduced. The organism shows a collapse of its adaptive energy. This last phase is called the *stage of exhaustion* (SE). Resistance in this final stage falls well below *normal* levels.

Specific and Nonspecific Stressor Agents

When ACTH stimulates the adrenal cortex the corticoid content of the bloodstream rises. The original message causing the pituitary to produce ACTH apparently is sent directly from the affected organ(s). In Fig. 2-11 this sequence of events is shown in diagrammatic form, for a specific agent acting upon the organism [Fig. 2-11(a)] and nonspecific agent [Fig. 2-11(b)].

When a specific agent (a burn, laceration, an infectious germ, etc.) acts upon any one bodily target, both a specific reaction and general alarm signals occur. This is shown in Fig. 2-11(a). Each of the four cells of this diagram represents a specific target (muscles, kidney, skin surface, etc.). The specific response of that organ or other bodily target is represented by discharge of the distinct patterns (circles, triangles, etc.) from the lower part of the square affected by the agent. The upper part of each

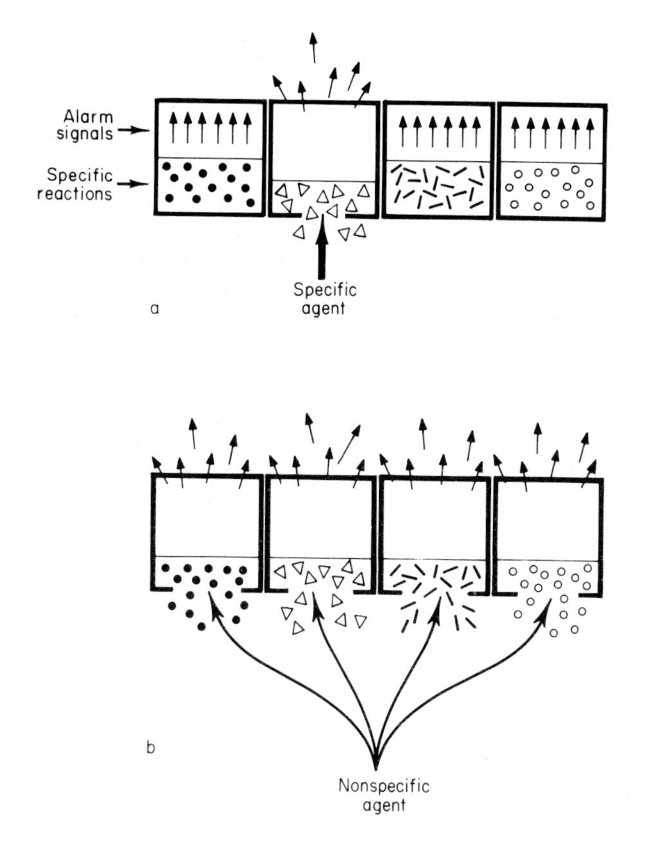

FIG. 2-11. The action of specific (a) and nonspecific (b) stressor agents. The alarm signals are additive, while the specific reactions are not. (Selye, 1960.)

square (the identical little arrows) represents a nonspecific bodily response, the alarm signals.

Figure 2-11(b) shows what happens when a nonspecific agent acts on the body. Suppose a nonspecific agent acts simultaneously upon all four targets. As shown in the lower part of the diagram, it will elicit a specific response in the case of each organ. These specific responses cannot be thought of as additive. On the other hand, the upper part of the diagram indicates how the nonspecific responses—the alarm signals—are discharged from each of the four cells. Since these reactions are identical, they may have a very strong additive effect. Hence, a nonspecific agent, for example, extremes of cold, fatigue, or psychological tension, may produce a much greater stress response than a specific agent acting on any one organ.

Bodily Effects of Alarm Signals

What exactly are the target organs, and just what is the nature of the alarm signals? At the present time we must be satisfied with only a partial answer to these questions. When the stressor is nervous tension, anxiety, frustration, and the like, it is known that virtually every organ and every chemical constituent of the body is involved. We can also trace a great deal of the endocrine and neural activity.

In general, acting through the nervous system, stressors produce adrenalines and they produce acetylcholine. Considerable tension exists as a result of this, because these two types of hormones often function to antagonize each other. At the same time the pituitary has been stimulated to secrete ACTH. Consequently antiinflammatory hormones are released from the adrenal cortex. They destroy white blood cells, influence local inflammation, and may act upon the kidneys so that an increased constriction of blood vessels throughout the body results. The proinflammatory hormones, which are simultaneously stimulated, also appear to have a damaging effect upon the kidneys.

The results of these complex events can be illustrated by a comparison of normal animals with those subjected to a nonspecific psychological stress situation, such as frustration by forceful immobilization. Figure 2-12 shows the adrenals, thymuses, lymph nodes, and inner surfaces of the stomach for a normal and a frustrated rat. The organs of the normal animal are shown on the left, those of the experimental animal, on the right.

In Fig. 2-12A the enlargement and dark discoloration of the adrenals is shown. B and C reveal shrinkage of the thymus and lymph nodes, respectively. D illustrates the numerous blood-covered stomach ulcers in the alarmed rat as compared to the normal animal.

Individual Differences in Response to Stress

Temperamental and maturational differences among individuals that would predispose them to different responses to stress have come under little scrutiny. On the other hand, a large number of experiments with animals have been conducted in recent years on the effects of stimulation in infancy on later adult behavior, particularly adult behavior in response to stress. Hunt and Otis (1955), for example, ran an experiment in which different groups of rat pups were systematically removed from the mother, systematically removed from the mother and shocked, shocked only, and received daily handling only. Although no differences were found among these groups during later testing, all of them differed from a control group of nontreated animals. The experimental animals were found to be much less timid about emerging from their living cages

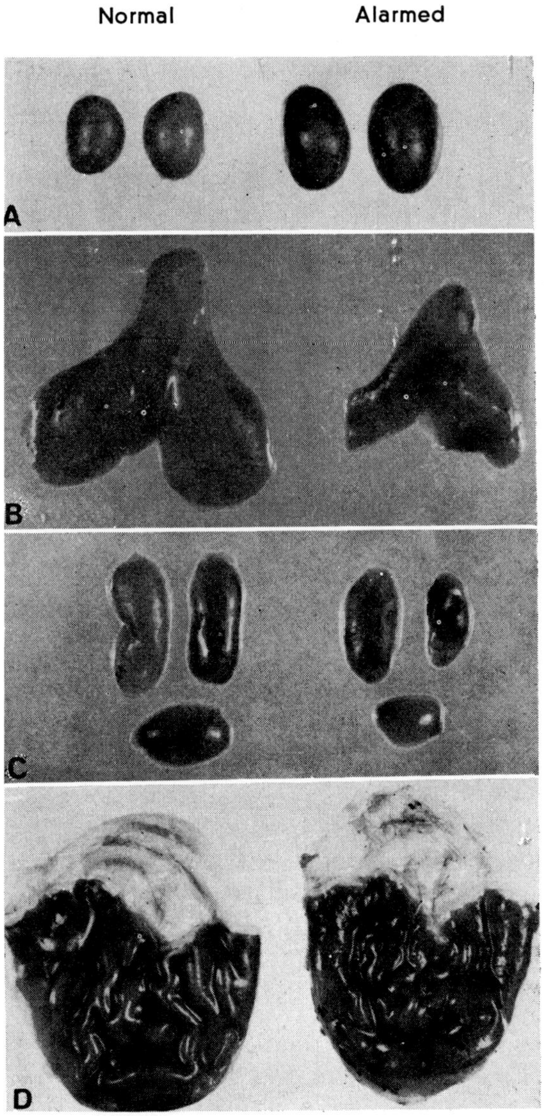

Normal Alarmed

A

B

C

D

FIG. 2-12. A comparison of organs of a normal rat (left) and one exposed to the stress of forceful immobilization (right). A. Adrenals. B. Thymus. C. Lymph nodes. D. Inner surfaces of the stomach. Note the enlargement and dark discoloration of the adrenals, the shrinkage of the thymus and lymph nodes, and the blood-covered stomach ulcerations of the alarmed rat. (After Selye, 1960.)

onto a runway placed in front of their open cage door when they were hungry.

Other experiments by Seymore Levine and his co-workers have shown that stimulated animals differ from normal control animals in terms of their emotional responsiveness (see Levine, 1962). Stimulated animals have been found to be significantly more aggressive as adults. They also perform better than nontreated animals in avoidance-learning situations. Animals lacking a history of early stimulation show a marked tendency to "freeze." Hence, they are much slower at learning to avoid a noxious stimulus.

Figure 2-13 shows the adrenal steroid output of stimulated vs. nonstimulated subjects following adult shock. The stimulated animals are represented by the solid line. Notice that the adrenal steroid output is considerably higher for the stimulated animals throughout. Stimulated animals are more responsive both physiologically and behaviorally to noxious stimulation such as electric shock. Nonstimulated animals do not respond appropriately to noxious stimulation, whereas they are over-responsive to stimulation that objectively appears to involve much less threat. A number of carefully designed experiments have led Levine to conclude that stimulation in infancy results in the capacity for the orga-

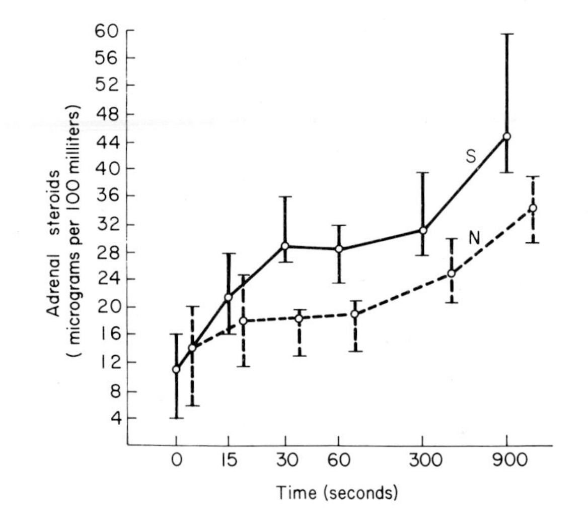

FIG. 2-13. Circulating corticosteroids following electric shock in stimulated (S) and nonstimulated (N) rats. Animals with a history of stimulation show a consistently higher corticosteroid response. (Levine, 1960.)

nism to respond more effectively to its environment: to exhibit an appropriate and motivating emotional response to threat and a diminished emotional response to novel stimuli.

Research in this area is active. At the present time certain factors have been ruled out as relevant variables. Others remain unexplored. Thus, the mere removal of the mother for varying periods of time has been demonstrated to have little or no effect on subsequent emotionality. It has also been found that experimenter contact, or handling, is not a critical variable, nor is shaking the living cage. Constant stimulation with a bright light during infancy produces no differences, and auditory stimulation does not appear to be a relevant variable.

Levine believes that tactile and kinesthetic stimulation in early infancy are crucial to the production of later individual differences in response to stress. However, he has expressed amazement that such relatively mild stimulation, for such a relatively small time period of the animal's life, can have such a potent effect. Certainly, the nontreated animals are not without natural sources of stimulation. What may be involved is not stimulation *per se*, but the dynamic quality of stimulation. It may be that relatively sudden and drastic change from whatever level of stimulation the animal normally has will eventually be found to be crucial.

Experimental Studies of Stress

There have been a variety of demonstrations of the physiological and behavioral effects of stress. Researchers attempting to study stress in the laboratory have been handicapped, however, by the fact that for both practical and ethical reasons human subjects cannot be experimentally exposed to any but the mildest varieties of stressors. Even in instances where the experimenter has had the subject's consent, it has been quite difficult to control the experiment, for the subject still knows that he is in an experiment; that he will not be harmed in any way; and that he is expected to act scared, frustrated, exhausted, or whatever.

This problem was essentially bypassed in a series of experiments reported by Berkun, Bialek, Kern, and Yagi (1962). They exposed military personnel—Army Basic Trainees and experienced soldiers—to situations that were, from the point of the view of the subjects, indistinguishible from actual danger or threat.

In one situation the subjects were aboard a plane on a routine shuttle flight when they heard one of the engines sputter and stop, felt the plane buck, and were given immediate ditching instructions. In another series subjects were taken to an isolated outpost and then their communications were broken as a forest fire in one instance, and artillery shells in another, were directed toward their position. Still another type of situation, apparently the most stressful of all, involved the subject being led to believe that he was responsible for a serious accident, and when he attempted to call for help the communications went dead.

A number of measures were used to evaluate the stress response. As one physiological index of stress the output of urinary 17-hydroxycorticosteroids (17-OH-CS) was used. The evidence is that this steroid group is sensitive to stress effects, and changes in the 17-OH-CS level can be used to define the presence of the stress response.

A second physiological measure was an eosinophil count. It will be recalled that the number of circulating eosinophil cells is reduced with the introduction of stress, but after prolonged stress the count increases. Further fluctuations may occur over a long period of time.

A subjective self-report measure called the subjective stress scale was used to assess the subjects' feelings during stress. It is a scale composed of 15 words. The words range from "wonderful," "fine," . . . through "indifferent," . . . "nervous" . . . to "in agony," and "scared stiff." The subjects were asked to estimate their feelings, which were then converted into a scale score value.

Berkun *et al.* found that the output of 17-OH-CS and the eosinophil counts were sensitive to the presence of stress. When the various situations were ranked in terms of the mean subjective stress scale scores, a systematic relationship appeared: with increasing stress the average number of urinary corticosteroids increased, and then decreased. At the extremes of subjectively felt stress, the corticosteroid level decreased. Berkun *et al.* inferred that under severe stress an initially very high rate of steroid output could not be maintained, with a resulting low overall average rate reflected in their measure of it. On the other hand, slightly less intense stress appeared to produce a rate which was not as high initially, but was capable of being sustained so that it was reflected in the average overall rate measure. These findings with human beings in simulated real-threat situations tend to support and extend the data from the animal laboratory. Incidently, Berkun *et al.* also found that although lesser degrees of stress enhance performance, performance begins to break down as the subjective experience of stress increases.

Stress and Relaxation

All of us have had the experience of being "keyed-up" from "nervous tension." Some of us may experience a slight amount of tension as pleasant, because we know it allows us to function at our best. However, too much excitement is usually experienced as being unpleasant. Under its influence we may behave in thoughtless ways, and find ourselves unable to be well organized in our daily activities. The sensation of being keyed-up is a very real one. It has physiochemical correlates, primarily the adrenalines and corticoids. The feeling can be reproduced by the artificial injection of these substances into the bloodstream.

Just as there are wide individual differences in the body's physiochemical reactions, there are wide individual differences in the kinds of events that will reduce tension. Many people tend to be helplessly influenced by their own stressful activities. They live in a chronic state of frustration. Others have learned to find recreation tension-reducing, whether it be reading, travel, or sports. On the other hand, for some people rest would increase, rather than decrease, tension. Some find that plunging themselves into their work is actually relaxing for them.

At the present time little is known about the physiochemical correlates of states of relaxation. However, there are some data relevant to this. Handlon (1962) conducted a study in which he used the 17-OH-CS count as a measure of arousal; he also included a relaxation, or bland condition. He showed two kinds of motion pictures to normal subjects. One group of films was selected for their arousal value. The films shown in this condition were *Ox-Bow Incident, High Noon,* and *A Walk in the Sun.* Those of the bland condition were a series of nature-study films by Walt Disney. The subjects were asked to make self-ratings of their emotional mood as it changed during the pictures.

The results of this experiment are shown in graphic form in Fig. 2-14. The mean changes in self-ratings of aggression, depression, and anxiety are plotted on the ordinate at the right of the figure. The mean changes in plasma hydrocortisone count are plotted on the left-hand ordinate. The effects of the arousing films are shown by the white bars. Those of the bland film condition are shown by lined bars. The results indicate unmistakably that the bland movies produced marked lowering of the 17-OH-CS count. They were more effective in lowering the 17-OH-CS count than the arousing films were in raising it. As in the case of previous data we have discussed, a high correlation between the subject's appraisal of his emotional mood and the 17-OH-CS count was evidenced.

Stress and Psychosomatic Disorders

We have seen that many stressors are not biological, or even physical. Stress may be produced by many events that do not, in fact, represent any real danger, aside from the harm they cause because the body must mobilize its energies to resist stress. Much stress is purely psychological in origin. The same events may cause stress-producing anxiety, worry, tension, and frustration for one individual and not for another.

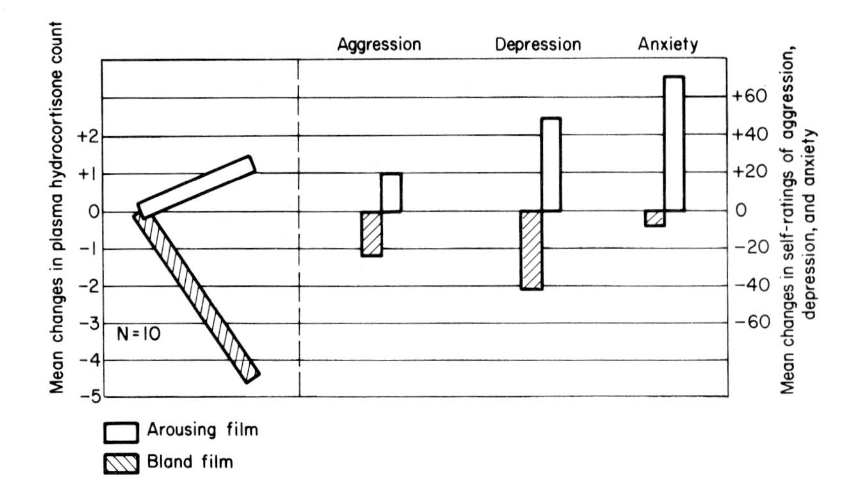

FIG. 2-14. Biological and psychological changes following exposure to "bland" and "arousing" movies. (Adapted from Handlon, 1962.)

Some bodily disorders can be shown to be related to events in the lives of individuals that do not innately produce pathology. The individual's inherited bodily endowment may make some contribution, but the reason some people develop bodily disorders in response to psychological events may also have to do with their life experiences. These are called *psychosomatic disorders*. They will be discussed more fully in Chapter 10.

At this point, however, let us note how certain psychogenic physical pathology has been produced in the laboratory by the arrangement of special conditions of stress. Porter, Brady, Conrad, Mason, Galambos, and Rioch (1958) restrained experimental monkeys in a chair and subjected them to a series of conditioning procedures. They had to push a bar periodically in order to avoid punishing shock. Control monkeys received the same number of shocks that the experimental monkeys did, but they did not have to anticipate, and do something about avoiding the shock. The experimental animals were required to perform for a 6-hour period, alternating with 6 hours of rest. Figure 2-15 shows this experimental arrangement. An experimental monkey, sometimes referred to as the executive monkey, is shown on the left; his control partner on the right. The control monkey had the same avoidance lever that his experimental partner did, but since it had no function and there was nothing the control monkeys could do to avoid the shock themselves, it was soon ignored. Eleven of the 19 executive monkeys used in the study died or became moribund and had to be killed after 2 to 17 weeks. All of them showed gastrointestinal pathology such as hemorrhaging and erosion of the stomach lining, duodenal ulceration, and colitis. There were no deaths among the control monkeys who were shocked equally as often—nor did they show gastrointestinal pathology. Similar results have been found by other investigators employing severe conflict (food and water vs. shock) (e.g., Sawrey, Conger & Turrell, 1956; Sawrey & Weiss, 1956).

SUMMARY

The structure and functioning of the central and autonomic nervous systems, and of the endocrine system have been discussed. Following this review of the fundamentals of biology and endocrine functioning,

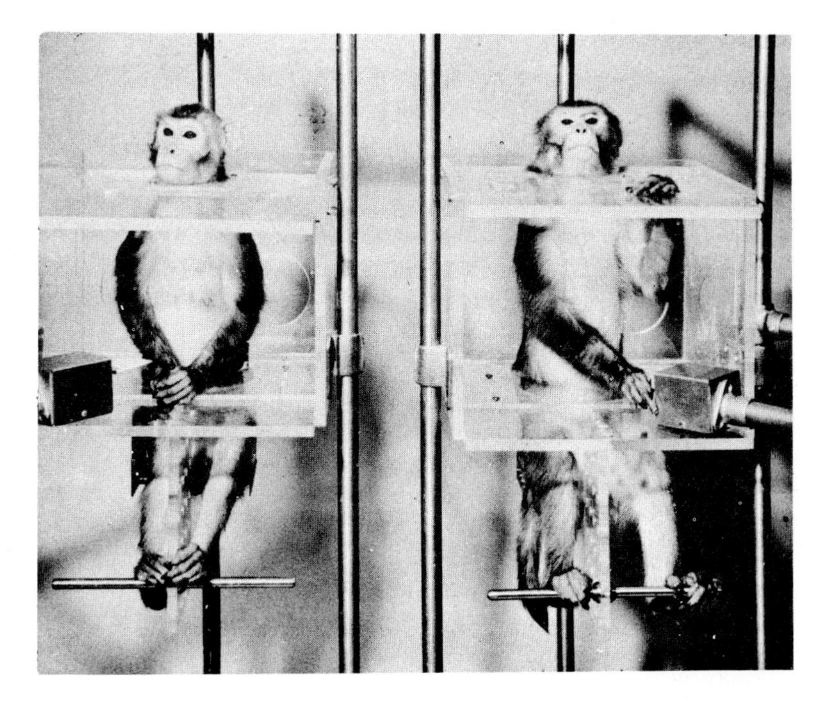

FIG. 2-15. Experimental situation employed for the production of ulcers in the executive monkey. The monkey at the left could postpone an electrical shock for itself and its partner by pressing a lever. The lever of the monkey at the right had no function and was ignored. (Courtesy of Dr. J. V. Brady, and the Walter Reed Army Institute of Research, Washington, D.C.)

various aspects of the topic of stress are considered, along with some relevant research.

The fundamental unit of all living organisms is the cell. The nucleus of the cell contains giant DNA molecules that control the metabolic processes in the surrounding cytoplasm and carry the genetic code for the reproduction of the cell. Polarization may occur across the membrane that marks the outer boundary of the cell. Cells are highly specialized within receptor, adjustor neuron, and effector structures. Although the body's adjustive machinery is exceedingly complex, in principle it is quite simple. By means of neurochemical transmission, sensory messages are carried from receptors into the central nervous system, which consists of the brain and spinal cord. They may either be translated directly into an effector impulse, reach effectors over connector neuron pathways, or be sent to the brain for interpretation. There is considerable localization in the brain. How an impulse is interpreted depends upon the projection pathway followed by it. After messages are received in the appropriate projection areas, motor impulses may be initiated from the great motor areas of the brain. The cerebellum serves to coordinate and integrate these impulses, which then activate specific groups of effectors, causing an adjustive response.

Control of glands, smooth muscle, and cardiac muscle is exerted by the autonomic nervous system. It also exerts an influence over many of the major organs of the body. In general, the sympathetic branch of the autonomic nervous system functions to prepare the body for emergency reactions, whereas the parasympathetic branch acts to conserve bodily resources.

The endocrine glands maintain the hormonal balance of the body. Body metabolism involves thousands of intricate processes. Blood-sugar metabolism is regulated by insulin. A decrease of the blood-sugar level results in a coma and possibly in death; an increase results in widespread metabolic disturbance, anxiety, depression, and personality disorganization. Epinephrine and norepinephrine are secreted by the adrenal medulla. They raise blood pressure and raise blood-sugar level. The adrenal cortex produces steroids that have many functions, including regulation of bodily sodium–potassium balance and carbohydrate metabolism. An overproduction of these hormones during stress reduces local infection, produces an exaggerated sense of well-being, and sometimes, hyperactivity. The thyroid gland also affects behavior. A thyroid deficiency produces sluggishness; oversecretion produces restlessness.

The anterior pituitary produces several hormones that stimulate other glands and organs: the gonadotropic hormone group, the adrenocorticotropic hormone (ACTH), and somatotropin, among others. These stimulate sexual functions, the production of steroids from the adrenal cortex, and growth-regulating mechanisms. The posterior pituitary secretes pituitrin. It causes constriction of blood vessels and inhibits urination.

The parathyroid glands secrete parathormone, which controls the calcium concentration of the blood. A decreased calcium concentration results in nerve tissue irritability. Among the sex hormones, produced by the gonads, are the androgens, progestins, and estrogens. Testosterone stimulates the development of secondary sex characteristics in the male and affects drive and aggressiveness. The estrogens maintain normal primary and secondary sexual characteristics in the female.

Prolonged extreme exposure to painful stimulation, infection, heat and cold, fatigue, frustration, anxiety, and many other conditions produces a syndrome of bodily responses referred to as *stress*. Local tissue injury produces a reactive inflammatory pattern of swelling, pain, heat, and reddening. It also induces a neural and hormonal alarm response, part of the overall pattern of stress. When the stressor is nonspecific, the bodily alarm signals may combine to produce an intense stress syndrome. The body's response to such conditions is triphasic. It involves first an alarm phase. During this phase emergency autonomic, adrenal, and cholinergic activity occurs, antiinflammatory hormones are stimulated, blood sugar is lowered, white blood cells are disintegrated, and the body's resistance drops below normal. However, during a well-defined second phase that occurs somewhat later, most of these processes are reversed. The body shows a superior resistance during this phase. Finally, after prolonged exposure to the stressful conditions, the body's adaptive energy collapses. During this final stage of exhaustion resistance falls well below normal levels.

There are wide individual differences in bodily responses to stressors, and in the body's general ability to withstand conditions of stress. Results of both human and animal experiments to date indicate that the amount and quality of stimulation in early infancy may have a profound affect on the ability of individuals to cope with their environments as adults. Experiments with human subjects have substantiated the fact that bodily neurochemical reactions to stress-producing conditions closely parallel those found in laboratory animals. Hormonal and other bodily responses, subjective feelings of stress, and ability to perform complex tasks are all systematically related to the intensity and quality of the actual conditions producing stress. Changes from stress to nonstress are clearly accompanied by changes in body chemistry. Finally, experimental analogs of human psychosomatic ailments in response to stress have been produced using laboratory animals as subjects.

REFERENCES

Berkun, M. M., Bialek, H. M., Kern, R., & Yagi, K. Experimental studies of psychological stress in man. *Psychol. Monog.* 1962, **76**, No. 15 (Whole No. 534).

Goldfarb, W. Infant rearing and problem behavior. *Amer. J. Orthopsychiat.*, 1943, **13**, 249–265.

Handlon, J. H. Hormonal activity and individual responses to the stresses and easements in everyday living. In N. Greenfield & R. Roessler (Eds.), *Physiological correlates of psychological disorder.* Madison, Wis.: Univer. of Wisconsin Press, 1962.

Hunt, H. H., & Otis, L. S. Restricted experience and timidity in the rat. *Amer. Psychologist,* 1955, **10**, 432. (Abstract)

Katz, B., The nerve impulse. *Sci. American*, 1952, Nov., p.61.

Levine, S. The effects of infantile experience on adult behavior. In A. J. Bachrach (Ed.), *Experimental foundations of clinical psychology.* New York: Basic Books, 1962. Pp. 139–169.

Levine, S. Infantile stimulation. *Sci. American,* 1960, **202**, pp. 81–86.

Maher, B. A. *Principles of psychopathology.* New York: McGraw-Hill, 1966.

Morgan, C. T. *Physiological psychology.* New York: McGraw-Hill, 1965.

Munn, N. L. *Introduction to psychology.* Boston: Houghton, 1962.

Penfield, W. *The excitable cortex of man.* Thomas, 1958.

Porter, R. W., Brady, J. V., Conrad, D., Mason, J. W., Galambos, R., & Rioch, D. Some experimental observations on gastro-intestinal lesions in behaviorally conditioned monkeys. *Psychosom. Med.*, 1958, **20**, 379–394.

Rogers, T. *Elementary human physiology.* New York: Wiley, 1961.

Roth, J., Glick, S. M., Yalow, R. S., & Berson, S. A. Hypoglycemia: A potent stimulus to secretion of growth hormone. *Science*, 1963, **140**, 987–988.

Sawrey, W., Conger, J., & Turrell, E. An experimental investigation of the role of psychological factors in the production of gastric ulcers in rats. *J. comp. physiol. Psychol.*, 1956, **49,** 457–461.

Sawrey, W., & Weiss, J. D. An experimental method of producing gastric ulcers. *J. comp. physiol. Psychol.*, 1956, **49**, 269–270.

Selye, H. *The story of the adaptation syndrome.* Montreal: Acta, Inc., 1960.

Weiss, W. P., & Sokoloff, L. Reversal of thyroxine-induced hypermetabolism by puromycin. *Science*, 1963, **140**, 1324–1326.

Chapter 3

Motivation and Emotion

INTRODUCTION

When the body is exposed to certain external conditions, there is an activation of the chemicals of neural transmission and changes in hormone activity, as described in the preceding chapter. The bodily mechanisms that respond to environmental conditions that function as stressors, complex as they may be, are not the only bodily mechanisms serving an arousal function. Psychologists have found it useful to infer a state of bodily arousal on the basis of measurement operations other than just indexes of hormonal activity, and to study the relationship between them and behavior.

It is possible to operationally define points along a continuum of bodily arousal, ranging from deep sleep, through light sleep, wakefulness, and vigilance, to extremes of excitement and tension. Some bodily mechanisms and processes serve a general arousal function. Others have the dual function of contributing to general arousal and giving rise to particular patterns of behavior and experience. Among these latter mechanisms are those related to the arousal of primary drive states and the mechanisms underlying emotional experience.

In this chapter we will discuss general arousal (sometimes called drive), the bodily mechanisms and processes associated with it, and its relation to behavior. We will then turn our attention to the primary drives and to the bodily processes and mechanisms underlying emotions. Because we will have frequent reference to the lower brain structures, and to the upper spinal cord, we will begin our discussion of arousal with certain anatomical features of the nervous system, namely, those structures comprising the limbic system.

THE LIMBIC SYSTEM

A longitudinal cross section of the human brain is shown in Fig. 3-1. The outer cerebral stratum (neocortex) encapsulates a substratum that is more primitive in an evolutionary sense. It is known as the paleocortex, or "old brain." It is sometimes also called the *rhinencephalon.*

Evidence has accumulated in recent years that important mechanisms underlying arousal function as integral parts of a system called the *limbic system.* Most of the structures of the limbic system are located within the paleocortex. They are shown in Fig. 3-1. The reader should familiarize himself with their location so that he can visualize it during our discussions of them in this chapter. The structures comprising the limbic system include the cingulate gyrus, located in the longitudinal fissure between the neocortex and the paleocortex; the septal area, amygdala, and hippocampus within the paleocortex; and the mammillary bodies. These structures are connected with one another by various pathways, or circuits, of neural transmission.

Complete removal of the cerebral cortex is called *decortication.* Experimental animals that have been decorticated are extremely sensitive to environmental stimulation. The slightest touch or pressure applied to the animal results in a full-fledged rage reaction. Also, unlike normal

FIG. 3-1. Side view of the right hemisphere and brain stem. The location of bodies within the limbic system, including the reticular formation.

animals who show an after-discharge in their emotional reactions, decorticated animals tend to recover very quickly.

On the other hand, when animals are selectively decorticated so as to spare the limbic system, they behave quite differently. It is very difficult to arouse an emotional response of any kind. The general picture they give is one of placidity. However, if lesions are made, severing either one of two major circuits within the limbic system of these same animals, they become ferocious and show the extreme sensitivity of totally decorticated animals. Apparently the neocortex inhibits paleocortical arousal unless certain pathways within the limbic system are severed or stimulated.

The Reticular Activating System

Within the paleocortex there is a highly diffuse, yet concentrated mass of neural fibers and cell bodies known as the *reticular formation*, shown in Fig. 3-1. The reticular formation contains both ascending and descending systems. Reticular projections go into the thalamus and down the cord. They are also relayed diffusely to all parts of the cortex.

It is now a commonplace demonstration that animals can be aroused into a state of alertness by mild electrical stimulation of the ascending branch of the reticular formation, whereas stimulation of adjacent areas has no effect (Segundo, Naquet, & Arana, 1955). Consequently, damage to this part of the reticular formation results in the animal becoming unresponsive to environmental stimulation. When the damage is extensive, a coma, and subsequently death, ensue (Field, Magoun, & Hall, 1960). Hence, one of the major functions of the ascending branch of the reticular formation appears to be the maintenance of a normal state of wakefulness. Insofar as this system also appears to mediate specific behavioral arousal, it has been called the *ascending reticular activating system* (ARAS).

There has been some evidence suggesting that the reticular formation may constitute a switching mechanism determining which impulses will, or will not, arrive at the sensory projection areas of the brain. Thus, if an experimental animal is attending to some other stimulus in the environment when, let us say, an auditory stimulus is activated, no impulses occur in the auditory cortex (Hernández-Peón, Sherrer, & Jouvet, 1956; Hubel, Henson, Rupert, & Galambos, 1959). Impulses coming from the cochlear nucleus to the cortex can be shunted by stimulation of certain parts of the reticular formation. The inhibition of incoming sensory impulses under natural circumstances may similarly be the result of reticular activity.

More in line with its functioning as an alerting, rather than an inhibitory mechanism, the ARAS has been shown to enhance learning in experimental monkeys. Fuster (1958) trained monkeys to discriminate between two white three-dimensional objects, different in shape. The choice of one of the objects, the position of which was randomly varied, was always rewarded by food. When a mild electrical stimulus to the ARAS preceded the animal's choice response by a short time, more correct choices were made compared to a control situation. An increase in the speed of response during the stimulation trials also occurred in this experiment.

Finally, the reticular formation appears to be activated by descending fibers from the cortex. This fact is of considerable interest to behavioral scientists. Later we will see that learned drives may be mediated by symbolic associations, words, and other cue-producing responses associated with the higher mental processes. Because of our experiences with the world, certain thoughts may arouse tension and/or vigilance—exactly what would be expected from cortical stimulation of the ARAS.

AROUSAL

Electroencephalography

One operational definition of arousal is given by the technique called electroencephalography. This technique involves attaching electrodes to the scalp of human subjects in order to record fluctuating voltage. When these currents are amplified and connected to a pen that writes on a continuously moving chart, waveforms such as those illustrated in Fig. 3-2 are obtained. Normal subjects in the resting state produce two main frequencies: an alpha rhythm and a beta rhythm. Alpha waves are of a frequency of between 8 and 13 cycles per second. Beta waves range from 14 to 25 cycles per second. Ordinarily, alpha rhythms are found when electrodes are placed in the occipital and parietal lobes of the brain, whereas beta rhythms are found in the frontal and temporal areas. However, generally, in stages of relaxed wakefulness the record is likely to consist of synchronized alpha waves of moderate voltage. Beta waves, on the other hand, are more characteristic of alertness. With excitement the waves become desynchronized, and tend to flatten, or lose the wave shape.

The transition from a state of relaxed wakefulness to a state of drowsiness is accompanied by occasional low-amplitude slow waves, known as delta waves. During sleep the alpha waves disappear entirely and the slow delta waves of 1 to 4 cycles per second predominate and become larger. Spindle bursts of about 14 per second also make an appearance. Finally, during deep sleep the spindle bursts disappear, leaving large,

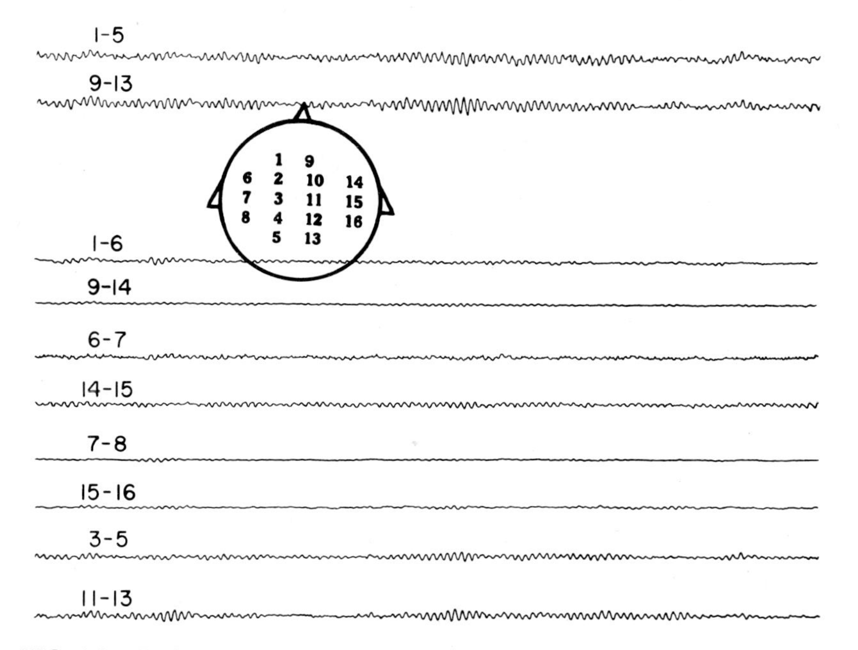

FIG. 3-2. Typical EEG recordings. The numbers in the diagram indicate the placement of the electrodes on the skull. Each tracing is from a pair the electrodes identified by the numbers. (Rogers, 1961.)

slow trains of delta waves. Typical records illustrating these changes from excitement through deep sleep are shown in Fig. 3-3.

The Orienting Response

Any unusual or novel stimulus in the environment, when it is first presented, elicits an attentive response. The actual receptor, skeletal, and postural adjustments vary, of course, with the type of stimulus, the sensory modality, and the circumstances of the moment. This response has been called succinctly the "What is it?" response. Russian psychologists call it the *orienting response.*

When this response occurs the electroencephalograph (EEG) record shows a disruption of the normal alpha rhythm, referred to as an *alpha block.* This desynchronization of the alpha wave is correlated with reticular activity. The occurrence of the alpha block indicates, or defines, arousal.

When a stimulus is repeated a number of times, it loses its power to evoke the alpha block. This process is called *adaptation.* The alpha

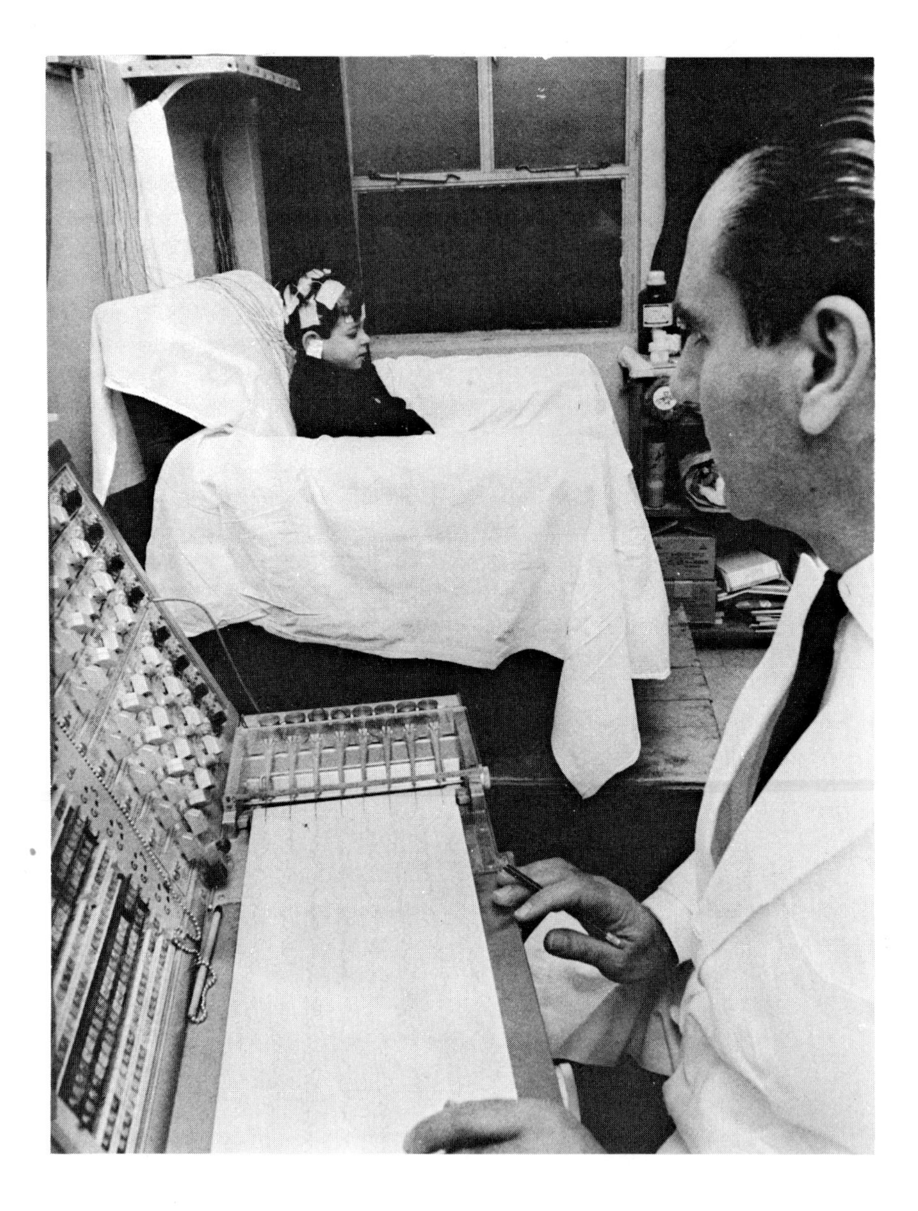

Electroencephalographic (EEG) recordings being obtained from a young patient. (Photograph by Tor Eigland.)

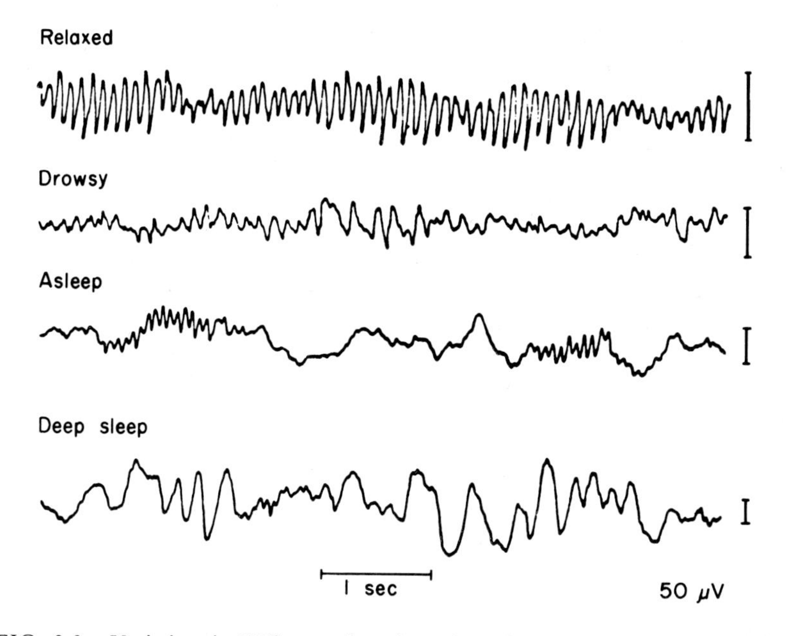

FIG. 3-3. Variations in EEG recordings from deep sleep through excitement. The top tracing is the alpha rhythm. The bottom tracing is the delta rhythm. (Jasper, 1941.)

block can be learned. After repeated pairings of a stimulus that produces the alpha block with one that is distinctive, but does not evoke it, the latter stimulus comes to elicit the alpha block.

Other Indexes of Arousal

We have seen that the desynchronization of the EEG wave can be used as an index of arousal. As excitement increases, more and more complex bodily response systems are activated. We will see, at least briefly, how complex these processes are when we discuss the mechanisms underlying emotions. Many states of activation involve overlapping response systems, with only slight differentiation among them. (See pp. 84–87.) Thus, although there may be slightly different processes and mechanisms involved in different states of arousal, the general activation syndrome invariably involves many of the same response patterns. Because the autonomic nervous system is so often involved in the general arousal pattern, measures of autonomic functioning are frequently used as indexes of arousal. We will discuss some of the more commonly used measures of autonomic functioning separately. In actual practice in the laboratory many different measures are taken simultaneously, giving a polygraphic account of the responses occurring in the subject.

Some measures of autonomic functioning are familiar to almost everyone. Pulse rate and blood pressure are examples. Blood pressure has been found to be particularly useful in conjunction with the mecholyl test (described on pp. 85–86). Respiration rate is another common autonomic index. Simple techniques such as skin temperature, pupillary dilation, and blood volume measures are less well known. Blood volume changes are usually recorded from the finger with a special instrument called a plethysmograph.

Polygraphic recordings usually involve measures of electrical activity in various parts of the body. One of these indexes is the *galvanic skin response* (GSR), sometimes called the *psychogalvanic reflex* (PGR). There is almost invariably some increase in sweat gland activity correlated with increasing sympathetic activation. Sweating lowers electrical resistance on the surface of the skin. Resistance changes can easily be recorded by placing an electrode on the palm of the hand, and putting it in circuit with a galvanometer. Any change from the basal resistance level can easily be measured by a needle deflection or a recording pen. Typical GSR responses to external stimulation are shown in Fig. 3-4.

Muscle tension can also be used as a measure of autonomic arousal. Electrodes placed on the body's surface can record electrical changes produced by the activation of a muscle, or of a muscle group. This is called an *electromyographic* response (EMG). Electromyographic recordings show considerable variability from very low frequencies when the muscle is inactive, to very high frequencies and periodic bursts during activation. These records are called electromyograms. A similar measure is the *electrocardiogram* (EKG). This involves a measure of the electrical activity produced by the contractions of the heart beat. A relatively stable sequential pattern is typically obtained, with periods of

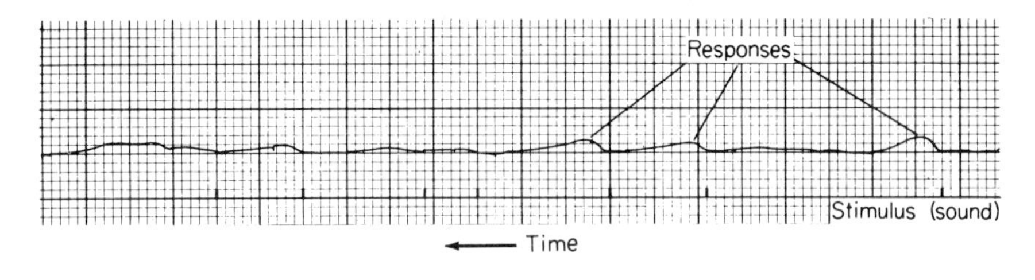

FIG. 3-4. GSR responses of a normal subject to random loud sound stimuli. (Courtesy of Maher, 1966.)

peak activity. On the usual EKG recording one beat of the heart is measured from peak to peak of the wave. This automatically gives a measure of the heart rate in terms of time elapsed, or in terms of distance on the paper between peak responses. A typical EKG record is shown in Fig. 3-5.

It might be noted in passing that a system known as *radiotelemetry* can be used to transmit the impulses from electrodes on the surface of the skin to an amplifier and recording pen, thus eliminating the constraints involved in the use of wires. There are a number of practical limitations on the use of this technique, however.

Having noted that a number of different kinds of indexes of autonomic responsiveness are possible, and that they are often used together in order that the investigator may obtain a total picture of the autonomic activity occurring in the subject, let us turn our attention to the way people actually respond autonomically.

Autonomic Response Specificity

Suppose the psychologist creates conditions in the laboratory that he believes ought to produce arousal, and suppose that he uses various autonomic measures to find how the subjects are affected by arousal. It might be expected that each measure, being an index of autonomic activity, would show a significant change during arousal. This is not what is found. One individual may show a marked increase in heart rate, but no appreciable change in sweat gland activity. Compared on these two measures, another individual may react in exactly the opposite way, showing little increase in heart rate, but a rapid rise in GSR. Laboratory analysis of autonomic responsiveness has led investigators to the conclusion that individuals tend to respond with a *pattern* of autonomic activation. Maximum activation occurs only in certain physiological functions for any given individual. Furthermore, the pattern of autonomic activity appears to be quite stable over time, and tends to occur in response to all stress situations, irrespective of their exact nature. In other words, individuals show relatively stable characteristic autonomic response patterns. This has been referred to as *autonomic response specificity* (Lacey, 1950; Lacey, Baterman, & VanLehn, 1953).

It is instructive to look at data from an actual experiment. Lacey (1950) introduced stress by requiring female subjects to respond as rapidly as possible by naming all the words they could think of beginning with a given letter. This might seem like an extremely simple task. It is not. After only a minute or so subjects begin to become frustrated and embarrassed. Lacey took measures of systolic and diastolic blood pressure, pulse pressure, galvanic skin response, heart rate, and heart rate

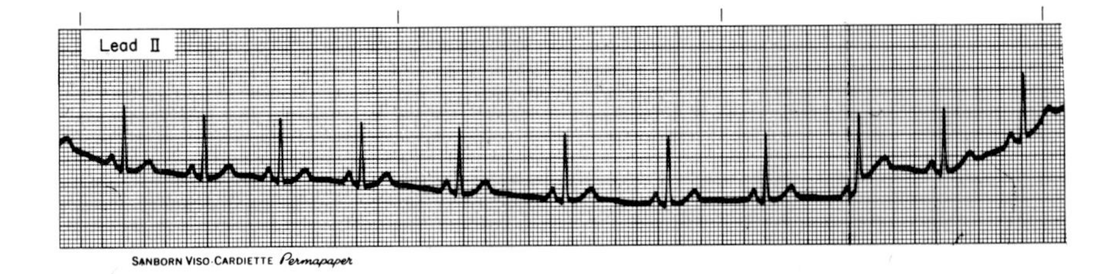

FIG. 3-5. A typical EKG recording.

variability. Figure 3-6 shows the amount of change that occurred in these measures for each individual. Rank, from 1 to 12, is plotted on the ordinate. It indicates how the individual ranked in the amount of change when her score on any given measure was compared with that of the other 12 subjects. The types of autonomic response measures are shown on the abscissas.

Look at Case FY11. She exhibited the greatest amount of change of anyone in heart rate—a rank of 12. However, her systolic blood pressure and diastolic blood pressure both only showed small amounts of change in comparison to the others. Each had a rank of 2. Quite in contrast, subject FN11 showed the most change of anybody on diastolic blood pressure, but ranked only third in the heart rate measure. Inspection of these profiles will give the reader an excellent appreciation of how each individual shows a different pattern of autonomic responsiveness.

There are further complications in the measure of autonomic response functions. There is some evidence that the lower the basal, or initial, value of a physiological system, the greater its response (see Wilder, 1957). Thus, a "ceiling effect" is possible in the measurement of autonomic functions. The individual's response level might be so high that further increases in it could hardly be produced. It appears that this principle does hold for some systems, and not for others. Respiration rate and heart rate, for example, do not appear to function according to this principle, whereas GSR does (Kaelbling, King, Achenbach, Branson, & Pasamanick, 1960).

There is also some evidence that individual differences in the response of different physiological systems may be related to psychiatric classification. Malmo and Shagass (1952) found that patients who were diagnosed as neurotic differed from normal controls in their blood pressure response to stress. The normal subjects as a group showed a rise in

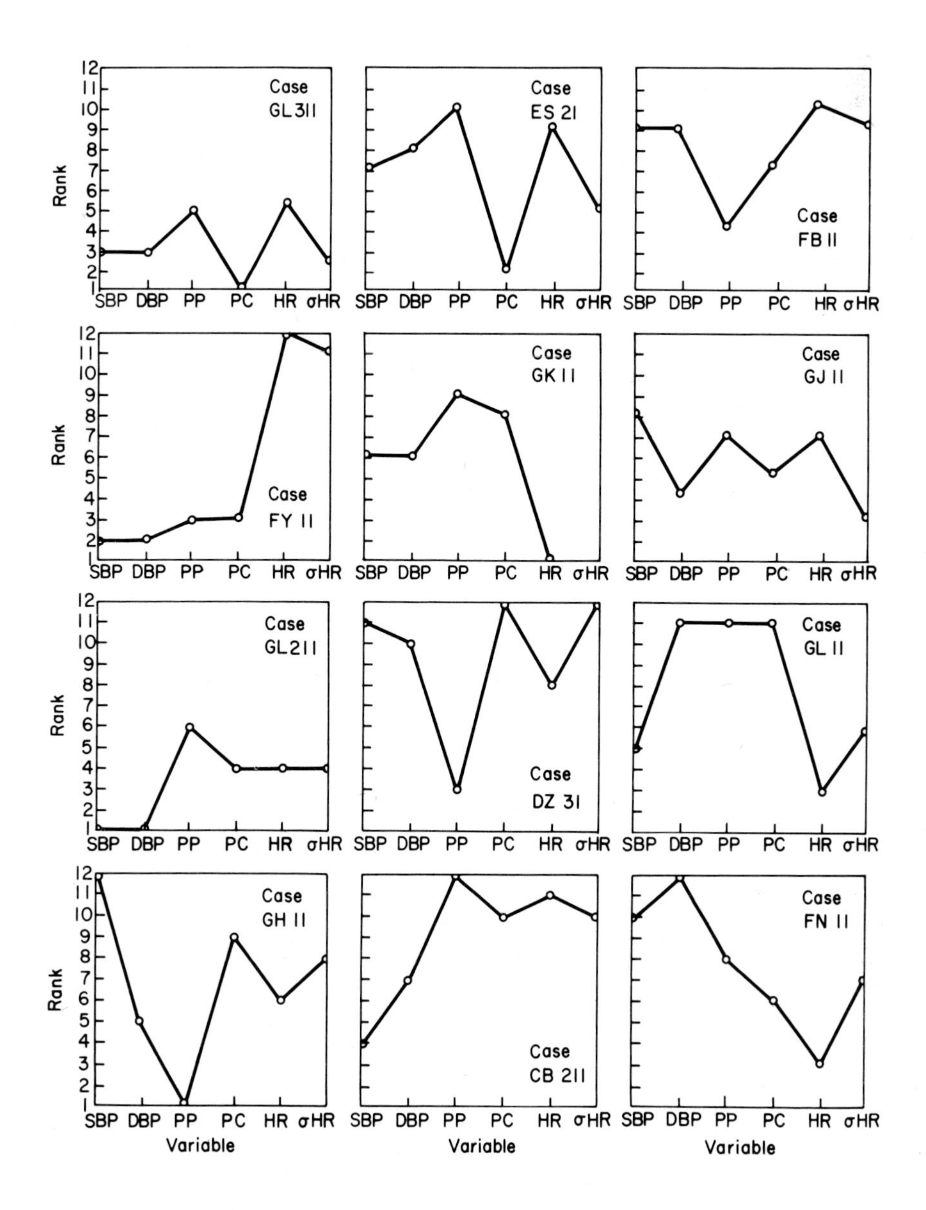

FIG. 3-6. Individual profiles of autonomic response. SBP, systolic blood pressure; DBP, diastolic blood pressure; PP, pulse pressure; PC, palmar conductance; HR, heart rate; σHR, heart rate variability. (Lacey, 1950.)

blood pressure with the onset of stress, followed by a leveling off or decline as stress continued. The neurotic group, in contrast, showed a continuing rise in blood pressure as stress continued. It is as if the psychiatric group lacked some regulatory mechanism that normally functions to check a rise in blood pressure; or perhaps such a regulatory mechanism is overridden by excitatory processes for the psychiatric group.

We have noted that there are stable individual differences in autonomic responsiveness to stress, and within an individual's pattern of autonomic activation specific physiological systems tend to show a maximum response. There is some evidence that this response specificity can be related to psychiatric symptoms. Malmo and Shagass (1949) found that psychiatric patients having cardiovascular symptoms tend to show heart rate changes in response to pain, whereas patients having frequent complaints of headache and related symptoms tend to show a muscular response to pain.

Arousal and Performance

The relationship between arousal and performance was discussed to some extent in Chapter 1. There it was noted that laboratory animals had often been used to determine the relationship between intensity of stimulus conditions and performance, on the basis of operations performed by the experimenter. We pointed out that extrapolations to complex human situations are possible and that as excitement increases, performance, or behavioral efficiency, also increases. However, as it mounts beyond a certain point there is a sharp decrease in performance and in effectiveness. This same trend was noted in Chapter 2, when we discussed the influence of stress on performance. For the most part in our previous discussions of this topic we have been concerned with stimulus conditions on the one hand, and with performance on the other. The level of excitement of the organism has been inferred, except in the one instance where we discussed the studies involving the use of the 17-OH-CS counts as an index of the stress response. We have now seen that the state of arousal is measurable in certain ways. It remains for us to observe that when a measure of autonomic or reticular activity is used, the same U-shaped relationship between arousal and performance that we have noted previously between stimulus conditions and performance occurs.

We will not belabor the point. A single animal experiment by Malmo (1959) exemplifies this relationship quite well. Malmo subjected rats to water deprivation for varying periods of time, ranging from 12 to 72 hours. He used heart rate as a measure of arousal, and rate of bar-pressing as a measure of performance. The results of this experiment are shown in Fig. 3-7. Notice the typical U-shaped function for the performance measure. As thirst increases performance becomes highly accelerated to a mean number of 120 responses per minute at the 48-hour deprivation point. Thereafter, response rate decreases rapidly. In contrast to this, heart rate shows a steady increase throughout the experiment. In this experiment deprivation days were spaced between several days of *ad lib* drinking, in order to control the possibility that any performance decrement would be the result of physical emaciation.

On the basis of data such as these, and human data as well, clearly it is possible to have an extreme state of bodily arousal, and yet for it not to be reflected in behavior. Research in behavior pathology will be discussed at greater length in later chapters, but with reference to the topic of arousal, it is important to note that when behavior is characterized by extreme apathy, withdrawal, and unresponsiveness to the environment it does not follow that the individual is little aroused. In some cases of

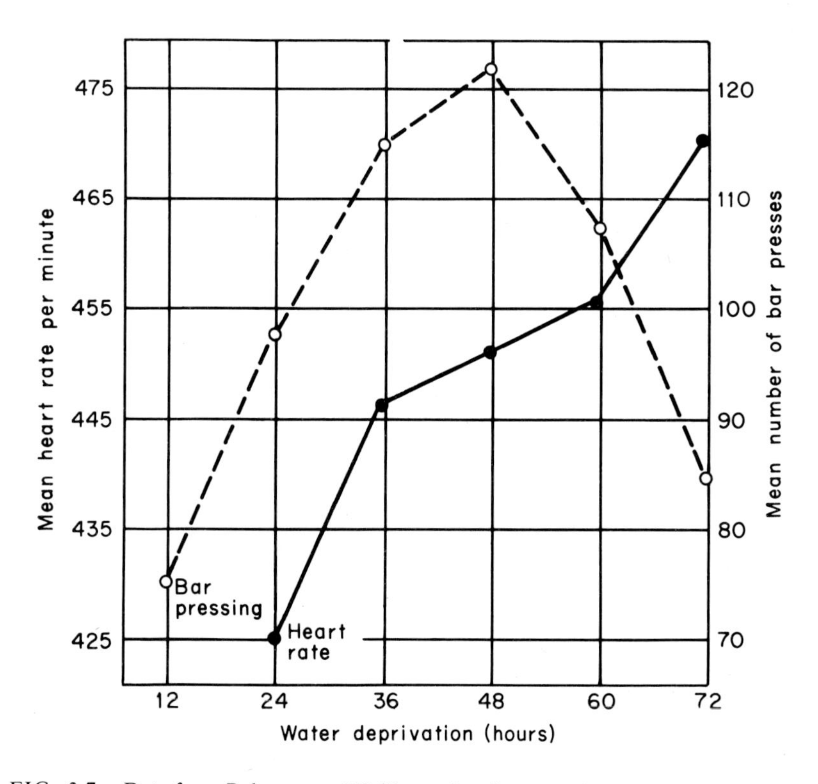

FIG. 3-7. Data from Belanger and Feldman showing mean heart rate and bar-pressing responses plotted as a function of amount of water deprivation. (Cited by Malmo, 1959.)

severe behavior pathology there is evidence that extremes of withdrawal and unresponsiveness are related to extremes of arousal.

This is worth illustrating at this point because of its relevance to the U-shaped function relating performance and arousal that we have been discussing. Figure 3-8 shows data from Venables and Wing (1962), who classified hospitalized chronic schizophrenic patients in terms of extent of social withdrawal. Figure 3-8 shows a plot of arousal as measured by skin potential as a function of the withdrawal classification. It can be seen that the more a group is rated as socially withdrawn, the higher its skin potential rating. The state of arousal appears to intensify whatever responses have been learned as a result of the individual's past experiences. It will be seen in Chapter 4, but it should be noted briefly here, that extreme states of apathy may be produced in other ways. Notably, when the individual finds little or no reward for the way he behaves, eventually little behavior will be found in strength. He may experience a lack of "will" to do much of anything. This state of apathy may or may not be characterized by anxiety or emotional arousal.

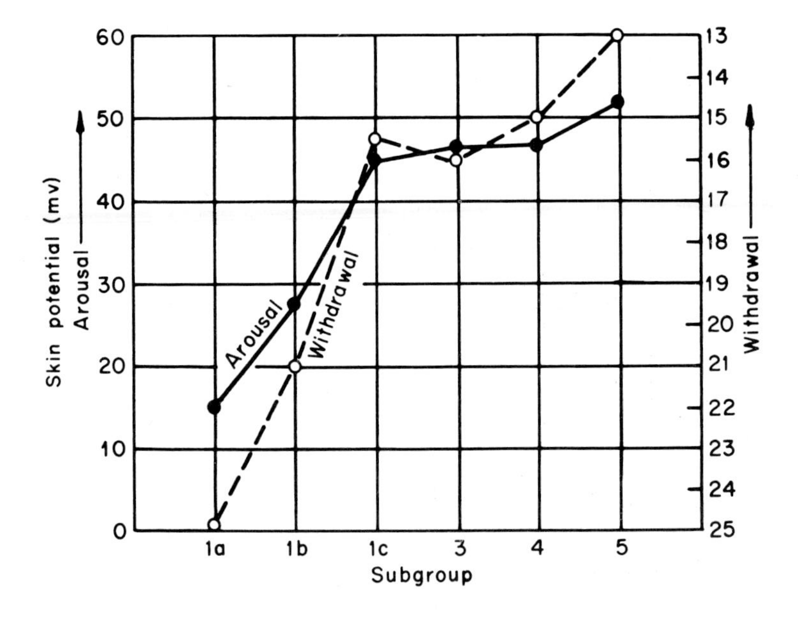

FIG. 3-8. The relationship between withdrawal and arousal as measured by the skin potential for different clinical subgroups of schizophrenics. (Venables & Wing, 1962.)

PRIMARY DRIVE STATES

Traditionally hunger, thirst, and sexual arousal have been referred to as *primary drive states.* All primary drive states contribute to bodily arousal. In terms of their activation function, primary drives are indistinguishable from each other. Either food or water deprivation, for example, increases the general activity level of laboratory animals roughly in proportion to the amount of deprivation. In laboratory animals activity level has also been found to be related to the administration of certain hormones serving a sexual arousal function.

Along with general activation, the primary drives tend to stimulate particular kinds of activity. For the hungry animal or human, specific food-seeking activity is greatly intensified. Energy is spent seeking water when the organism is thirsty. The organism appears to be steered toward the satisfaction of the drive: toward conditions which alter or remove the source of drive stimulation. This directionality of behavioral arousal comes about because the pattern of internal stimulation involved in such states is distinctive. It can serve as a cue, termed a *drive stimulus*, which can become associated both with external stimuli and with specific patterns of behavior.

The drive stimuli may become conditioned stimuli for very specific and highly complex behavior. That is, they may serve as a cue. Even laboratory animals such as the rat have been able to differentiate among these cues, and among different intensities associated with the same drive. For example, they have learned to turn right in a T maze when hungry and left when thirsty; or to turn in one direction under relatively weak hunger, and in the opposite direction under stronger hunger (Leeper, 1935). In the case of the human being the stimulus of hunger may elicit a complex chain of sensory-motor adjustments. A civilized man may stop his work, go to a restaurant, order from the menu, etc.

The Biology of Hunger and Thirst

When the body is deprived of food or water a complex series of biochemical changes and neural excitations occur. These changes are said to constitute a state of physiological *need*, a term which, when used in this context, is synonymous with the term "drive." Biochemical and neural responses have drive-stimulus value. As the state of food deprivation increases, "hunger pangs" become a major part of the feeling of hunger. These sensations are correlated with contractions of the stomach and duodenum (Cannon, 1929; Carlson, 1916). They may occur even if the stomach is denervated (Quigley, 1955).

Some investigators have suggested that the driving force of hunger may be the increase in metabolic activity in certain hypothalamic nuclei (Larsson, 1954; Miner, 1955). Both Larsson and Miner have reported that prolonged eating can be induced by stimulation of the lateral hypothalamic nucleus. Forssberg and Larsson (1954) also found that deprivation increases the uptake of glucose and phosphorus by hypothalamic nuclei. Although the exact mechanisms by which increased hypothalamic activity occurs are not established, it is known that the hypothalamus can detect and respond to a number of chemical stimuli. Hence, there may be hypothalamic receptors that register the number of blood constituents related to the depletion of food, water, salt, etc.

The various experiences occurring during hunger may be accounted for by relays from the hypothalamic nuclei to other areas. Arnold (1960) has pointed out that connections exist to the vagus nerve which could bring about stomach contractions and their corresponding sensations; to the premotor and motor cortex, which could bring about discharges from the adrenal cortex and activation; and to the amygdala. Part of the function of this latter body appears to be imagination and hallucinatory activity. Electrical stimulation of part of this body produces licking, sniffing, and chewing in cats—as if they detected an imaginary odor or taste in the mouth.

Pathways from the hypothalamus affect the secretion of gastric hydrochloric acid (French, Longmire, Porter, & Movius, 1953). Stimulation of certain association areas of the brain is also known to induce secretion of gastric juice and salivation (Fulton, 1951). Interestingly enough, a hypnotized subject secretes gastric juice and bile when he imagines eating food (Barber, 1958).

The hypothalamic detection and response system also appears to operate in the case of thirst. Thus, Andersson and McCann (1955) found that electrical stimulation of various hypothalamic areas in the goat produced increased drinking and the licking of salt, and destruction of certain other parts resulted in a refusal to drink. The release or inhibition of antidiuretic hormone seems to be a part of this circuit as well.

Dryness of the mouth and throat is induced by water deprivation. The presence of water in the mouth has some affect upon the cessation of drinking. The stopping of eating and subsequent relaxation may be brought about by the ventromedial hypothalamus. Miner (1955) found that when this nucleus is destroyed, animals overeat. In contrast, Grossman (1950) has reported that hunger pangs cease when food containing sugar, fat, or salt leaves the stomach and enters the duodenum. It is possible that the subsequent release into the bloodstream of the hormone enterogasterone activates the ventromedial hypothalamus and inhibits the lateral hypothalamus.

The ventromedial hypothalamus has relays to the premotor and motor regions, which may bring about relaxation of both voluntary and involuntary muscles. Lesions in this nucleus cause animals to become vicious and irritable; unable to relax, except during sleep, which comes on abruptly (Spiegel, Miller, & Oppenheimer, 1940; Wheatly, 1944). The ventromedial hypothalamus also may function to activate "imagination" (in this case ideas associated with the arrest of action), because connections exist between it and the amygdala (Arnold, 1960). Finally, secretion of the growth hormone, somatotropin, promotes utilization of the food eaten by stimulating secretion of insulin and glucogen from the pancreas (Bullough, 1955). Relays from the ventromedial hypothalamus may induce this secretion.

Hunger, Personality, and Behavior

There have been many anecdotal descriptions of personality and behavior during states of starvation. They range from diaries kept by individuals who have been marooned in various ways to accounts of massive starvation such as the Russian famine of 1918–1922. For a summary of many of these anecdotal accounts the reader is referred to Keys, Brozek, Henschel, Mickelsen, and Taylor (1950).

Accounts of the starvation of whole peoples during famine typically document progressive personal and social deterioration. The majority of people take on an emaciated appearance. There is an increase in the incidences of mental illness, crime, and prostitution. Children may be abandoned by their parents, who are unable to provide food for them. Cannibalism may be practiced. Famine victims are often reported to show a heightened irritability and, at the same time, extreme apathy. The symptomatic picture may include resignation, indifference, a loss of sexual desire, and an overall apathy. The later stages of hunger involve an almost complete immobility of the starvation victims.

Reports from marooned victims of natural starvation, prisoners of war, and the like tell of many physical and psychological symptoms. These include irritability, exhaustion, faintness, chilliness, vertigo, unsteady and effortful movements, a dulling of intellectual functioning, a highly nervous state, and startle reactions to novel stimuli. In the early stages of hunger the individual may daydream about food. Thoughts and images of food and of food-seeking tend to dominate his life. This may not be so during the later stages of hunger. As the victim's subjective life becomes less well-organized, he may suffer from frequent hallucinations involving food, but eventually he becomes the victim of a semi-conscious state of awareness in which conscious thoughts of food tend not to be present.

One of the few instances of scientific study of hunger in human beings was an experiment carried out at the University of Minnesota during World War II (see Schiele & Brozek, 1948). This experiment was on the effects of semistarvation. Actually, one of the major purposes of the experiment was an exploration of different types of diet in rehabilitation. In this study 36 volunteers were first subjected to a 3-month period during which they were maintained on a good diet, averaging 3492 calories per day. This was followed by 6 months of semistarvation, with the average daily intake reduced to 1570 calories. The experimental design also included a third period, a rehabilitation period, when they received an additional 1200 calories.

During the semistarvation period subjects took on the physical appearance of hunger. There was an average 25% loss in body weight. There was also a steady loss in physical strength, and a lowering of body metabolism and of pulse rate. Many physical symptoms appeared, such as tiredness, weakness and dizziness, an increased sensitivity to cold or noise, visual difficulties, and a loss of sexual desire. They also reported a loss of intellectual ability, but this was not substantiated by the investigators. That they should feel less intellectually capable was probably attributable to their physical disabilities and to emotional factors.

Psychological symptoms were also numerous. They became irritable and indecisive. For some there were marked fluctuations in mood. Apathy, depression, a loss of sociability, and a complete preoccupation with thoughts of food were all common symptoms. The personality effects of semistarvation were strikingly revealed by changes in the Minnesota Multiphasic Personality Inventory (MMPI) profiles of the group. These are shown in Fig. 3-9. During the control period their scores on all of the scales except one, the interest scale (Mf), were very close to the theoretical norm of 50. Semistarvation resulted in a sharp, systematic increase in their scores on the neurotic scales—hypochondriasis (Hs), depression (D), and hysteria (Hy). There were only minor fluctuations on the other scales. The group profile for the follow-up at the end of the experiment, after 33 weeks, was almost identical with the original profile (see Fig. 3-9). The heightened neurotic tendencies disappeared after the semistarvation period had passed.

In addition to the common psychological reactions of the group, some subjects developed unusual or severe neuropsychiatric disturbances. One subject tried to injure himself, and eventually chopped off his fingers in order to get out of the experiment. Another developed hysterical

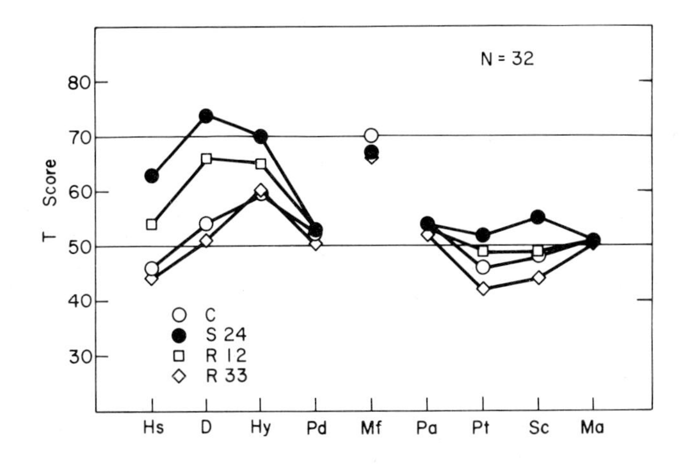

FIG. 3-9. Minnesota Multiphasic Personality Inventory results for subjects in the semistarvation experiment. The data points represent average values for the 32 subjects who completed the experiment. The administration during the control period is designated C. The administration at the end of the semistarvation period is designated S24; after 12 weeks of controlled rehabilitation, R12; and the R33 values were obtained from 20 subjects after 12 weeks of controlled rehabilitation and 21 weeks of unrestricted diet. Hs, hypochondriasis; D, depression; Hy, hysteria; Pd, psychopathic deviate; MF, masculinity–femininity; Pa, paranoia; Pt, psychasthenia; Sc, schizophrenia; Ma, hypomania. (Schiele & Brozek, 1948.)

sensory symptoms. A few subjects broke the dietary regimen that was essential to the experiment. Typically, however, this was done in inefficient and neurotic ways. Instead of going to a restaurant and ordering a meal, they would eat garbage, or steal small amounts of food in an attempt to expiate or minimize their guilt.

Biological Aspects of Sex

Conditions that produce sexual excitement in the absence of learning are varied. Genital responses have been observed in male infants, 3 to 20 weeks old. Similarly, tumescence may be evoked in infants by such factors as frustration of feeding responses, bowel and bladder tensions, and mild fear-arousing situations (e.g., Halverson, 1940).

A great variety of sensory stimuli can become conditioned to arouse sexual excitement. Conversely, inhibitory behavior may be learned in response to conditions which would otherwise be sexually exciting. For a discussion of learning and sexual behavior see pp. 278–279.

Although learning plays a major role in human sexual behavior, much of the aggressive and active behavior of the adult male is the result of the activity of the sexual hormones. Human female menstrual cycles are also roughly comparable to the estrus cycles of animals lower in the phylogenetic scale. Several studies have reported that human females show a heightened sexual urge just before and just after the menstrual period (e.g., Davis, 1929).

The artificial administration of sex hormones to various infrahuman organisms has complex consequences which depend upon the type of hormone, the amount, the sex of the organism, and the age of the organism at the time of administration. If a very high dosage of the male hormone testosterone is injected in the mother during pregnancy females are born as hermaphrodites. The only way to determine that they were generic females is by surgery, which shows the presence of internal female tissue. The neural tissues underlying the performance of sexual behavior may also be affected. That is, as adults these animals (guinea pigs) display the typical pattern of normal male mounting behavior (Phoenix, Goy, Gerall, & Young, 1959). The administration of maternal testosterone appears to have no affect on male fetuses, nor on their adult sexual behavior. Heterotypical (opposite-sex) sex hormones administered in early infancy suppress the development of normal sexual behavior in both the male and the female (Harris & Levine, 1962).

The administration of homotypical (same-sex) sex hormones to organisms that have not attained maturity, but are not in the early stages of infancy, has the effect of making them sexually active at an early age. Similarly, in many species that breed seasonally sexual behavior has been induced during off-season by the injection of gonadotrophic or gonadal hormones. The sexual potency of aged animals has also been restored in this manner (Minnick, Marden, & Arieti, 1946). Few systematic data are available on the effects of hormone administration to humans. Mild hormone extracts are employed occasionally to foster the development of secondary sex characteristics in individuals with various mild glandular abnormalities. Human adults have reported dramatic increases in sexual desire following hormonal injections.

For animals low on the evolutionary scale, removal of the ovaries of the female or of the testes of the male abolishes sexual behavior. In female primates and human beings the effects of ovariectomy are much more variable (Beach, 1947). Most women report a loss of sexual desire, but some maintain the same level of sexual activity as prior to the operation, whereas the sexual activity of others diminishes rapidly. Similarly, the effects of castration on primate and human males appears to be variable, and to depend to a large extent on preoperative experience. Apparently, experience can be a factor even in lower animals. Rosenblatt and

Aronson (1958) found that male cats which had had sexual experience prior to castration manifested sexual performance after castration which was considerably superior to that of cats with little or no precastration experience, even though in both cases there was a decline of activity following castration. In both male and female animals the effects of removal of the gonads can be reversed by injecting the appropriate hormones. Again, in primates, and especially in humans, the results of such replacement therapy are highly variable.

The hypothalamus appears to be crucial in sexual behavior. Analogous to its functioning in hunger, hypothalamic receptors may serve to detect the presence of sex hormones in the blood. In female cats the implantation of small amounts of solid estrogen in the hypothalamus has induced sexual receptivity, or heat, lasting for as long as 50 or 60 days (Michael, 1962). In this case the effect appears to be directly on the hypothalamus, because the vagina and uterus remain in anestrual condition. These animals became sexually receptive behaviorally, yet the stimulation of increased production of gonadal or gonadotrophic hormones was not involved.

Arnold (1960) has argued on the basis of data of Rothballer and Dugger (1955) that the impulse to mate involves excitation of the hypothalamus which activates relays in the same structures that receive relays during hunger, the amygdala and the hippocampus. Part of this neural circuit appears to involve relays which induce the motor and autonomic changes associated with sexual arousal, as well as stimulation of the pituitary gland for the discharge of gonadotrophic hormone.

Several studies have indicated that the amygdala plays some role in the sexual arousal pattern. Faure (1957) reported that electrical stimulation of part of the amygdala and hippocampus aroused sexual behavior. Wood (1958) reported that bilateral lesions of part of the amygdala produced a hypersexed state 8 to 10 weeks following the operation. Others, e.g., Koikegami, Yamada, and Usui (1954), have produced ovulation in the rabbit by electrical stimulation of the amygdala. As in the case of other bodies within the limbic system, the amygdala appears to have a diversity of function. It appears to be implicated in certain aspects of hunger, thirst, and sex.

It appears that the experience of pleasure in orgasm is a reaction to sensations arising from the rhythmic contraction of the clitoris, vagina, and uterus of the female, and of the muscles of the prostate, seminal vesicles, and ejaculatory ducts in the male (cf. Fulton, 1951). The orgasm is followed by quiescence and relaxation, which necessitates a pathway mediating the inhibition of muscular activity. Arnold believes that this inhibition is mediated by the ventromedial hypothalamic nucleus (Arnold, 1960). Data from Schreiner and Kling (1953) tend to support this idea.

Intracranial Stimulation

One of the most interesting areas of research on the limbic system is that dealing with the rewarding and aversive effects of electrical stimulation. Investigators who had been studying the effects of electrically stimulating the ARAS unintentionally placed an electrode near the septum in a rat. When the animal was stimulated in the open field situation, it tended to focus its behavior on the spot where it happened to be at the time of stimulation (Olds, 1955). Furthermore, the animal's behavior could be brought under control of the electrical stimulus. Small stimulations were made contingent upon responses of the animal which the experimenter wanted him to make. No stimulation was given for unwanted responses. In this manner the animal could be led successfully through a maze. It was soon discovered that animals could learn to turn right or left in a T maze, or learn more complicated behavior, with electrical stimulation serving as the sole reward.

Figure 3-10 illustrates the powerful effect of such mild electrical stimulation, when the electrode is optimally placed. In this instance the apparatus was designed so that the animal could stimulate itself by pressing the bar of a Skinner box. The animal whose record is shown made about 2000 bar-pressing responses an hour for a period to 26 hours. After the animal had slept for 19 hours it continued to stimulate itself at the same rate. The reinforcing potency of intracranial stimulation has been demonstrated in many other instances. Olds and Sinclair (1957) showed, for example, that hungry rats will cross an electrically charged grid at greater shock intensities to obtain brain stimulation than to obtain food. Olds (1956) reported similar results when intracranial stimulation is pitted against such drives as thirst and sex.

Exact replication of some of these demonstrations is often difficult. The loci of stimulation at which positive results are obtained are scattered widely throughout the paleocortex and adjacent areas. For the most part, the best results are obtained within the limbic system. It appears that there are two distinct positively reinforcing areas (Olds, Travis, & Schwing, 1960): an area from the septum, extending back toward the dorsal part of the thalamus; and an area involving the hypothalamus, amygdala, and hippocampus (see Fig. 3-1). The picture is further complicated by the fact that locus combines with other stimulus variables such as wave form, frequency, and intensity to determine the results.

The problem of how the action of electrical stimulation of these loci is related to the motivation of animals under natural circumstances has

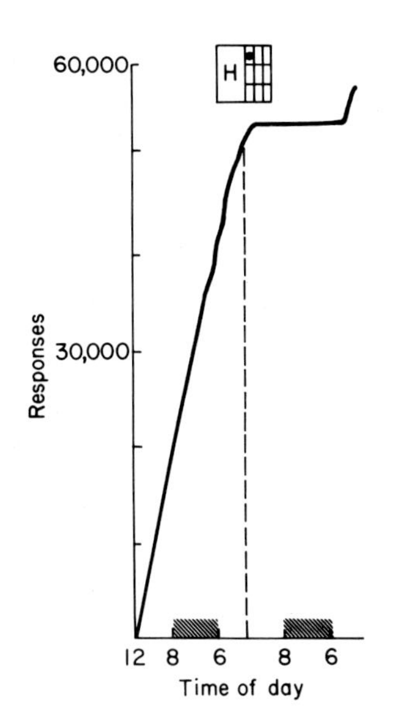

FIG. 3-10. Continuous mild electrical self-stimulation for a 48-hour period. The cross hatching on the abscissa indicates darkness (8 p.m. to 6 a.m.). This animal stimulated itself at a rate of more than 2000 responses per hour for 26 hours, slept, and then continued self-stimulation at the same rate. (Olds, 1958.)

scarcely been approached as yet. Several studies have jointly varied primary drive deprivations and intracranial stimulation (Brady, Boren, Conrad, & Sidman, 1957; Olds, 1958). Although some negative results have been obtained (e.g., Hodos & Valenstein, 1960), under optimal placement rats press more for self-stimulation when a primary drive is activated simultaneously. This effect may depend upon whether the electrode is placed in an area which has some functional relevance to the particular primary drive involved (Hoebel & Teitelbaum, 1962).

Some clarification as to the motivational effects of intracranial stimulation will be expected as more human data are made available. Two studies have employed psychotics who had failed to respond satisfactorily to any existing therapies (Bishop, Elder, & Heath, 1963; Heath, 1954). The data from these studies have only been suggestive of positive effects. Sem-Jacobsen and Torkildsen (1960), however, employed patients suffering from Parkinson's disease as subjects. They found several loci which appeared to have a rewarding effect. The subjects reported that it "tickled," or "felt good." They were observed to smile, grin, and request further stimulation.

Conditioning is a relatively permanent affair, except under special circumstances. One set of such circumstances involves the arrangement of special conditions whereby extinction can take place. The animal must be required to respond in the same stimulus complex, except that the original reward is absent. Howarth and Deutsch (1962) found that animals conditioned by intracranial stimulation *and not permitted to respond* during the period following the last stimulation, extinguished just as rapidly as animals who were permitted to respond without reward. Thus, behavior conditioned by employing brain stimulation as reinforcement appears to be rapidly lost unless the reinforcing stimulus is in effect. Since nerve conduction is basically chemical in nature, rather than electrical, it may be that such discrepancies will be resolved by future research employing chemical intracranial stimulation. A slight re-placement of the electrode in some areas results in stimulation not being rewarding at all. It may even be aversive. Animals have learned to avoid stimulation when the electrodes are placed in certain loci. In one experiment a tone was paired with stimulation of the anterior hypothalamus. The animals learned to run upon presentation of the tone, and thus avoid electrical stimulation (Cohen, Landgren, Strom, & Zotterman, 1957). These animals also displayed behavioral symptoms associated with emergency reactions, such as piloerection and hissing (Roberts, 1958). Although it is possible that in some instances known pathways are being stimulated, for the most part this has not been the case. What appears to be involved is the elicitation of fear.

Roberts (1962) reported alarm and flight reactions in cats upon stimulation of the posterior hypothalamus. There are pathways from the hypothalamus which connect with sympathetic fibers, and with the adrenal medulla (Folkow & von Euler, 1954). The existing evidence has led Arnold (1960) to postulate that fear is mediated by a special pathway within a general "action circuit" (see Fig. 3-12). Such hints of the causal chain of physiological mechanisms represent an advance over older conceptions of relatively isolated reward and aversion "centers." Anger and rage have also resulted from stimulation of various limbic areas (e.g. Kaada, Jansen, & Anderson, 1953; Spiegel, Miller, & Oppenheimer, 1940). As in the case of fear, there are suggestions of specific pathways, and of specific connections with the autonomic nervous system.

THE DIFFERENTIATION OF EMOTION

On a subjective or on a verbal basis a wide range of emotional experience can be differentiated by the normal adult. However, differentiating emotions scientifically is very difficult. When people are actually put in laboratory situations designed to elicit anger, fear, disgust, delight, etc., their descriptions of their experiences immediately afterwards are meager. Many words in our language describing emotional experience tend to be vague, and to have an overlap in meaning. An emotional event may be described by a number of more or less synonymous terms, e.g., distasteful, repulsive, disgusting, or unpleasant. In spite of the many shadings of possible emotional experiences, it appears to be relatively easy for people to classify them into "pleasant" and "unpleasant." Psychologists have found it necessary to study emotion operationally: in terms of the approach (adient) or avoidant behavior related to them. We will have a great deal to say about response-defined emotion in later chapters. Now, let us turn to the physiological differentiation of emotion.

Anger and Fear Differentiation

Some success has been reported in differentiating autonomic functions in anger and fear. Ax (1953) stimulated anger and fear in the laboratory by arranging conditions that would normally be thought to produce them. He devised a response index based on the measurement of heart rate, GSR, EMG, facial temperature, etc. It consisted of profile difference scores obtained from maximal and minimal changes in these measures. He was able to show that the obtained response patterns differed under anger vs. fear conditions. He also found greater physiological integration when anger was present than when fear was present.

The response patterns obtained for fear and anger in this experiment were suggestive of those produced by injections of the hormone epinephrine (adrenaline) and a combination of epinephrine and norepinephrine (noradrenaline), respectively. It will be recalled (see Chapter 2) that both epinephrine and norepinephrine are secreted by the adrenal medulla. Epinephrine elicits profound changes in almost all bodily systems, whereas the effect of norepinephrine is more circumscribed. Specifically, norepinephrine stimulates the contraction of small blood vessels and increases the resistance to blood flow.

High blood pressure may result from either sympathetic or parasympathetic activity. In the former it is produced by the direct increase in heart rate. In the latter, vasoconstriction may have the same effect. Funkenstein (1955) and his group found that the drug mecholyl, which stimulates the parasympathetic system, produced two different reactions in psychotic patients with high blood pressure. Some patients (N group) showed a slight drop in blood pressure, followed by a return to its initial level within 3 to 8 minutes. Others (E group) showed a marked drop in blood pressure, which remained below its preinjection level for 25 minutes or more.

These patients also had different responses to electric shock treatment. The majority of E group patients improved in response to electric shock treatment, whereas only a small percentage of patients in the N group showed improvement. When clinicians rated the predominant emotional patterns of these patients without knowing to which group they belonged, it was found that almost all those who generally reacted with depression or fright belonged to the E group, and almost all those who generally got angry at other people belonged to the N group. Evidence was later obtained (Funkenstein, King, & Drolette, 1957) that the N group type of mecholyl reaction is associated with excessive secretion of norepinephrine, and the E group type of reaction is associated with excessive secretion of epinephrine.

In another series of experiments the immediate reaction of students during a stress-inducing situation was interpreted as indicating their habitual acute emergency reaction to stress. The reactions were graded on the basis of the subject's reports. The most frequent categories were (1) anger-out, anger directed toward an external object or person; (2) anger-in, anger directed toward the self; (3) anxiety. The anger-out group showed a low intensity of physiological response and excessive secretion of norepinephrine. Both the anger-in and anxiety groups showed a high intensity of physiological response and an excessive secretion of epinephrine. Under stress, these latter groups also performed experimental tasks more poorly.

Subjects of these experiments were also graded on their ability to cope with stress over a period of time. These behavioral evaluations were divided into four categories: mastery, and delayed, unchanged, and deteriorated mastery. The type of acute emergency reaction subjects had (anxiety, anger-in, or anger-out) appeared to have little relation to their eventual mastery of continued stress. The individual's acute emergency response may be counteracted by his previous learning experiences, although the initial physiological and psychological responses to stress may be predictive of the type of reaction the individual would have under conditions producing extreme behavioral disorganization.

THEORIES OF EMOTION

The development of instrumentation and of research techniques, along with a growing interest in psychobiological research, has led behavioral scientists to certain reinterpretations of theories of emotion. An understanding of the development of theoretical work in this area requires a brief survey of the systematic revisions and changes which have taken place since the early James-Lange and Cannon-Bard theories.

The James-Lange Theory

Both James (1884) and Lange (1885 as cited by James, 1920) separately proposed that emotion is the *perception of physiological changes*. James felt that visceral changes were of major importance, whereas Lange thought changes in the circulatory system were more important. James is often quoted as saying, "We are afraid because we run; we do not run because we are afraid." James held that the perception of physical responses gives rise to emotion and that these responses arise because certain situations have a "physical effect upon the nerves."

The Cannon-Bard Theory

Cannon (1927, 1931) felt that the perception of bodily sensations cannot adequately explain emotional experience and expression. He showed that appropriate stimulation after removal of the sympathetic nervous system results in the same signs of anger and fear in laboratory cats that occur preoperatively (Bard, 1928; Cannon, 1927). He further pointed to evidence that visceral changes were too slow to be a source of emotion, and that artificially induced visceral changes do not produce emotion. At this time it appeared that emotion might be generated cortically by thalamic processes. Decorticated animals showed exaggerated rage reactions which disappeared when the thalamus was removed. Cannon and Bard thus hypothesized a thalamic theory of emotion. Emotion was postulated to result from disinhibitory impulses from the cortex which released the appropriate thalamic pattern. Disinhibited thalamic discharges were thought to excite both the peripheral nervous system and the cortex, giving rise to emotion. Actually, the evidence for this theory was extremely weak. Such evidence as there was could either be opposed by other findings or interpreted differently.

The Papez-MacLean Theory

The first theory to suggest definite brain mechanisms activated during emotion was proposed by Papez (1937). According to Papez, emotional experience can be aroused by sensory impulses which are relayed over a definite limbic circuit (hippocampus–fornix–mammillary bodies of the hypothalamus–cingulate gyrus) to the cortex (see Fig. 3-1). It can also be aroused from the cortex (from the frontal lobes or from the hippocampal gyrus) and relayed by way of the mammillary bodies to the cingulate gyrus. Thus, Papez proposed that the cingulate gyrus is like the somatosensory projection areas, in the sense that neural impulses are translated into emotional experience when they are received there. Papez also outlined a motor system, mediating the expression of emotion. It consisted of the descending hypothalamic tracts which connect with the sympathetic motor neurons of the spinal cord, and of the hypothalamic fibers which mediate parasympathetic excitation.

MacLean (1949) supported and extended Papez's theory. He postulated that the hippocampal area was a correlation center for sensations from the eyes, ears, oral region, viscera, sexual organs, and body wall. He called this area the *visceral brain* which he felt functioned to integrate primary drive stimuli and affect autonomic functioning through the hypothalamus. MacLean also suggested that the hippocampal system was capable of crude, nonverbal symbolism.

Psychologists were slow to give the Papez–MacLean theory consideration, primarily because of its highly speculative content. At the same time they retained the old Cannon theory which, in comparison to the Papez–MacLean theory, lacked the neuroanatomical and experimental evidence. As speculative as the original Papez theory was, the weight of experimental evidence has now substantiated the fact that the limbic system is of central importance in emotion.

Lindsley's Theory

In 1951 Lindsley proposed an *activation theory* of emotion which emphasized the role of the reticular formation in the motivation of behavior.

He proposed that the reticular formation plays a central role in motivation and emotion. He argued that it has precisely those functions required: (1.) It is an activator which is sensitive to all types of internal and external activity and to cortical events. (2.) Both the ARAS and limbic structures activated by it can manifest a persistence, or preservation, of arousal. (3.) At least in some instances, motivation may require a direction, or goal.

The perception and definition of distinctive stimulus cues associated with goal-directed behavior requires the elimination of routine sensory influx and specific alerting. The reticular formation appears to be capable of both of these functions. As has been pointed out by Morgan (1965), Lindsley's theory is no doubt correct in its emphasis on the activation aspects of emotion. However, more recent evidence indicates that limbic structures, especially the hypothalamus, have complex activation functions also. Hence, Lindsley's activation theory merely serves to indicate that the reticular system is an integral part of the total activation pattern.

Arnold's Theory

Arnold has tried (1960) to integrate existing neurophysiological and psychological evidence to account for the physiological mechanisms of

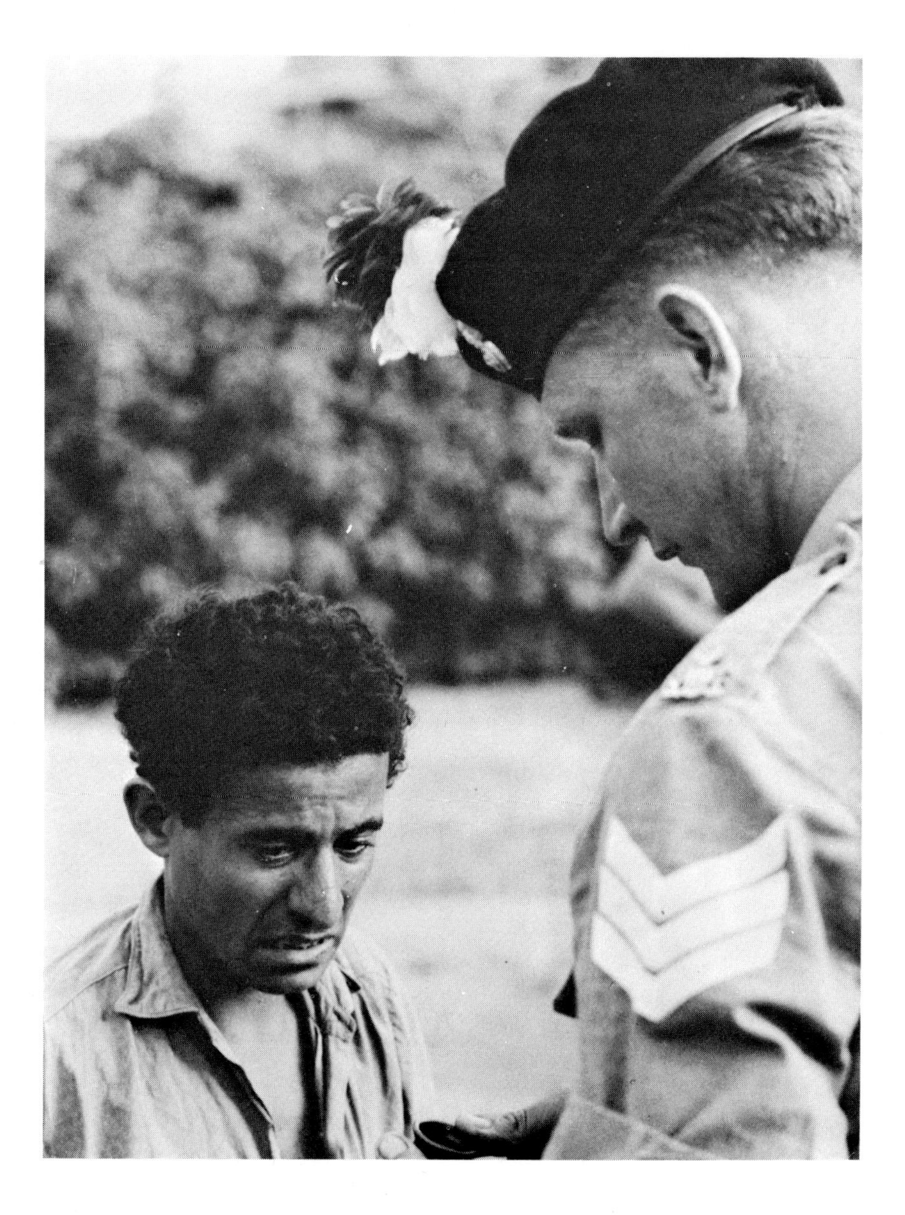

Events within the limbic system mediate appraisal and emotion. This person is appraising and clearly having an emotional reaction to the stimulus situation—interrogation by the soldier. (Photograph by Tor Eigland.)

emotion. It is an ambitious undertaking. Arnold has clearly recognized that certain shortcomings must exist in integrative theory at our present stage of knowledge. She has also recognized the importance of, and necessity for, responsible scientific theorizing:

> Since research activity in this field is lively, our integration cannot be final and may prove wrong in many details. At the same time, the attempt to map out the sequence of neural events in emotion in as much detail as possible should prove fruitful because details can be checked, proved or disproved, while general models and theories, though brilliant and stimulating, may be in the end a barrier to knowledge because they are based on analogies instead of verifiable facts. [Arnold, 1960, p. 30]

Arnold attempts to specify the underlying mechanisms and complex neuroanatonomical events which mediate both the experience and expression of emotion. The theory begins with simple propositions. Thus, a stimulus must be perceived and its effects *appraised* before an emotional state can occur in reference to it. The appraisal of what is perceived in many instances depends upon the memory of earlier experiences involving the same or similar stimulation. Further, the qualitative and physiological response differences underlying our many different emotional experiences must depend upon different appraisals. Accordingly, sensory impulses must be received and integrated in the cortex before the mechanisms involved in emotions become operative. Once the sensory input is integrated, a system is required to translate an awareness of its significance. The mechanisms involved in this must also provide access to somatic and autonomic motor systems.

Arnold postulates that appraisal and the consequent experience of liking or disliking are brought about by a special system which she refers to as the *estimative system,* illustrated in Fig. 3-11. The dotted lines in Fig. 3-11 show the projection pathways of the estimative system. The estimative system includes the afferent connections from sensory receptors to the brain-stem reticular formation, certain nuclei of the thalamus, and the limbic cortex. The evaluation of sheer sense impressions may occur at the thalamic level. However, the experience of liking or disliking occurs when the impulses coming in over the projection pathways of the estimative system reach the posterior cingulate gyrus, as Fig. 3-11 indicates.

Arnold defines emotion as "a *felt tendency* [author's italics] toward something appraised as good or away from something appraised as bad." The system that serves to correlate impulses from sensory areas and relay them to the motor cortex where the "felt tendency" can be experienced is proposed to be the *hippocampal system.* Impulses from the

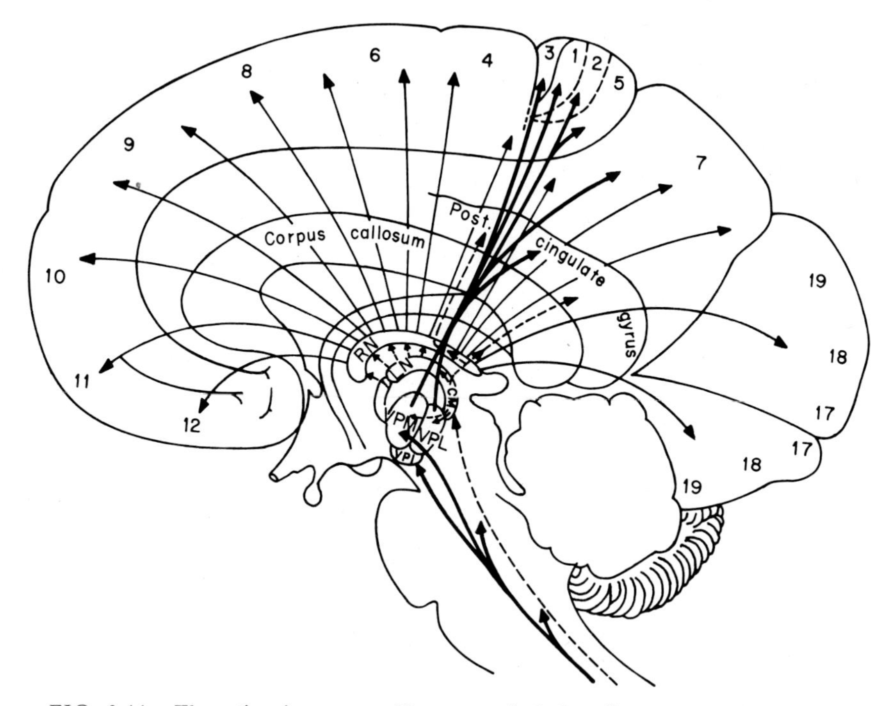

FIG. 3-11. The estimative system. The system includes afferent connections from the sensory receptors to the brain-stem reticular formation, thalamic nuclei, and the limbic cortex. The heavy lines show the sensory projection pathways. The actual projections of the estimative system are represented by the dashed lines. The solid, light lines indicate diffuse projection to all parts of the cortex. (Arnold, 1960.)

sensory areas are relayed to the adjoining limbic region, and thence to the hippocampus, where they are collected and sent to the hypothalamus, midbrain, and motor cortex (frontal lobes).

The hippocampus also has the necessary connections to function for the recall of relevant memories. It is thought to provide a circuit from the cortical and limbic sensory areas connecting with sensory nuclei of the thalamus in such a way that impulses can be accompanied (primarily via the midbrain reticular formation) back to the sensory projection areas. Sensory projections are thought to carry impulses simultaneously to primary and secondary sensory areas, to the association areas, and to the limbic areas serving the respective functions of recognition, registration (for recall), and evaluation.

Recall alone does not appear to be a sufficient condition for expectation of the same or similar effects to follow. Such expectation, or "imag-

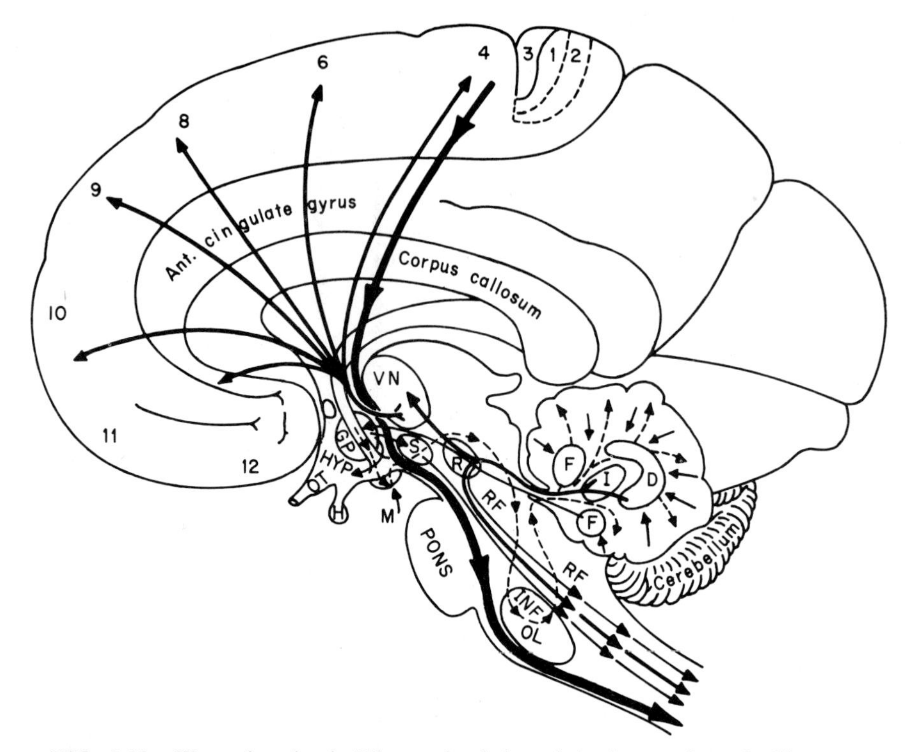

FIG. 3-12. The action circuit. When action is intended relays go from the hippocampus to the cerebellum (dotted lines). From there an organized pattern of action is relayed to the frontal lobe (medium line); down the corticospinal tract, mediating muscle movement (heavy line); and at the same time relays are sent to extrapyramidal and hypothalamic effectors (thin line). (Arnold, 1960.)

ination" is postulated to be mediated by the amygdaloid complex. This complex receives afferents from sensory and limbic areas allowing for its excitation by any perceived object. It receives afferents from hypothalamic nuclei, allowing for its excitation by primary drive tensions. Moreover, it sends projections to association and to limbic areas.

Some impulses to action are an inseparable part of the foregoing dynamics of appraisal. Thus, even the first thalamic appraisal appears to initiate relays to motor pathways. However, these action impulses may be held in check until appraisal has occurred and the impulse to act has itself "been appraised." Arnold believes this latter appraisal involves a special circuit (hippocampus — mamillary bodies — thalamus — cingulate gyrus).

Following this final evaluation, impulses are postulated to be sent from the hippocampus to the midbrain, cerebellum, and frontal lobe to connect with cells in the motor cortex. The cerebellum appears to amplify and coordinate impulses for action and to relay them to the ventral thalamus and frontal lobe. This system of relays, called the *action circuit,* is illustrated in Fig. 3-12. Arrival of these impulses in the premotor area appears to give rise to an urge to action. The hippocampus is thus visualized as functioning as a transformer, translating the sensory influx eventually into motor outflow. The motor outflow includes the activation of both sympathetic and parasympathetic systems, and subsequent vasoconstriction, vasodilation, etc.

Generally speaking, the same circuit that mediates action is also thought to mediate the inhibition of action. The cerebellum coordinates inhibitory impulses from the hippocampus into the action pattern and relays them appropriately.

This is the general picture which Arnold proposes. Insofar as it attempts specification of the complex functional systems of the human cortex, it constitutes a valuable step toward a more complete understanding of the physiological basis of behavior.

PSYCHOACTIVE DRUGS

There are a great variety of chemical agents which affect behavior. Most people are familiar with the effects of alcohol and of sedatives. In recent years tranquilizers and activators such as Dexedrine and Benzedrine have also come to have widespread use.

Alcohol

When alcohol first enters the bloodstream, it acts as a mild stimulant. Following this early stage, it becomes a depressant. With continued consumption speech, motor coordination, judgment, and discrimination become increasingly impaired. Under the influence of extreme amounts of alcohol, vascularcirculatory collapse (fainting) occurs. Alcohol is well known as a releaser of inhibitions. Experimental work with animals clearly indicates that it has the effect of reducing fear. Its addictive power can be attributed mostly to anxiety-reduction, and hence, to existing emotional disturbance, rather than to any habit-forming effects *per se.*

Although there are very broad, common classes of behavior in our culture which are normally inhibited, such as the expression of sex or aggression, the effects of alcohol may differ markedly from individual to individual because of the idiosyncratic learning history peculiar to any given one. Commonly alcohol allows people to relax and lose some of their inhibitions, to become more sociable and friendly. In contrast, some individuals may become depressed. Ordinarily there are social sanctions against openly showing certain feelings, such as the fact that one is miserable and depressed. Under the influence of alcohol, the individual may not care about concealing feelings the open display of which is mildly punished under normal circumstances. The intoxicated person may become disinhibited to the point where he will do things which would normally generate a large amount of anxiety for him. When the individual actually begins doing things that he may have a desire to do but which are actually quite frightening he may suddenly sober up.

Chronic alcoholics undergo acute withdrawal reactions. The most severe are those associated with the delirium tremens: tremor of the muscles of fine motor coordination, vivid hallucinations, extreme startle reactions, and a disorientation for time and place. In rare cases pathological intoxication occurs: some individuals become disoriented, hallucinated, confused, and even violent with a moderate amount of alcoholic consumption. Korsakoff's psychosis, a brain disorder resulting from acute deficiency of vitamin B, may also appear as part of chronic alcoholic deterioration. Of an estimated seventy million users of alcohol in the United States only about 6% can be classified as alcoholics. Generally, alcohol may be used to promote a lessening of tension and increased sociability. It may constitute a first-line defense against anxiety and thus serve the same function that television viewing or excessive reading does for hundreds of thousands of people.

Sedatives

Sedatives are sometimes called hypnotics. Like alcohol in mild doses they function to take away the individual's defenses against anxiety. The most commonly used sedatives are barbiturates such as sodium Amytal and pentobarbital. Intravenous injection of barbiturates can be used to promote a temporary remission of symptoms in some severe anxiety and borderline psychotic patients. It may enable them to report information they would otherwise be unable to give because the thoughts in question are too frightening to cope with. Sodium Pentothal (commonly known as "truth serum") has been effectively used to speed up psychotherapy, especially in cases of combat neuroses.

Since the sedatives depress the threshold of neurons and generally reduce nervous activity, they have also been frequently used as anticonvulsants. Barbiturates such as sodium Pentothal are sometimes used in minor surgery. Frequently, while the drug is taking effect, before the individual loses consciousness completely, violent emotional reactions in speech may be manifested and careful physical restraint is necessary. Following surgery the patient may awaken with marked feelings of vigor, relaxation, and emotional lability.

Tranquilizers

The tranquilizing drugs have a calming effect without seriously interfering with the individual's normal mental functioning. They vary widely in chemical structure and in their effect on the nervous system. Most tranquilizers have a depressing effect on the reticular formation. Depending upon their chemical structure, they may also either stimulate or depress hypothalamic activity and the limbic system generally. Reserpine actually activates the reticular system and hence its value as a tranquilizer must have some other basis.

Tranquilizers have now proven to be very effective in ameliorating depression and anxiety states in psychotics and neurotics, and in curbing violent behavior. There are wide individual differences in reaction to any given tranquilizer. However, usually one can be found which has the desired effect for a given individual. An enormous amount of research has already been instituted in an attempt to determine the exact physiological changes which accompany the administration of various tranquilizers. Another vast area of research deals with the effect these drugs have on behavior. Firm conclusions are not available at the present time. However, this general area of research appears to hold a great deal of promise.

Activators

There is another class of drugs generally referred to as antidepressants or activators. They may produce general hyperexcitability of the nervous system, as is the case of the commonly used Benzedrine (amphetamine) and Dexedrine (dextroamphetamine). Many drugs in this class have specific effects on body metabolism or on certain aspects of neural functioning. The cholinergic drugs are representative of the latter category. They increase the level of acetylcholine in the brain and nervous system. Although the activating drugs may enhance alertness and motor responses, they have not been shown to have any appreciable effect on learning.

Psychotogens

The effects of certain drugs are primarily psychological. They are referred to as *psychotogens,* or *psychotomimetic drugs.* Certain tranquilizers and stimulants may fall into this class, but it is especially composed of drugs which produce symptoms of psychosis to some degree. This includes the drugs commonly associated with narcotic addiction, such as opium and its derivatives (morphine, heroin, paregoric, and codeine). Generally opium produces euphoria, a microscopic sense of time and distance, contentment, daydreaming, a decrease in sexual desire, and inhibition of voluntary movement. Once addiction is established, acute and severe withdrawal symptoms occur if the opiates are not taken continually. Certain other narcotics such as cocaine and marijuana produce a euphoric state, feelings of confidence and well-being, and may stimulate sexual behavior, but they do not involve increased tolerance or withdrawal symptoms. Strictly speaking, they do not cause addiction. In some cases individuals who are predisposed may develop acute psychotic symptoms under the influence of these drugs.

A psychotomimetic drug which has recently attracted a good deal of popular interest is lysergic acid diethylamide (LSD-25). Another is bulbocapnine. These drugs, in particular, appear to produce a schizophrenic-like state. Actually, they do not differ greatly from several other drugs in this respect. Interest in these drugs has been stimulated by the search for biochemical correlates of psychosis. Although there is a large and increasing literature in this area no definite generalizations are warranted at the present time.

SUMMARY

In this chapter we have discussed general arousal, or drive, and the mechanisms associated with it; the primary drives of hunger, thirst, and sex; and then the biological mechanisms mediating them, and their relationship to arousal and behavior; the differentiation of emotion and theories of the biological mechanisms associated with emotion; and finally, the relationships among certain drugs, arousal, and behavior.

It has been found from studies of decortication that emotional arousal depends upon the limbic system. The reticular formation functions as an arousal mechanism, and also determines whether or not impulses arrive in the projection areas and hence, whether or not they are accessible to consciousness. Reticular arousal can be operationally defined by the alpha block. Degree of arousal can be indexed by electroencephalography. There are also a number of ways to measure autonomic functions. Under conditions of autonomic arousal individuals show relatively stable autonomic response patterns, with maximum activation occurring only in certain physiological functions for any given individual. This is called *autonomic response specificity.* The relationship between arousal and performance is a U-shaped function. Behavioral efficiency increases and then decreases again with increasing increments of arousal. The body may be in an extreme state of arousal, and yet the individual may appear withdrawn and apathetic.

Hunger, thirst, and sexual arousal have distinctive biochemical and neural responses associated with them. These responses have drive-stimulus value. The drive stimuli may come to elicit chains of responses, and hence serve a steering function for behavior. Hypothalamic mechanisms are important in the elicitation and regulation of bodily responses that occur during hunger, thirst, and sexual arousal. There is clear evidence that semistarvation can result in emotional disturbance, which disappears when the normal diet is reinstated. Learning plays a major role in human sexual behavior, but hormones also have an arousal function for humans.

The stimulation of certain limbic areas has been shown to have both pleasurable and aversive effects. It appears that specific pathways are being stimulated in these experiments, rather than reward and aversion "centers." The response patterns of anger and fear can be differentiated. Greater physiological integration occurs under anger-producing conditions than under fear-producing conditions. Excessive secretion of epinephrine and of norepinephrine under stress is associated with fright vs. anger, and may be predictive of response to certain kinds of therapy. The acute emergency response may later be counteracted as a result of previous learning experiences. Theories of emotion have gradually become more sophisticated so that at the present time systems, mechanisms, and pathways within the brain can be specified to account for the bodily responses and for the experiences that occur during emotional arousal.

A great many chemical agents affect arousal and behavior. There are many well-known and commonly used drugs. Some relatively new tranquilizers and activators have enjoyed widespread use, but more research is required before firm conclusions can be reached as to their effects.

REFERENCES

Andersson, B., & McCann, S. M. Drinking, antidiuresis and milk ejection from electrical stimulation within the hypothalamus of the goat. *Acta Physiol. Scand.*, 1955, **35**, 191–201.

Arnold, Magda V. *Emotion and personality.* Vol. II. New York: Columbia Univer. Press, 1960.

Ax, A. F. The physiological differentiation between fear and anger in humans. *Psychosom. Med.*, 1953, **15**, 433–442.

Barber, T. X. The concept of hypnoses. *J. Psychol.*, 1958, **45**, 115–31.

Bard, P. A. A diencephalic mechanism for the expression of rage with specific reference to the sympathic nervous system. *Amer. J. Physiol.*, 1928, **84**, 490–515.

Beach, F. A. Evolutionary changes in the physiological control of mating behavior in mammals. *Psychol. Rev.*, 1947, **54**, 297–315.

Brady, J. V., Boren, J. J., Conrad, D., & Sidman, M. The effect of food and water deprivation upon intracranial self-stimulation. *J. comp. physiol. Psychol.*, 1957, **50**, 134–137.

Bullough, W. S. Hormones and mitotic activity. *Vitamins & Hormones*, 1955, **13**, 261–292.

Cannon, W. D. The James-Lange theory of emotion. *Amer. J. Psychol.*, 1927, **39**, 106–124.

Cannon, W. D. *Bodily changes in pain, hunger, fear, and rage.* New York: Appleton, 1929.

Cannon, W. D. Again the James-Lange and the Thalamic theories of emotions. *Psychol. Rev.*, 1931, **38**, 281–295.

Carlson, A. J. *The control of hunger in health and disease.* Chicago: Univer. of Chicago Press, 1916.

Cohen, M. J., Landgren, S., Ström, L., & Zotterman, Y. Cortical reception of touch and taste in the cat. *Acta Physiol. Scand.*, 1957, **40**, Suppl. 135.

Davis, K. B. *Factors in the lives of twenty-two hundred women.* New York: Harper, 1929.

Faure, J. Modifications of the electrical activity of the rhinencephalon: Behavioral changes induced by hypophyseal gonadotropic hormones in the female rabbit. *EEG clin. Neurophysiol.*, 1957, **9**, 361–362.

Field, J., Magoun, H. W., Hall, V. E. (Eds.) *Handbook of physiology.* Vol. 3. Washington, D.C.: American Physiological Society, 1960.

Folkow, B., & von Euler, U.S. Selective activation of the noradrenaline and adrenaline producing cells in the cat's adrenal gland by hypothalamic stimulation. *Circulation Res.*, 1954, **2**, 191–195.

Forssberg, A., & Larsson, S. On the hypothalamic organization of the nervous mechanism regulating food intake. II. Studies of isotope distribution and chemical composition in the hypothalamic region of hungry and fed rats. *Acta Physiol. Scand.*, 1954, **32**, Suppl. 115, 41–63.

French, J. D., Longmire, R., Porter, R. W., & Movius, H. J. Extravagal influences on gastric hydrochloric acid secretion induced by stress stimuli. *Surgery*, 1953, **34**, 621–632.

Fulton, J. F. *Physiology of the nervous system.* (3rd ed.) London & New York: Oxford Univer. Press, 1951.

Funkenstein, D. H. The physiology of fear and anger. *Sci. Amer.*, 1955, **192**, 74.

Funkenstein, D. H., King, S. H., & Drolette, Margaret E. *Mastery of stress.* Cambridge, Mass.: Harvard Univer. Press, 1957.

Fuster, J. M. Subcortical effects of stimulation of brain stem on tachistoscopic perception. *Science*, 1958, **127**, 150.

Grossman, M. I. Gastrointestinal hormones. *Physiol. Rev.*, 1950, **30**, 33–90.

Halverson, H. M. Genital and sphiencter behavior of the male infant. *J. genet. Psychol.*, 1940, **56**, 95.

Harris, G. W., & Levine, S. Sexual differentiation of the brain and its experimental control. *J. Physiol. (London)* 1962, **163**, 42–43.

Heath, R. G. (Ed.) *Studies in schizophrenia.* Cambridge, Mass.: Harvard Univer. Press, 1954.

Hernández-Péon, R., Sherrer, H., & Jouvet, M. Modification of electrical activity in the cochlear nucleus during "attention" in unanesthetized cats. *Science*, 1956, **123**, 331–332.

Hodos, W., & Valenstein, E. S. Motivational variables affecting the rate of behavior maintained by intracranial stimulation. *J. comp. physiol. Psychol.*, 1960, **53**, 502–508.

Hoebel, B. G., & Tietelbaum, P. Hypothalamic control of feeding and self-stimulation. *Science*, 1962, **135**, 375–376.

Howarth, C. I., & Deutsch, J. A. Drive decay: The cause of fast "extinction" of habits learned for brain stimulation. *Science*, 1962, **137**, 35–36.

Hubel, D. J., Henson, C. O., Rupert, A., and Galambos, R. "Attention" units in the auditory cortex. *Science*, 1959, **129**, 1279–1280.

Hull, C. L. *Essentials of behavior.* New Haven: Yale Univer. Press, 1951.
James, W. On some omissions of introspective psychology. *Mind,* 1884, **9**, 1–260.
James, W. *The varieties of religious experience.* New York: Longmans, Green, 1920.
Jasper, H. H. Electroencephalography. In Wenfield, W., & Ericksen, T. (Eds.) *Epilepsy and cerebral localization.* Springfield, Ill.: Thomas, 1941.
Kaada, C. R., Jansen, J., Jr., & Andersen, E. Stimulation of the hippocampus and medial cortical areas in unanesthetized cats. *Neureology,* 1953, **3**, 844–857.
Kaelbling, R., King, F. A., Achenbach, K., Branson, R., & Pasamanick, B. Reliability of autonomic responses. *Psychol. Rep.,* 1960, **6**, 143–163.
Keys, A., Brozek, J., Henschel, A., Mickelsen, O., and Taylor, H. L. *The biology of human starvation.* Vol. II. Minneapolis: Univer. of Minnesota Press, 1950.
Koikegami, H., Yamada, T., & Usui, K. Stimulation of amygdaloid nuclei and periamygdaloid cortex with special reference to its effects on uterine movements and ovulation. *Folia Psychiat. Neurol. Japon.,* 1954, **8**, 7–31.
Lacey, J. I. Individual differences in somatic response patterns. *J. comp. physiol. Psychol.,* 1950, **43**, 338–350.
Lacey, J. I., Baterman, D. E., & VanLehn, R. Autonomic response specificity. *Psychosom. Med.,* 1953, **15**, 10–21.
Larsson, S. On he hypothalamic organization of the nervous mechanism regulating food intake. *Acta Physiol. Scand.,* 1954, **32**, Suppl. 115, 1–40.
Leeper, R. W. The role of motivation in learning: A study of the phenomenon of differential motivational control of the utilization of habits. *J. Genet. Psychol.,* 1935, **46**, 3–40.
Lindsley, D. B. Emotion. In S. S. Stevens (ed.), *Handbook of experimental psychology.* New York: Wiley (Interscience), 1951. Pp. 473–516.
MacLean, P. D. Psychosomatic disease and the "visceral" brain. *Psychosom. Med.,* 1949, **11**, 338–353.
Maher, B. *Principles of psychopathology.* New York: McGraw-Hill, 1966. p. 87.
Malmo, R. B. Activation: A neuropsychological dimension. *Psychol. Rev.,* 1959, **66**, 367–386.
Malmo, R. B., & Shagass, C. Physiological study of symptom mechanisms in psychiatric patients under stress. *Psychosom. Med.,* 1949, **11**, 25–29.
Michael, R. P. Estrogen-sensitive neurons in sexual behavior in female cats. *Science,* 1962, **136**, 322–323.
Miner, R. W. (Ed.) The regulation of hunger and appetite. *Ann. N.Y. Acad. Sci.,* 1955, **63**, 1–144.
Minnick, R. S., Marden, C. J., & Arieti, S. The effect of sex hormones on the copulatory behavior of senial white rats. *Science,* 1946, **103**, 749–750.
Morgan, C. T. *Physiological psychology.* (3rd ed.) New York: McGraw-Hill, 1965.
Olds, J. A preliminary mapping of electrical reinforcing events in the rat brain. *J. comp. physiol. Psychol.,* 1956, **49**, 281–285.
Olds, J. Physiological mechanisms of reward. In N. R. Jones (Ed.), *Nebraska symposium on motivation.* Lincoln: Univer. of Nebraska Press, 1955.
Olds, J. Self-stimulation of the brain. *Science,* 1958, **127**, 315–323.
Olds, J., & Sinclair, J. Self-stimulation in the obstruction box. *Amer. Psychologist,* 1957, **12**, 464.
Olds, J., Travis, R. D., & Schwing, R. C. Topographic organization of hypothalamic self-stimulation. *J. comp. physiol. Psychol.,* 1960, **53**, 23–28.
Papez, J. W. A proposed mechanism of emotion. *A.M.A. Arch. Neurol. Psychiat.,* 1937, **38**, 725–743.
Phoenix, C. H., Goy, R. W., Gerall, A. A., & Young, W. C. Organizing action of pre-

natially administered testosterone proprionate on the tissues mediating mating behavior of the guinea pig. *Endocrinology,* 1959, **65**, 369–382.

Quigley, J. P. The role of the digestive tract in regulating the ingestion of food. *Ann. N.Y. Acad. Sci.,* 1955, **63**, 6–14.

Roberts, W. W. Both rewarding and punishing effects from stimulation of posterior hypothalamus of cat with same intensity. *J. comp. physiol. Psychol.,* 1958, **51**, 400–407.

Roberts, W. W. Fear-like behavior elicited from dorsomedial thalamus of cat. *J. comp. physiol. Psychol.,* 1962, **55**, 191–197.

Rogers, T. A. *Elementary human physiology.* New York: Wiley, 1961. P. 338.

Rosenblatt, J. S., & Aronson, L. R. The decline of sexual behavior in male cats after castration with special reference to the role of prior sexual experience. *Behavior,* 1958, **12**, 285–338.

Rothballer, A. B., & Dugger, G. S. Hypothalamic tumor. *Neurology,* 1955, **5**, 160–177.

Schiele, B. C., & Brozek, J. Experimental neurosis resulting from semi-starvation in man. *Psychosom. Med.,* 1948, **10**, 31–51.

Schreiner, L., & Kling, A. Behavioral changes following rhiencephalic injury in the cat. *J. Neurophysiol.,* 1953, **16**, 643–659.

Segundo, J.P., Naquets, R., and Arana, R. Subcortical connections from the temporal cortex of monkey. *Arch, Neurol. Psychiat.,* 1955, **73**, 515–524.

Sem-Jacobsen, C. W., & Torkildsen, A. In E. R. Ramey and D. S. O'Doherty (Eds.), *Electrical studies on the unanesthetized brain.* New York: Harper (Hoeber), 1960. Pp. 280–288.

Spiegel, E. A., Miller, H. R., & Oppenheimer, M. A. Forebrain enraged reactions. *J. Neurophysiol.,* 1940, **3**, 538–548.

Venables, P. H., & Wing, J. K. Level of arousal and the subclassification of schizophrenia. *Arch. Gen. Psychiat.,* 1962, **7**, 114–119.

Wheatley, M. D. The hypothalamus and affective behavior in cats. *A.M.A. Arch. Neurol. Psychiat.,* 1944, **52**, 298, 316.

Wilder, T. The law of initial value in neurology and psychiatry. *J. nerv. ment. dis.,* 1957, **125**, 73–86.

Wood, C. D. Behavioral changes following discrete lesions of temporal lobe structures. *Neurology,* 1958, **63**, 215–220.

Chapter 4

Principles of Learning I. Conditioning, Reinforcement, and Extinction

INTRODUCTION

Some of the principles that have been established by experimental psychologists are relevant to a scientific interpretation of human behavior. Frequently these principles have been established by using lower animals as subjects. By far the most important generalizations based on animal experiments apply to the learning process.

A change in the probability that a given kind of behavior will occur in the presence of specific stimulus conditions is called *learning.* Many variables influence the likelihood of the occurrence of behavior. Some function to maintain the status quo; others give rise to, and strengthen, new behavior; still others modify behavior by weakening it, and hence decreasing the likelihood that it will occur under the same or similar circumstances. These are all aspects of the same process—the learning process.

Human learning is an exceedingly complex topic. The purpose of this chapter and the following chapter is to review and summarize some of the most important basic principles of learning. In this chapter we will cover respondent and operant (instrumental) conditioning, the processes of generalization, extinction, stimulus discrimination, and the delay of reinforcement. Some attention will be given to the effects of programming reinforcement, and to the relevance of programming to human behavior. Chapter 5 will deal with the effects of aversive stimuli, escape and avoidance conditioning, and punishment and conflict. Subsequent chapters will deal separately with other aspects of learning.

RESPONDENT CONDITIONING

The procedures and findings of the experiments of Ivan Pavlov and his associates are the basis for the definition of what has traditionally been called *classical conditioning.* This type of learning is now more frequently referred to as *respondent conditioning.* The terms are synonymous.

While studying the activity of the digestive glands and of certain aspects of the cerebral hemispheres, Pavlov noted that the salivary glands secreted at times even though there was no food present in the mouth. He called these *psychic secretions* and set out to discover their cause.

First, it was necessary to control all conditions which might possibly influence the experimental animal, a dog. The amount of contact between the dog and the experimenter was reduced as much as possible. A special chamber was designed so that external stimuli such as light, temperature, and sound could be controlled. The dog was secured in a harness and a glass tube was inserted into its cheek by a minor surgical operation in such a way that the exact amount of saliva flow could be measured. This apparatus is illustrated in Fig. 4-1.

In the early experiments psychic secretions were obtained by sounding a tuning fork and 7 or 8 seconds later presenting food to the animal. This procedure was carried out three times daily, with intervals from 5 to 35 minutes separating the paired presentations of food and tone. After approximately 10 presentations the tone alone evoked some salivation. Following 30 paired presentations, the tone alone evoked 60 drops of saliva. Thus, a stimulus which had previously been "neutral," in the sense that it had no power to reflexly elicit the salivary response, had come to elicit salivation when it repeatedly preceded a stimulus which was naturally adequate to elicit the salivary response.

The response to the substitute stimulus also showed characteristic changes in latency as the pairing of the natural and the previously neutral stimuli continued. During early tests salivation did not occur in response to the tone alone until the tone had been sounded for a considerable time (e.g., 18 seconds), whereas, as the number of paired presentations increased, salivation occurred in response to the tone alone after only 1 or 2 seconds.

The natural, or adequate, stimulus for the response is referred to as the *unconditioned stimulus* (UCS). The response to the UCS is referred

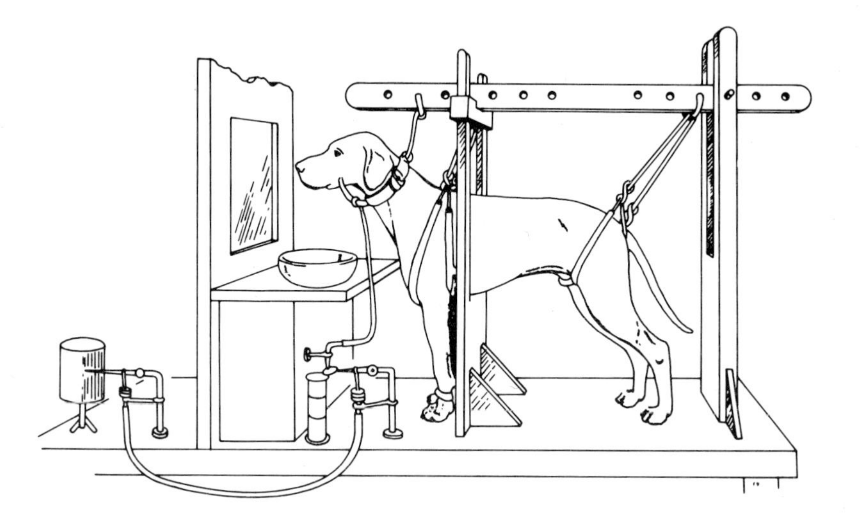

FIG. 4-1. Apparatus used by Pavlov to study respondent conditioning. (After Yerkes & Morgulis, 1909.)

to as the *unconditioned response* (UCR). The neutral stimulus which comes to act as a substitute for the original (unconditioned) stimulus is referred to as a *conditioned stimulus* (CS). Finally, after conditioning has taken place the response to the CS is called a *conditioned response* (CR).

During conditioning the CS–UCS pairing and the response are related as follows:

CS (tone)
UCS (food) ——————————————→ UCR (salivation)

In Pavlov's experiments meat powder, or some other food placed in the animal's mouth served as a UCS for the salivary response (UCR). Of course, the tone which was paired with it initially had no power to evoke the salivary response. As conditioning proceeds, the CR shows an increased probability of occurrence to the CS alone. At the end of conditioning, the CS regularly and automatically elicits the CR. The relationship is simply as follows:

CS (tone) — — — — — — — — — — — — — — — — → CR (salivation)

The salivary gland response is not a voluntary response. It is mediated by the autonomic nervous system. The word *respondent* refers to bodily reactions which occur automatically and involuntarily in the presence of certain stimuli—those designated as UCSs. The basis of respondent behavior is primarily the smooth muscle and glandular responses which we discussed in connection with the autonomic nervous system in Chapters 2 and 3. It is perhaps unfortunate that the conditioned salivary gland response has become so well known, for its use as an illustration of respondent conditioning overshadows the more important fact that conditioned respondents may involve major bodily organs and systems. Although the parasympathetic branch of the autonomic nervous system may act in piecemeal fashion, the sympathetic branch tends to stimulate every organ it innervates. It is only for the purpose of controlled experimentation that the response of a single organ, gland, or muscle is isolated for study.

The UCR may be a defensive response, such as the rapid and automatic withdrawal of a finger when shock is applied to it, or an eye blink in response to a puff of air applied to the cornea of the eye. Respondent conditioning in which an aversive UCS is employed is referred to as *respondent defense conditioning.* The Pavlovian experiment in which the salivary gland response was conditioned, or any conditioning in which the UCS is pleasant, satisfies an appetite, or is of a rewarding nature is called *respondent reward conditioning.*

Respondent defense conditioning need not involve highly circumscribed, simple reflexes, any more than the respondent reward conditioning discussed above must necessarily involve them. Respondent defense conditioning may involve the entire syndrome of responses associated with the sympathetic branch of the autonomic nervous system. It may also involve emotional circuits and pathways of the limbic system. When a distinctive stimulus such as a sharp clicking noise is used as a CS with shock as a UCS, an emotional response to the clicking noise can be built up very rapidly (i.e., in from three to six pairings). Animals conditioned in this way show a rapid acquisition of the freeze response, or other behavior associated with severe fear reactions. Emotional responses conditioned in this manner are given a special designation. They are called *conditioned emotional responses* and designated CER(a). The small letter (a) is used to distinguish this type of conditioned emotional response from other varieties to be discussed later (see p. 143). Many responses of the autonomic nervous system have been conditioned separately in the laboratory, including respiration, heart rate, changes in blood-sugar level, and vasoconstriction (see Kimble, 1961).

In view of the fact that so many of the bodily respondents that are known to be conditionable also have cue value (e.g., the person can Feel his heart beating, or may be aware of increased breathing), it is an interesting question whether conditioning can take place with the cues produced by internal bodily responses serving as conditioned stimuli. The usual conditioned stimulus is exteroceptive. Could an internal bodily cue, an interoceptive stimulus, also serve as a CS? The answer is that it certainly can. An electrical stimulus applied to practically any area of the cortex, for example, can serve as a CS for various motor responses. Even motor responses elicited by electrical stimulation may serve as conditioned stimuli for other motor responses (Loucks, 1936).

Scientists in the Soviet Union have been particularly active in experimenting with either conditioned stimuli or unconditioned stimuli, or both applied directly to various organs of the body. For example, in one study, by means of appropriate surgical procedures the amount of bladder distention of hospital patients could be controlled externally by the experimenter. He could pour liquid through a tube which had been inserted directly into the patient's bladder, and hence control increases or decreases of pressure in the bladder. During conditioning a dial placed in front of the patient indicated the bladder pressure and served as a CS. Following conditioning the experimenter could control the patients' urge to urinate by manipulating the dials appropriately. The patients reported an intense urge to urinate and displayed associated agitation at the sight of a high dial reading even though there was no liquid in their bladders. In this experiment the unconditioned stimulus was interoceptive and the conditioned stimulus was extroceptive.

Conditioning experiments which employ words or nonsense syllables as conditioned stimuli have particularly important implications for human adjustment. Vasoconstriction can be produced by a variety of UCSs, including electric shock or submerging the subject's hand in ice water. If either of these two unconditioned stimulus events are paired with a spoken nonsense syllable as a conditioned stimulus, vasoconstriction can be conditioned to the spoken nonsense syllable (e.g., Roessler & Brogden, 1943). A further refinement of this procedure is to have the subject think, or form an image of the nonsense syllable. Vasoconstriction can become conditioned to the subject's thought or image of it.

Higher Order Conditioning

Once the CR is well established the CS can be used to condition other responses. The CS may then be employed as if it were a UCS. A new neutral stimulus paired with it can come to elicit the response. This represents a second-order conditioned response. The process is referred to as *higher order conditioning.*

The second-order CR is appreciably weaker than the first-order CR, because during the conditioning of the second-order CR the first-order CS is undergoing extinction (see below). In the dog Pavlov found third-order conditioned responses almost impossible to establish. Detailed data on higher order conditioning in humans are not available. It might be expected, however, that the capacity of the human to acquire higher orders of conditioned responses is considerably greater.

Higher order conditioning demonstrates that neutral stimuli can acquire the capacity to strengthen stimulus–response associations. Consider the experiment in which the bodily response-produced cues of bladder distention were conditioned to dial readings. A second-order CR might be conditioned by pairing running tap water with high dial readings. The patients' urge to urinate would then occur in response to the sight and sound of running water. Before the original experiment, the high dial readings were a neutral stimulus. By virtue of the fact that they became a strong CS for the urge to urinate after first-order conditioning, they could be used as if they had the properties of a UCS in order to strengthen new associations.

Extinction of Conditioned Responses

Conditioned responses are relatively permanent. They show little or no dimunition with the passage of time. For example, retention of the conditioned eyelid reaction in man for 19 months has been demonstrated (Hilgard & Humphreys, 1938). Conditioned salivation in man has been demonstrated for 16 weeks (G.H.S. Razran, 1939). Conditioned flexion reflexes in the dog have been found to be retained after 30 months (Wendt, 1930). Some investigators have postulated that traumatically strong shocks may produce respondently conditioned fears which remain fixed even under special conditions that are known normally to have a weakening effect on CRs (e.g., Solomon, Kamin, & Wynne, 1953).

Conditioned stimuli do not lose their power to evoke CRs unless the individual remains in the situation, and the CS is presented repeatedly without the UCS being presented with it. When a CS is repeatedly presented without the UCS there is a progressive weakening of the CR. As the number of presentations of the CS alone increases, the probability of the CR occurring in response to it decreases. This process is called *experimental extinction.*

Many variables influence the rate of experimental extinction. Experimental psychologists talk about "resistance to extinction," referring to the fact that under certain conditions CRs are much more difficult to extinguish than under other conditions. Some of the more important of these conditions will be discussed on p. 122 and in connection with the programming of reinforcement (see pp. 120–122).

Suppose a student is taken into the psychological laboratory and seated in a barber chair with his head held in a fixed position by apparatus designed for that purpose. Then a small low intensity light is repeatedly turned on, followed each time by a sharp 4-pound puff of air delivered through a reduction fistula to the cornea of his eye. After conditioning the student will respond with a short-latency, involuntary eye-blink to the onset of the light CS. Let us suppose further that it took 25 trials, or CS–UCS pairings, to condition the response to full strength, and that extinction was begun immediately, with the probability of a CR occurring reduced to approximately 0 after 12 or 13 extinction trials. If these specially arranged extinction trials had not been given to the subject he probably would respond with a CR if he were brought back into the laboratory a week, a month, or perhaps even a year later.

During his daily life outside the laboratory the subject would certainly be exposed to low intensity lights of one variety or another. Couldn't it be expected that he would respond to them with a CR? The answer is probably not, because all of the other cues associated with the experimental situation would not be present. Just as the experimenter isolates one of the many responses made by the individual in the experimental situation in order to study it systematically, he also arranges for a particular, isolated CS to occur. However, the subject is affected by more than just that one cue. He responds to the CS, but he responds to it within the context of the entire stimulus complex. The CS–CR association is learned within a particular context. The CR may not extinguish unless the CS is presented repeatedly without the CR, within the context in which the original learning took place.

Spontaneous Recovery

If the response to the conditioned stimulus is tested some time after it has been extinguished, it will be found to have recovered. However, the CR rarely recovers in full strength with the passage of time, although it may temporarily almost equal the response strength which existed at the end of conditioning. Sometimes spontaneous recovery of the response occurs after only a few minutes rest. In other experiments spontaneous recovery does not reach its full strength until several hours have elapsed. In spite of this phenomenon it is possible to completely extinguish a conditioned response. After a number of reextinction sessions, the CR no longer shows spontaneous recovery following rest.

Spontaneous recovery undoubtedly operates in complex human behavioral situations. For example, in the psychotherapeutic process, where conditions are arranged specifically to produce behavioral change, such change is associated with many relapses, or the "spontaneous" return of maladaptive patterns of behavior, in spite of all the best intentions of the client. When the therapeutic process is successful these maladaptive behavior patterns eventually diminish in strength and finally disappear. Minor habits which have been extinguished (e.g., an incorrect grip on a golf club) also show spontaneous recovery with the passage of time, and a number of relearning sessions is frequently required before such habits can be completely extinguished.

Reinforcement

In respondent conditioning the UCS is referred to as a *reinforcer.* This term is used with reference to other kinds of conditioning as well. In the most generalized sense a reinforcer is any variable that strengthens behavior. When stimulus events are pleasant or rewarding, and when they occur immediately after a given behavior, they increase the likelihood that the same behavior will recur under the same or similar circumstances in the future. Those variables are said to reinforce the particular behavior in question.

Reinforcing events can be positive or negative, somewhat analogous to the distinction between respondent reward and respondent defense conditioning. Positive reinforcers can be thought of as rewarding events. A *positive reinforcer* is any variable whose *presentation* strengthens behavior. In contrast, a negative reinforcer can bc thought of as an aversive or punishing event. A *negative reinforcer* is any variable whose *withdrawal* strengthens behavior. As an example of the latter, the withdrawal of shock is a negative reinforcer in the laboratory situations we have mentioned in connection with respondent defense conditioning: when shock is withdrawn there is an increased probability that the individual will repeat that aspect of his behavior which immediately preceded the withdrawal of shock the next time it is presented.

Thus, suppose a tone is used as a CS and shock as an UCS in order to condition finger withdrawal to the tone. When the shock occurs the individual will reflexly withdraw his finger in response to it. Finger withdrawal is associated with release from pain. The shock which causes pain is a negative reinforcer. The tone paired with it will eventually elicit the CR of finger withdrawal, because finger withdrawal in response to the tone is anxiety-reducing. It avoids the painful stimulus that follows the onset of the tone. In effect the withdrawal or ending of shock is reinforcing. It strengthens the learning of the CR to the CS.

During respondent conditioning the UCS acts as a reinforcer, because without it the CS–CR connection could not be strengthened. However in higher order conditioning a stimulus which was previously a CS functions as a UCS. It has reinforcing properties. Neutral stimuli that have gained reinforcing properties through conditioning, through being paired with a UCS, are called *conditioned reinforcers.*

Neutral stimuli may be paired with not one, but a number of primary reinforcers. Therefore they acquire the capacity to serve as reinforcers for a wide variety of responses. Such stimuli are called *generalized reinforcers.*

A good deal of human behavior appears to be maintained in strength by stimulus variables which have acquired generalized reinforcing properties. Tokens such as money, degrees, awards, etc., are generalized reinforcers. Sometimes subtle behavioral cues have quite a strong generalized reinforcing capacity. For example, behavior on the part of another individual which radiates calmness, acceptance, understanding, warmth, and the like may have generalized reinforcing properties. Similarly, generalized negative reinforcers may consist of quite subtle behavioral cues indicative of detachment, hostility, or anxiety. They can be negatively reinforcing under a wide variety of behavioral circumstances.

Stimulus Generalization

When a response has been conditioned to a given stimulus it can be evoked by other similar stimuli even though these other stimuli have never been present during conditioning. That is, a CR may occur to stimuli which resemble the CS in some way. Stimuli which evoke a CR because they are similar to the CS, but which have not actually been used as CSs are called *generalized stimuli* (GSs). The process is called *stimulus generalization.*

The strength of the CR decreases with decreasing similarity of the GS to the original CS. That is, there is a gradient of stimulus generalization. The strength of the CR diminishes as the properties of the GSs resemble the original CS less. Some gradients of stimulus generalization are very steep. An individual conditioned to respond to a 1000-cycle tone may also respond to a 990-cycle tone and to a 1010-cycle tone, but thereafter show a sharp diminution in response strength as the stimulus tone (GS) deviates from a 1000 cycles in either direction. Other gradients may be quite flat: the organism may respond in almost full strength to stimuli which are very different from the original stimulus.

Before considering the variables which affect the slope of the generalization gradient, perhaps an example will make the phenomenon clearer. Figure 4-2 illustrates data from a generalization experiment by Hovland (1937a). In this study four tones were selected by psychophysical methods which were an equal number of just-noticeable differences (jnds) apart. The psychogalvanic skin response (GSR) was then conditioned to the highest tone for some subjects and to the lowest tone for others. The conditioned response of the subjects was then tested in extinction on the other tones. The design of the experiment made possible the pooling of results from all the subjects. Figure 4-2(a) shows a generalization gradient from this study. Response strength as measured in terms of the amplitude of the GSR decreased systematically with the difference between the training stimulus, designated 0 on the abscissa, and the various test stimuli, represented in terms of number of jnds from the training stimulus.

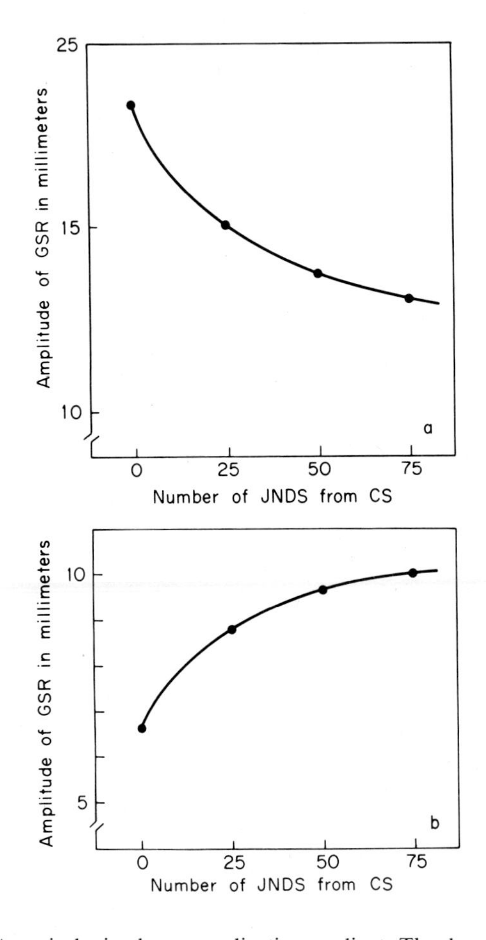

FIG. 4-2. (a) A typical stimulus generalization gradient. The dependent measure was the galvanic skin response; the conditioned stimulus is represented by the 0 point on the abscissa. The other tones are represented on the abscissa in terms of "just-noticeable differences" (JNDS) from it. (b) Typical gradient of generalization of extinction. The GSR was conditioned to all tones and then extinguished to an extreme tone only. The abscissa is the same as in (a). (Hovland, 1937.)

As a part of this same experiment Hovland conditioned another group of subjects to all four frequencies. He then extinguished half of the subjects on the highest tone alone and the other half of the subjects on the lowest tone alone. Zero on the abscissa of Fig. 4-2(b) represents the extreme tone to which the conditioned response had been extinguished. Notice that there was a gradient of *extinction.* That is, the extinction effects generalized to other tones on which the subjects had not been extinguished. Furthermore, the amount of extinction showed a systematic decrease as a function of dissimilarity between the test tone and the actual tone used in extinction. This part of the experiment illustrates that there is a *generalization of extinction.* When a response to a given stimulus is extinguished the same response to other similar stimuli shows a loss of strength.

A large number of variables affect the form of generalization gradients. Some, such as the number of stimulus dimensions changed, and drive, or level of arousal, influence generalization very strongly.

There is some evidence to support the expected fact that the more aspects of the stimulus which are changed the weaker is the CR to the GS (e.g., Fink & Patton, 1953; White, 1958). However, dimensions of stimulus similarity are frequently very difficult to define. It may also be difficult to determine the specific aspects of the stimulus complex to which the subject is responding. It would seem that the degree of generalization between two stimuli would serve as an index of the extent to which they are perceived as similar. But it is a fact that generalization frequently occurs in almost full strength to stimuli which, at least from an objective observer's point of view, appear to be widely dissimilar.

Human language habits mediate generalization in extremely complex ways. For example, in an experiment by Diven (1937) which has been replicated many times a GSR conditioned to the word "barn" was found to generalize to other rural words. The importance of symbolic and ideational mediation for human adjustment will be detailed in later chapters. Stimulus generalization is an integral aspect of human behavior patterns —those of major importance for human adjustment as well as minor habits.

Arousal flattens the generalization gradient. For human subjects arousal can be produced by words. This can easily be demonstrated in the laboratory. It has been found that highly anxious subjects show a flatter generalization gradient than less anxious subjects (e.g., Rosenbaum, 1953). The experience of anxiety is one of the central factors in human maladaptive behavior. Complex and strong generalization effects can readily be shown to be operative in human maladjustment. The chronically anxious person responds to many stimuli with anxiety that was originally learned in a circumscribed interpersonal context. The approval-dependent individual may try to seek the approval of almost everyone he meets, although such behavior was originally learned in response to parental demands.

OPERANT CONDITIONING

In laboratory respondent conditioning the reinforcing stimulus occurs irrespective of the individual's behavior. The subject is restrained so that the behavioral possibilities in the situation are highly limited. The reinforcing stimulus (UCS) elicits the UCR without the subject having much control over it. In contrast, operant, or instrumental, conditioning is so named because in this type of conditioning situation the behavior of the subject is instrumental in producing the reinforcing stimulus. The subject is far less restrained. The response to be learned is not forced to occur. In fact, it must be emitted before it can be reinforced.

One of the early pioneers in the study of instrumental conditioning was the American psychologist, Edward Lee Thorndike. One of his original experiments involved placing a cat in a puzzle box. It could release itself from the box in order to reach food only by a series of relatively complex manipulations. The dependent variable was the amount of time required for the animals to escape from the box and obtain reinforcement. Each time the animal escaped was counted as one discrete trial. On the first successful trial the animal would hit the release mechanism by accident. On successive trials its behavior would become more and more concentrated in the area of the release mechanism. After a number of successful trials it would release the mechanism immediately upon being placed in the box.

Generally in laboratory instrumental conditioning situations the response to be conditioned is not one that would be expected to occur very frequently. Instrumental conditioning consists of strengthening the response to the point where it occurs consistently and promptly in the presence of a particular stimulus complex. The paradigm for instrumental conditioning is as follows:

$$S \rightarrow R \rightarrow S^r$$

where S represents the stimulus complex, R represents the response to be conditioned, and S^r, the reinforcing stimulus.

The instrumental learning situation allows for a good deal of free responding. The animal's behavior is not controlled to the point where it is possible to specify the exact stimuli which elicit it, although, of course, there are certain definite stimuli at work. There is some stimulus generalization from other situations the animal has been in. Even the restless behavior which characterizes the presence of a drive is not entirely random, but it is, at least in part, determined by the animal's past learning in the presence of the particular drive stimulus. Nevertheless, since the stimuli cannot always be specified psychologists prefer to think of behavior in these situations as being emitted, rather than elicited. The emitted acts of the animal are referred to as *operants*, and instrumental conditioning is frequently referred to as *operant conditioning.* Sometimes this latter terminology is more appropriate, particularly in regard to human verbal behavior. It is often conceptually simpler to think of such behavior as being emitted and later to specify the variables controlling it.

There are a wide variety of laboratory instrumental situations. Each is designed for the observation and measurement of specific aspects of the conditioning process. The reader is probably already familiar with the T maze which is used to study simple choice behavior in animals, with the straight runway, and with the Skinner box. Whereas in the straight runway or T maze the animal's response may be measured in terms of speed or latency, in the Skinner box situation, rate, or frequency of emitted response serves as the dependent variable. A typical Skinner box conditioning apparatus is shown in Fig. 4-3. This situation allows for a good deal of free responding. The box is empty except for a manipulandum which the animal must press, or, when pigeons are used, a disc which must be pecked. The apparatus is wired so that appropriate bar-pressing or pecking activates a food mechanism.

Operant behavior in the Skinner box is typically measured by a device known as a cumulative recorder. A recording device is attached to the apparatus so that each time the disc is pecked or the lever pressed a recording pen moves one step upward, indicating the response on a continuously moving strip of paper. The paper moves under the pen at a constant rate. If no response is made a straight line is recorded.

Figure 4-4 illustrates cumulatively recorded extinction curves for four different experimental animals (rats). Each vertical movement of the pen indicates a response, whereas the length of the pauses between responses is indicated by the straight horizontal lines on the record. Notice that the pauses (horizontal lines) become longer as extinction proceeds.

The Respondent–Instrumental Conditioning Relationship

The relationship between respondent and instrumental conditioning can be seen most clearly in the simplest of laboratory learning situations: the straight runway. Here the animal must learn to traverse a

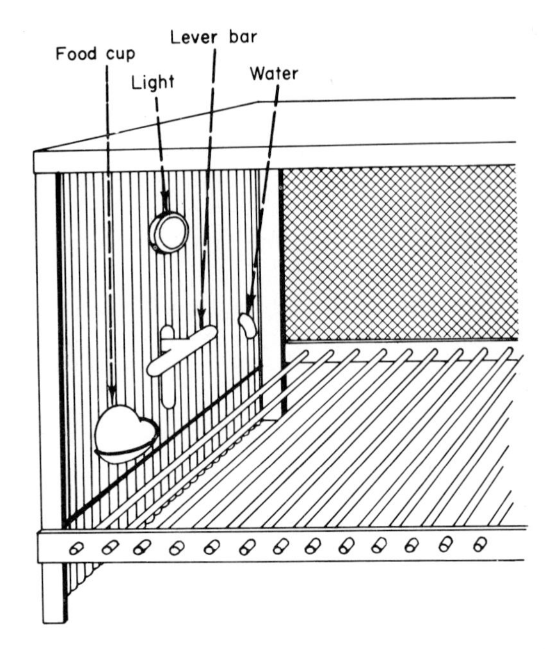

FIG. 4-3. A Skinner box. Depression of the lever bar causes a food pellet to drop into the food cup.

runway for food which is located in a goal box at the end of it. Starting and running speeds of the animal are determined by small photoelectric cells appropriately placed at various points in the alley. When animals are first placed in the starting box at the beginning of the runway and the starting door is raised they usually show random activity such as sniffing the walls, raising and biting the wire mesh at the top of the runway, or going a small distance and then returning. Eventually the animals will traverse the runway, and at the sight of food show a marked consummatory response which consists of seizing and eating. On subsequent trials competing responses are still evident. The animal may proceed half-way down the alley and double back again before going to the goal box. As the trials continue the speed of locomotion increases. Competing responses gradually drop out. At the end of conditioning the animal begins running immediately upon the raising of the starting box door, traverses the runway at full speed, and executes the consummatory seizing and eating response.

Figure 4-5(a) represents the end of the instrumental chain near the goal box. S_A and S_A' represent stimulus cues at various points in the alley. The locomotive response elicited at various points in the alley is

A rat in a Skinner box, used to study the operant behavior of lower animals. (Photograph by Tor Eigland.)

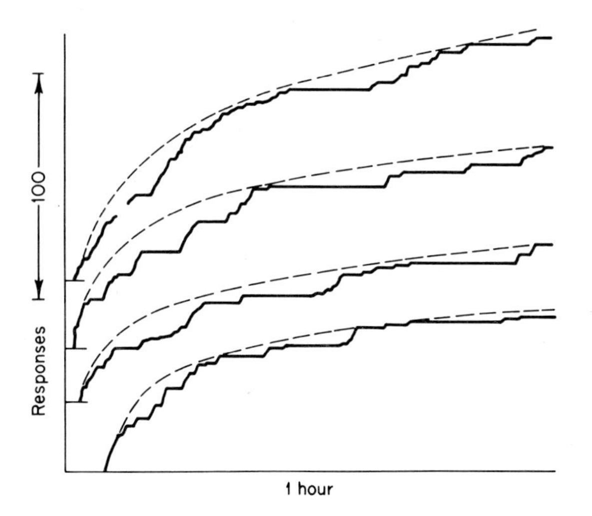

FIG. 4-4. Cumulatively recorded extinction curves for four experimental animals. Note that the pauses (horizontal lines) become longer as extinction proceeds. (Lundin, 1961.)

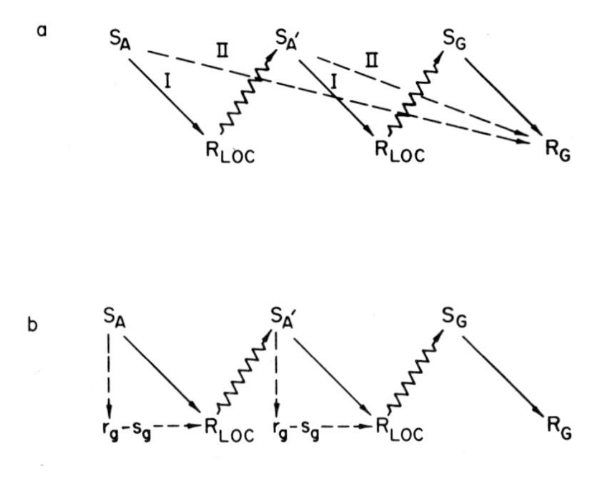

FIG. 4-5. (a) The relationship in a straight alley of stimulus cues in the alley ($S_{A,}$ $S_{A'}$) and in sight of the goal (S_G) to responses to them (R_{LOC}, R_G). (b) The operation of the r_g-s_g mechanism in a straight alley situation; r_g is the fractional anticipatory goal response and s_g, its associated stimulus cue. In (a) the arrows marked I represent the instrumental conditioning of the locomoting response. The arrows marked II show how the consummatory response (R_G) becomes an integral part of the instrumental learning. The respondent pairing of S_G, sight of the goal, and R_G, the goal response, generalizes to alley cues ($S_{A,}$ S_A) further and further back in the alley as learning continues. The dashed lines in (b) represent the hypothetical development of the r_g–s_g mechanism. See text for further explanation. (Spence, 1956.)

represented by R_{LOC}. The stimulus cues associated with the sight of food are represented by S_G, and the consummatory response, by R_G. Because the stimulus cues near the goal box (S_G) immediately precede the reinforcing stimulus, food, respondent conditioning would be expected to occur. That is, alley cues designated S_A' in Fig. 4-5(a) become conditioned stimuli, eliciting anticipatory chewing, salivary, or other related responses, because they have been paired with the sight and smell of food (UCS), which, in turn, reflexly elicits the consummatory response, R_G. The R_G response is analogous to a UCR.

In other words, the following respondent conditioning events are embedded within the instrumental behavior:

CS (end of alley)
UCS (food) ————————————→ UCR (eat)

After a number of trials respondent conditioning occurs:

CS (end of alley) — — — — — — — — — — — → CR (salivate)

Since these end of alley cues (S_A') are repeatedly paired with the sight of food (S_G) they would be expected to take on the properties of conditioned reinforcers. Therefore, they strengthen the connection between S_A' and stimuli further back in the alley. That is, higher order conditioning occurs because of the following pairing:

CS (middle of alley)
CS (end of alley) — — — — — — — — — — — — → CR (salivate)

This results in a higher order CR:

CS (middle of the alley) — — — — — — — — → CR (salivate)

Although as the number of trials increases this higher order CR continues to be strengthened, like all higher order CRs, it is weaker than the CRs formed at the end of the alley (first-order CRs). Finally, still higher order CRs occur as the number of experiences the animal has in the situation increases: eventually cues toward the *beginning* of the alley develop some power to elicit anticipatory CRs. With thorough conditioning the animal shows anticipatory chewing movements, salivation, and excitement as soon as it is placed in the start box at the beginning of the alley.

Learning psychologists call these anticipatory conditioned respondents *fractional anticipatory goal responses* (r_g). Presumably they have cue value for the animal (s_g), and because of this, they serve to increase the animal's general level of excitement. The r_g and the stimulus cue associated with it, s_g, is called the r_g–s_g mechanism (Spence, 1956). It is illustrated in Fig. 4-5(b).

Habit sequences similar to those represented by the r_g–s_g mechanism undoubtedly underlie gradients of complex human responses. Quite analogous to the increasing strength of the r_g–s_g mechanism as the animal approaches the end of the alley is the behavior of the man who, coming home for dinner, *walks* down the street and *runs* up the front steps.

The distance from the reinforcing stimulus that defines gradients of anticipation and excitement may be temporal as well as spatial, especially in the case of human behavior, where the conditioned stimuli involved may be symbolic. Because of generalization from other goal-gradients in our lives, the r_g–s_g may also be strengthened by additional anticipatory cues from an outside source. Thus, in the middle of the month of May a TV audience may be told, "There are 30 great shows coming on this network beginning September 11th;" and then they will be told the same thing from time to time during the summer, and with increasing frequency as the end of August and the beginning of September approach. By the time they hear the phrases, "There are 30 great shows coming on this network . . . next week," or " . . . tomorrow night," excitement among the TV audience may be high. The average man is likely to scoff at such "propaganda" insofar as he feels that it does not influence him, yet it can be shown that the probability of TV-viewing behavior in a population is increased by such techniques.

The straight alley situation was specifically chosen to illustrate the r_g–s_g mechanism because of its simplicity. Analysis of complex instrumental or operant situations shows the operation of respondent conditioning as an inherent part of the process. In later chapters it will be seen how anticipatory emotional cues operate in a similar manner. Cues of frustration, for example, become anticipatory, allowing for the operation of a mechanism similar to the r_g–s_g mechanism (r_f–s_f).

Delay of Reinforcement

In the straight runway, when the r_g–s_g is not strongly conditioned to cues far back in the alley, there are differential running speeds in various portions of the alley. The animals run faster the closer they are to the goal portion. There are many ways of conceptualizing this. An interpretation which is entirely congruent with the analysis of conditioning involving the operation of the r_g–s_g mechanism described above postulates the operation of a delay-of-reinforcement gradient in conditioning. That is, the fact that the greater the distance the animal is from the reinforcing stimulus, the less the strength of conditioning, might be explained by the fact that the further the animal is from the reinforcement, the longer the delay before he receives it. Delay of reinforcement is known to be an important variable in conditioning. Thus, in the Skinner box the longer the delay between the animal's bar-press and the delivery of reinforcement, the slower is the acquisition of the bar-pressing response. Conditioning is strongest when reinforcement is delivered immediately. Furthermore, performance can be improved or impaired by decreasing or increasing the delay of reinforcement (Spence, 1956). This is a widely generalizable principle of learning.

Under most regular life circumstances reinforcement seldom follows behavior immediately. On the other hand, most human situations are more complex, insofar as reinforcement can be mediated symbolically. Symbolic reinforcers can be delivered quickly. However, the reinforcement value of symbolic cues must itself be learned. With two equally bright children, one from a lower socioeconomic environment, and one with a middle-class background, the use of symbolic or of generalized reinforcers might be effective for one and not for the other. The middle-class child has typically learned to respond to symbolic rewards (e.g., a gold star on a chart), to value achievement, and to tolerate a delay of tangible reward.

Delay of reinforcement may be an important aspect of maladaptive behavior. The tenacity of a great deal of maladaptive behavior appears to be due to the fact that it is reinforced immediately (usually by prompt anxiety reduction), although in the long run the behavior is punishing. Punishment that occurs after a period of time is far less effective than punishment that occurs immediately. The powerful effects of immediate anxiety reduction, or of immediate positive gratification easily win out in competition with delayed punishment as determiners of how the individual will behave. The delinquent may steal, the alcoholic may drink, the neurotic may be inhibited because of the immediate consequences of such behavior.

Conditioned Discrimination

The response to a generalized stimulus can be weakened by presenting the stimulus continuously without reinforcement. If both the appropriate CS and a similar GS are presented, and if the response to the former is always reinforced, whereas the response to the latter is never reinforced, then a conditioned discrimination develops. That is, the individual learns a differential response to the two stimuli. He learns to respond to the CS, and not to respond to the GS.

Under the free-responding conditions of instrumental or operant conditioning, if an organism is rewarded for a particular behavior only when a certain stimulus is present, and never in its absence, then behavior can be brought under stimulus control. A stimulus which is always present when reinforcement occurs, and therefore marks the occasion when the occurrence of particular behavior will be followed by reinforcement, is called a *discriminative stimulus* (S^D). A stimulus which marks the occasion for behavior not being reinforced, because reinforcement never occurs when it is present, is called an S-delta (S^Δ).

The experimental psychologist may use a green light as an S^D, and a red light as an S^Δ. When the green light is on in the Skinner box, depression of the bar will activate the food magazine. On the other hand, a red light on, and the green light off, can be arranged to coincide with a circuit break so that the food magazine cannot be operated by depressing the bar. Bar-pressing behavior then becomes dependent upon the presence of a green light. A *discriminative function* is said to have been set up. Behavior has been brought under stimulus control.

People learn to respond differentially to a multiplicity of stimulus cues. Certain gestures, a smile, a pleasant expression, may be discriminated by the individual as marking the occasion when certain behavior will be rewarded. Other cues signal punishment or nonreward. The face of an employer or a parent wearing a frown may be discriminated out of a crowd of faces, and may serve as an S^Δ effecting the inhibition of certain behavior.

Shaping

Behavior may be shaped, or molded, by a process involving differential reinforcement. Once the reinforcing mechanisms of a given individual are known (once it is understood what is reinforcing for *him*), then reinforcement may be made contingent only upon evidence of behavior which is desired, whereas other behavior can be consistently not reinforced. The process of shaping behavior involves the administration of reinforcement for successive approximations to the final desired behavior. During the initial stages of the shaping of behavior, any behavior which even remotely resembles that which is eventually desired is reinforced. After the first approximate responses are made, reinforcement can then be withheld until a closer approximation to the desired response occurs. By gradually shifting the performance criterion, extremely complex behavior may be conditioned. Parents and agents of social institutions shape behavior all the time in this manner, although the behavioral control exerted under these circumstances is generally intuitive, lacking planning and a truly efficient arrangement of reinforcing contingencies.

VARIABLES AFFECTING RESISTANCE TO EXTINCTION

Experimental extinction was spoken of as if it were an invariant process, given the conditions that allow for its operational definition. That is, the individual must remain in the learning situation and the CS must be repeatedly presented without the reinforcing stimulus. However, behavior varies widely in its resistance to extinction. Some CRs extinguish very rapidly. Others are highly resistant to change. Generally, when behavior shows itself to be highly resistant to extinction, only a few reinforcements breaking the extinction process are enough to bring the behavior back up to full strength. It can be maintained at full strength indefinitely with a minimal number of intermittent reinforcing episodes. It is important, therefore, to look at some of the conditions affecting resistance to extinction.

Reinforcement Programming

Among the most important of these is the programming of reinforcement during learning. One common type of schedule of reinforcement is administration of reinforcement every time the individual makes the appropriate response. That is called a *continuous reinforcement schedule* (crf). Reinforcement programming other than the crf schedule involves partial, or intermittent, reinforcement. There are many ways that intermittent reinforcement can be programmed: every other response may be reinforced; reinforcement may occur at random, and so forth. The effects of various schedules of reinforcement during both acquisition and extinction have been studied extensively. Some of the more important of these are discussed below (see pp. 124–130).

Intermittent reinforcement results in greater resistance to extinction than does continuous reinforcement. This is true for laboratory respondent conditioning, for operant situations in which discrete trials are employed, and for free-responding operant situations.

Typical extinction curves for intermittent vs. continuous reinforcement are shown in Fig. 4-6. The total number of responses is plotted on the ordinate, and time is plotted on the abscissa. Compared to the curve of the continuously reinforced subject, the curve for the intermittently reinforced subject rises higher, and flattens out later. This indicates that the intermittently reinforced subject makes more responses and continues to respond for a longer period of time.

Continuously reinforced behavior does not show persistence. Compared to intermittently reinforced behavior it tends to show rapid extinction. Another important feature of behavior learned under a crf schedule is the fact that the course of its extinction is characterized by emotionality. This may be reflected by the presence of bursts and depressions in the record. It can be demonstrated if a measure is taken of the force with which the animal depresses the response bar in a Skinner box. In any event, subjects undergoing extinction who have been conditioned on

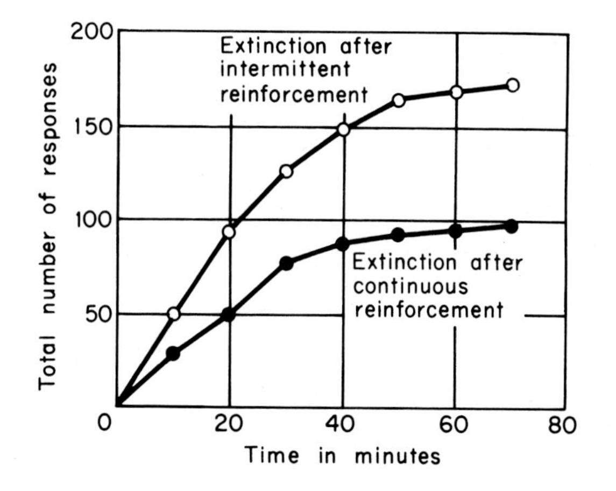

FIG. 4-6. Extinction curves following intermittent and continuous reinforcement. Notice that intermittent reinforcement produces greater resistance to extinction. (Adapted from Kendler, 1963.)

a crf schedule react emotionally. Animal subjects may become irritable, attempt to bite the experimenter, etc. Human subjects may show signs of restlessness, anger, or irritability.

Intermittently reinforced subjects show less signs of emotionality during extinction. Extinction conditions for them do not represent such an abrupt change because during learning, reinforcement did not occur every time they responded, even though the response was "correct."

Varied Reinforcement

A second set of conditions which increases resistance to extinction has come to be known as *varied reinforcement*. The term comes from studies in which various aspects of reinforcement or various conditions in the experimental situation associated with reinforcement have been systematically changed from trial to trial. For example, McNamara and Wilke (1958) conditioned five groups of rats to traverse a straight runway. For one group, the control group, the experimental conditions were kept constant from trial to trial. For the other four groups variations were systematically introduced. For one group changes in the goal box and in the illumination level were made from trial to trial; in another group 0 and 20 second delays of reward were alternated from trial to trial; in a third group the two previously mentioned procedures were combined; and in a fourth group the percentage of reinforcement, drive state, reward, environmental conditions, and delay of reinforcement were all manipulated simultaneously. Following 36 conditioning trials the animals were extinguished.

The results of this experiment are shown graphically in Fig. 4-7. In the legend in the upper left-hand corner of Fig. 4-7 "cue, response" refers to changes in illumination, the shifting of goal boxes, and the insertion of hurdles in the runway. "Delay" in the legend refers to the alternation of immediate reward and 20-second delay of reward. "Reward" refers to the manipulation of drive: animals were deprived of food on some days, of water on others, and of both on still other days. The curves of Fig. 4-7 illustrate an inverse relation between regularity of training conditions and resistance to extinction. The more variable and incidental conditions are during learning, the greater is the resistance to extinction. This is another generalizable principle. The variable, essentially "chance" character of most human learning undoubtedly contributes greatly to its persistence even when it is not rewarded.

Aversive Stimulation

Conditioned responses based upon unconditioned aversive stimuli which elicit the emotional responses of fear, anxiety, or frustration are

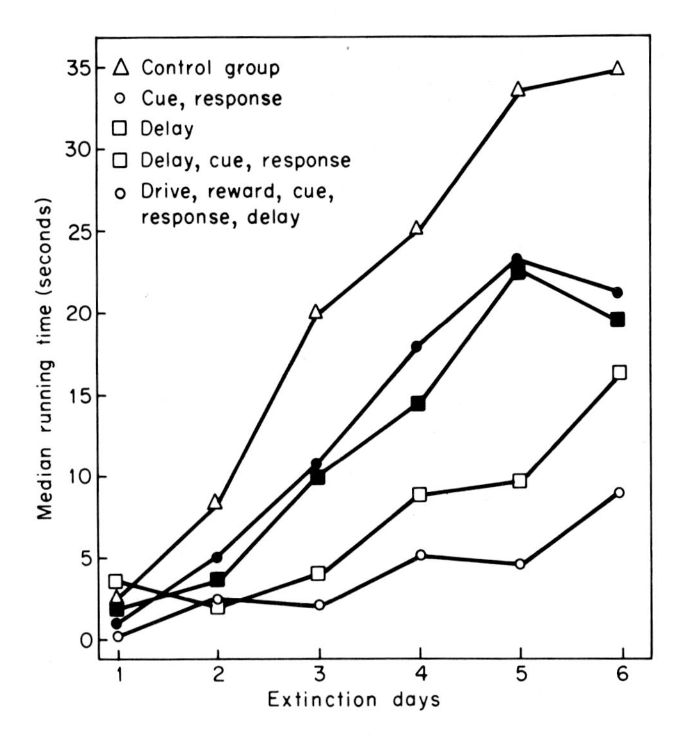

FIG. 4-7. A plot of the running time during extinction for groups trained under a variety of conditions of reinforcement. In general, the more variable the conditions the greater the resistance to extinction. See text for further explanation(After McNamara & Wilke, 1958.)

extremely resistant to extinction. Typically, since such aversive stimuli lead to escape and avoidance behavior, the individual does not allow himself to get back into the situation. Hence, the conditions necessary for extinction to take place simply never occur.

An experiment by Adelman and Maatsch (1956) illustrates the resistance to extinction that does occur when the appropriate conditions are arranged for experimental extinction to take place. They used escape from frustration as a reinforcer during learning. They then compared the extinction of these animals with another group for which food had been the reinforcer during learning. The food-reinforced animals required about 60 trials to reach the criterion of extinction. In contrast, the frustration-conditioned animals showed no indication of extinction after 100 trials.

The work of Maier (1939) on abnormal fixation in rats also illustrates the tenacity of frustration-produced responses. In Maier's experiments animals were forced to jump at one of two doors, both of which were marked by distinctive stimuli, e.g., a square and a triangle. The incorrect stimulus door was locked. When the animal jumped at it he sustained a blow and fell into a net below. When he jumped to the correct door he was rewarded by landing on a platform where there was food. The problem was then made insoluble by randomly varying the door which was considered to be "correct." Under these conditions animals developed very strong and persistent adjustments. They consisted mainly of right or left position habits. These response tendencies become highly fixated and compulsive in nature. Following the development of these strongly fixated responses, 75 percent of the animals were unable to learn when the problem was again made soluble. The fact that animals continued the wrong response was not due to their lack of ability to discriminate between the two stimuli. The fact that they were able to discriminate was indicated by their much longer latencies in response to the nonreinforced stimulus (Feldman, 1953). Nevertheless they persisted in the same stereotyped behavior.

Analysis of the precise contingencies involved in these experiments is quite complex. It is possible that these results can be explained theoretically on the basis of the partial reinforcement effect. Frustration-induced behavior will be more extensively considered in Chapter 6. At this point it should be noted that the rigid, compulsive behavior produced under these conditions does appear to be the analogue of similar human behavior under frustrative conditions. It is very likely that similar principles underlie human compulsive behavior.

SCHEDULES OF REINFORCEMENT

Interval Schedules

In the Skinner box type of experiment reinforcement may be programmed according to a number of different schedules. In some schedules reinforcement is administered after a certain amount of time has elapsed following the last reinforcement. These are referred to as *interval schedules.* They are of two varieties—fixed or variable. In *fixed-interval* schedules reinforcement is presented for the first response that occurs following a fixed interval of time. If the fixed interval is 5 minutes the apparatus is so arranged that the animal is reinforced for the first response made following a 5-minute interval since the previous reinforcement. In *variable-interval* schedules reinforcement is programmed to occur at varying intervals, having an arbitrary range and a fixed mean value. That is, a variable-interval 5-minute schedule might involve intervals as short as 1 minute or as long as 10 minutes, with the average interval being 5 minutes.

Fixed-Interval Reinforcement: First, let's turn our attention to the behavioral characteristics exhibited during conditioning on fixed-interval schedules. Fixed-interval conditioning involves three distinct stages. In the very beginning of conditioning on fixed-interval schedules organisms typically respond at a rapid rate following each reinforcement; then, particularly if the interval is long, the response begins to extinguish. When the next reinforcement is presented the process is repeated. In the middle phases of conditioning, however, a greater resistance to extinction develops and the rate becomes higher and more regular. Finally, in later stages of conditioning animals develop a conditioned time discrimination. That is, there is a pause following each reinforcement, and the animal begins to respond only as the time for the next reinforcement approaches.

Both the response rate (slope of the cumulative record) and the length of time required for the transition from one stage to the next vary with the length of the fixed interval. Generally, the shorter the interval the higher the response rate. With very long intervals the discrimination in the third stage may not develop.

The conditioned discrimination can be enhanced by adding an external stimulus which serves as a "clock." Typically, a small spot of light is shown in front of the animal; it gradually becomes larger until, when it reaches its maximum size, reinforcement is presented.

Figure 4-8 shows a portion of a cumulative record obtained from an animal operating on a fixed-interval schedule of reinforcement. Each vertical line (hash mark) on these curves indicates a single reinforcement. Notice the developing time discrimination from segment A to the later segment C.

In contrast to extinction under crf schedules, extinction under fixed-interval schedules is smoother and characterized by a more regular rate of responding. Behavior conditioned under fixed-interval schedules is also more resistant to extinction. Response rates show a slow steady decline.

Variable-Interval Reinforcement: Figure 4-9 shows a portion of a cumulative response record obtained from an animal operating on a variable-interval schedule of reinforcement. Behavior under variable-interval schedules does not show the characteristic time discrimination found in fixed-interval schedules. Under variable-interval schedules animals develop sustained steady rates of responding, high or low depending upon the size of the interval employed (the shorter the mean interval the

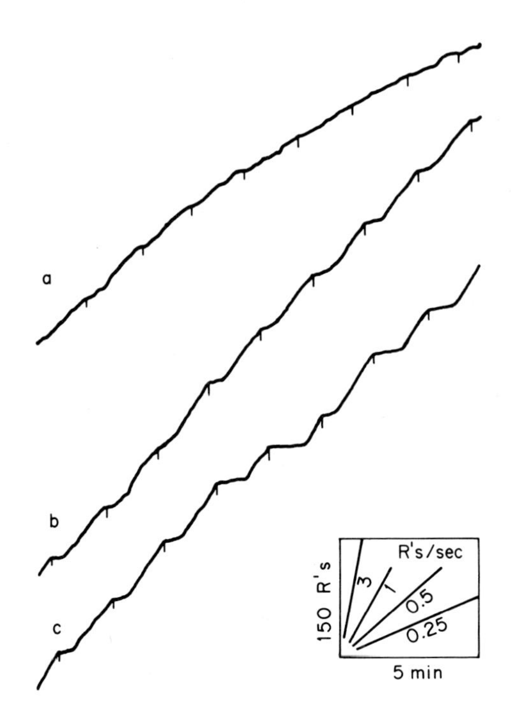

FIG. 4-8. Development of behavior on a fixed-interval schedule. Notice the developing time discrimination (longer pauses after each reinforcement) from the early segment A to the later segment C of the record. (After Ferster & Skinner, 1957.)

higher the response rate). Pigeons reinforced on a variable-interval schedule having a mean interval of 5 minutes have responded for as long as 15 hours at a rate of two responses a second without pausing for more than 20 seconds at any time (Skinner, 1953). During extinction on variable-interval schedules animals typically continue to respond for long periods of time at a sustained rate which declines very slowly as extinction continues.

Ratio Schedules

In ratio schedules reinforcement is made contingent upon the animal's making a certain number of responses. Reinforcement may be programmed on a *fixed-ratio schedule*, involving the delivery of reinforcement following a fixed number of responses, or it may be programmed on a *variable-ratio schedule* with the average ratio of responses to reinforcement fixed at some mean value.

Fixed-Ratio Reinforcement: Assuming that the response: reinforcement ratio is not extremely high, and that the amount of reinforcement is not too small, high response rates tend to develop under fixed-ratio schedules. If conditioning is begun using a very low ratio and the ratio is gradually increased, it is possible to develop very high ratios. The maintenance of a high rate of response from a pigeon operating on a ratio of 1000:1 is not uncommon. As conditioning proceeds a discrimination is established under fixed-ratio schedules. The animal is never reinforced immediately following the last reinforcement. Therefore, a characteristic pause following reinforcement soon develops. The length of the pause increases with the size of the ratio. Following this characteristic break the animal then responds very rapidly until the next reinforcement.

Figure 4-10 shows the final performance of two pigeons, birds A and B, conditioned on fixed-ratio schedules of 200:1 and 120:1, respectively. Bird A paused after each reinforcement for a period varying from a few seconds to more than a minute; then its rate shifted abruptly to 3.5 responses per second, which was maintained until reinforcement. Even

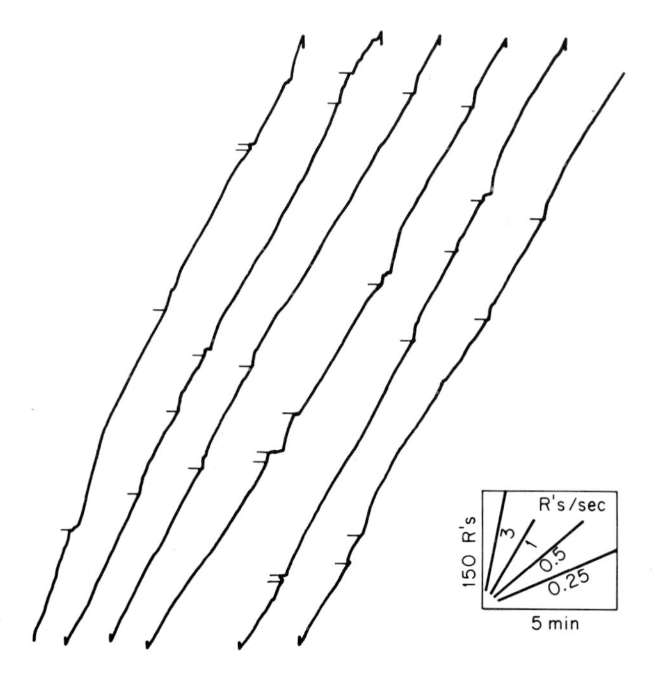

FIG. 4-9. Behavior on a variable-interval schedule. Note the steady, sustained, moderately high rate of responding characteristic of this schedule. (After Ferster & Skinner, 1957.)

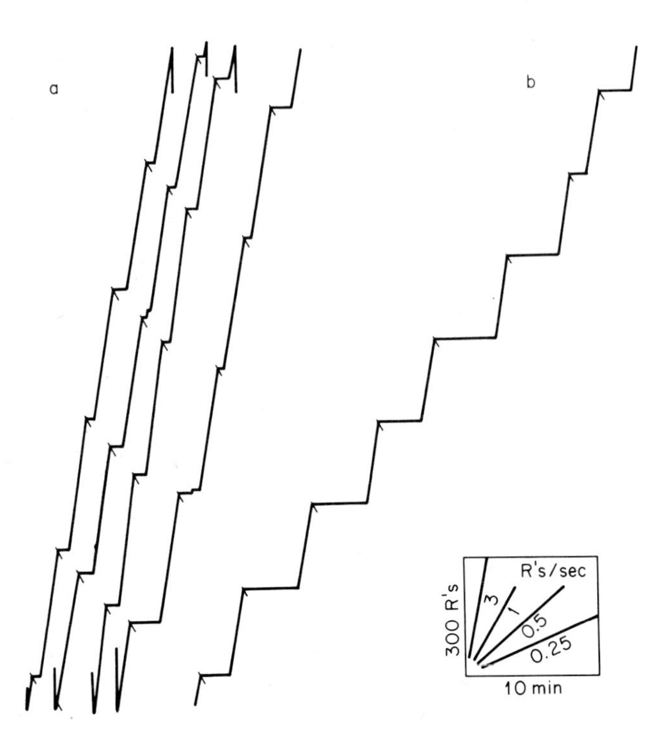

FIG. 4-10. Fixed-ratio behavior. These are the records of two different pigeons, bird A on a fixed-ratio schedule of 200 responses to 1 reinforcement and bird B on a ratio of 120 responses to 1 reinforcement. (After Ferster & Skinner, 1957.)

though it had a smaller ratio than bird A, bird B showed somewhat longer pauses and an instantaneous rate change from 0 to the terminal rate (Ferster & Skinner, 1957). Although in general the length of the pause increases with the size of the ratio, there are exceptions—individual differences among birds—at least when the two ratios being compared are as close as the 200:1 vs. 120:1 ratios of this particular example.

During extinction following fixed-ratio reinforcement there is a continuation of the response rate at a high level, followed eventually by an abrupt cessation of behavior. The occurrence of extinction on this schedule is typically precipitous and relatively complete; although a long pause may be followed by another burst of responding and, in turn, a sudden cessation of behavior.

Variable-Ratio Reinforcement: Under variable-ratio schedules, especially if a low ratio is employed in the beginning of conditioning, extremely high response rates may be established. Figure 4-11 shows the rate of responding of a bird under a variable-ratio schedule of 110:1 with a range of 0 (the very next response) to 500. After beginning briefly with crf, followed by relatively brief transitions to increasing variable-ratio schedules, a stable and extremely high overall rate of approximately 400 responses per minute was obtained under the variable-ratio reinforcement schedule of 110:1 for this animal.

As in the case with fixed-ratio schedules, the response rate during extinction does not show a continuous decline; rather, responding continues at a high rate which drops abruptly to zero. Sustained runs of responding at a high rate, separated by pauses, are characteristic of variable-ratio extinction.

Complex Reinforcing Contingencies

The four schedules outlined above only represent the basic schedules of reinforcement. In laboratory experiments on operant conditioning these four schedules have been manipulated in various combinations and combined with other reinforcing contingencies for the systematic study of more complex reinforcement programming. Elaboration of these complex schedules is beyond the scope of this text. Some of the more

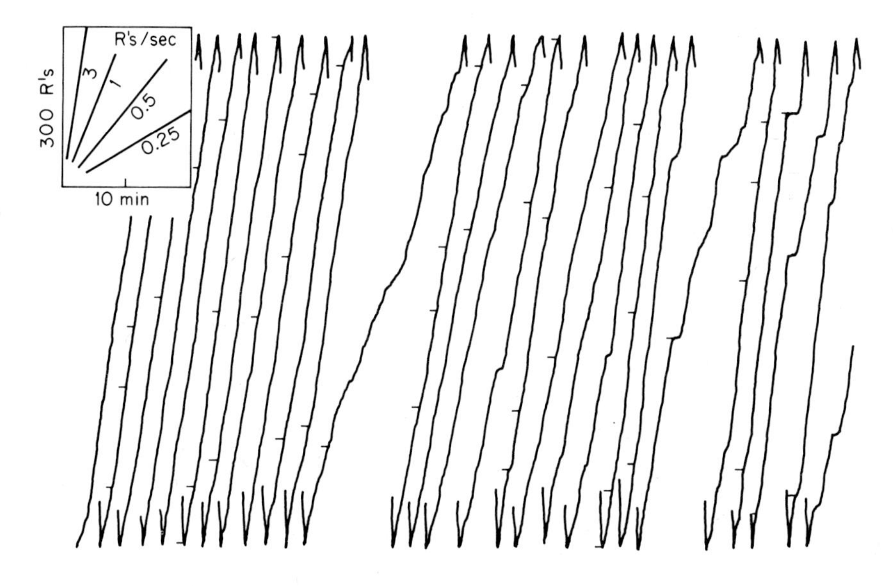

FIG. 4-11. Variable-ratio behavior. Most characteristic of this schedule is a high sustained rate of responding. This is indicated by the steep slope and lack of pauses in the record. (After Ferster & Skinner, 1957.)

common types of combined schedules are listed briefly below. In some instances the reader may recognize immediately the resemblance between a particular complex schedule and the "programming" of reinforcement which occurs under natural conditions.

Ferster and Skinner in their monumental volume *Schedules of reinforcement* (1957) defined the following types of schedules:

1. *Alternative.* Here, reinforcement is programmed by either of two schedules, ratio or interval. It is made contingent upon either of them, whichever is satisfied first.

2. *Conjunctive.* In this schedule reinforcement is programmed to occur when both a ratio and an interval schedule have been satisified.

3. *Interlocking.* Here, the organism is reinforced after a number of responses have been complete. However, following a reinforcement the number of responses required for the next reinforcement is changed.

4. *Tandem.* In this schedule reinforcement occurs when two schedules have been successfully completed.

5. *Chained.* This schedule is the same as a tandem schedule except that a conspicuous change in stimuli follows completion of the first component of the schedule.

6. *Adjusting.* In this schedule after a reinforcement the value of the interval or ratio is changed. The change is contingent upon some arbitrarily defined feature of the preceding performance.

7. *Mixed.* Reinforcement is programmed by more than one schedule, and the schedules employed are alternated, often at random.

8. *Multiple.* This schedule is the same as mixed, except that a distinctly different stimulus is present during each schedule.

OPERANT CONDITIONING AND HUMAN BEHAVIOR

Apparently, both simple habit systems and major patterns of behavior in humans are subject to lawful processes very similar to those demonstrated in lower animals. This is suggested by common observation as well as by laboratory experiments employing human subjects. Many very common life situations are highly suggestive of the operations of basic schedules of reinforcement. Everyone is familiar with the spoiled child whose behavior has been consistently reinforced. The frustration tolerance of such individuals is extremely low. Under certain stress situations, particularly those involving nonreinforcement, such individuals become tempermental and they tend to give up easily. On the other hand, those who appear to have a history of only intermittent reinforcement are more likely to be less emotional under stress and to persevere in their behavior even though reinforcement is not forthcoming.

On a relatively trivial behavioral level, the difference can be seen by comparing the behavior of an individual attempting to operate a malfunctioning soft drink or cigarette machine with that of individual playing a "one-armed-bandit." In the former case the individual is likely to get angry and not to deposit much more money in the machine (crf extinction behavior). In the latter case he may continue to respond even after very long periods without a payoff (variable-ratio extinction). The naive laymen will argue that the expectancies of the individuals differ in these two cases. From the psychological point of view, however, the cognitions referred to as "expectancies" are no more nor less than responses to conditioned stimuli.

Ratio and interval schedules are quite common under natural circumstances. Every college professor is familiar with the eager freshman who has read the first chapter before the first class meeting. In the middle of the semester the same eager freshman may work at a relatively high and steady rate. By the end of the freshman year, or perhaps at the beginning of the sophomore year, the chances are that the same indivdual has learned instead how to cram just before examinations. If all of the conditions surrounding the reward for studying contributed to the operation of fixed-interval schedules, such behavior in students would be even more evident and predictable. On the other hand, consider the difficulty the individual has in getting back to work after completing a long assignment or a paper (fixed-ratio schedule), or the individual who shows a great deal of zest and productivity and who accordingly finds his behavior rewarded in the most unexpected ways (variable-ratio schedule).

In the laboratory, the orderliness of behavior shown by lower animals is sometimes difficult to replicate with human subjects because of lack of adequate control over the source and accuracy of reinforcers. Sometimes socially mediated sources of reinforcement such as approval or expectations from the experimenter may affect the degree of control exercised by the reinforcer employed in the experiment (Bijou & Sturges, 1959). Also, certain schedules are very sensitive to variables other than the reinforcement programming itself, e.g., novel stimuli, deprivation, and quality and amount of reinforcement. Nevertheless, in numerous studies where the experimental conditions have been appropriately controlled even the fine-grain effects obtained in animal experiments have been replicated with children and adults. (e.g., Azrin, 1960; Lindsley, 1960; Long, Hemmack, May, & Campbell, 1958; Stoddard, Sidman, & Brady, 1960).

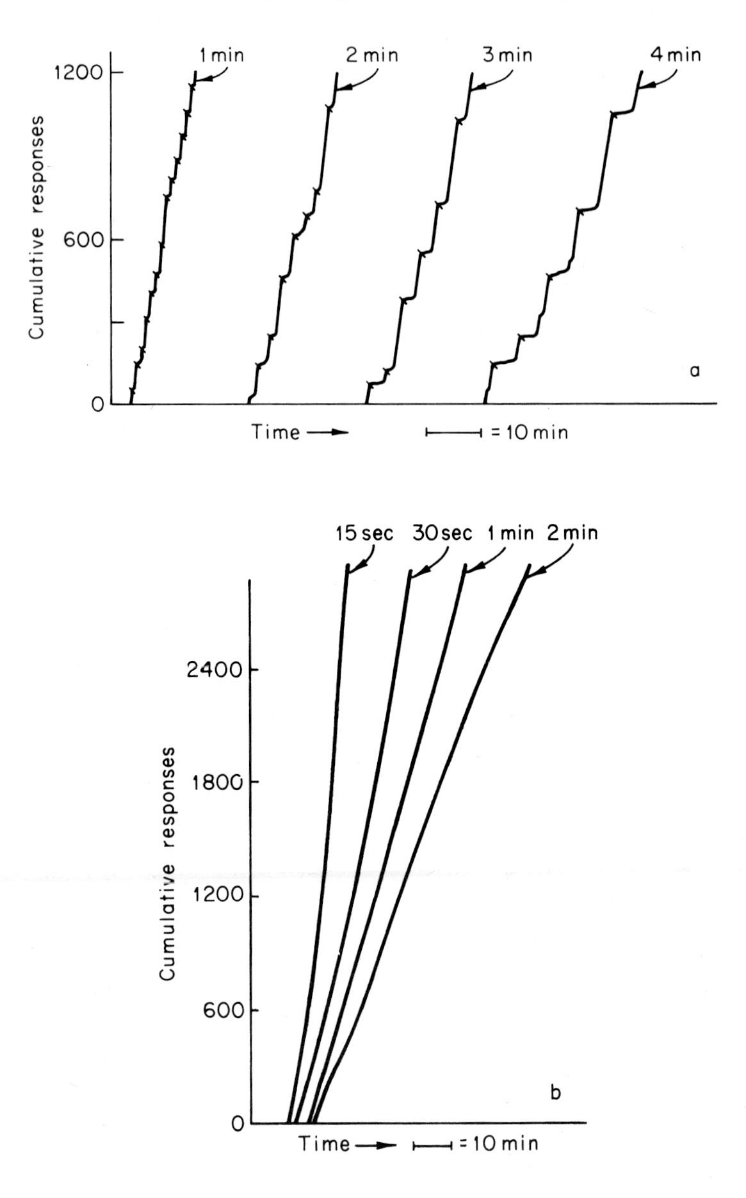

FIG. 4-12. Reinforcement programming applied to human behavior. These records show observing behavior (signal detection) under a fixed-interval schedule (a), a variable-interval schedule (b), and a combined fixed-interval–fixed-ratio schedule (c). In (c) the letters R and I show whether the interval or the ratio schedule is in effect. Note the similarity of these records to animal data shown in Figs. 4-8, 4-9, and 4-10 obtained using the same types of schedules. (Holland, 1958.)

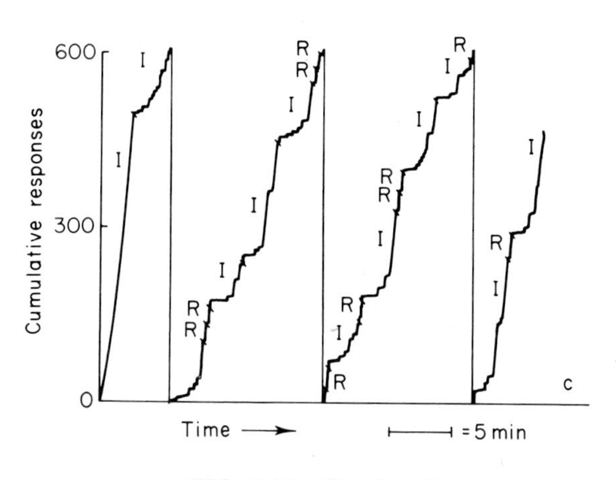

FIG. 4-12. (Continued)

Figure 4-12 illustrates data obtained from a well-controlled experiment by Holland (1958), employing Navy enlisted men as subjects. Holland was interested in certain aspects of signal detection. Figure 4-12(a) shows the observing behavior of a subject reinforced on fixed-interval schedules of 1, 2, 3, and 4 minutes. The diagonal marks indicate reinforcement. A comparison of these data with those shown in Fig. 4-8 reveals how closely these records resemble the animal data. A striking resemblance to curves obtained in lower organisms is also apparent in Fig. 4-12(b) (compare with Fig. 4-9). These are cumulative response records for variable-interval schedules with average intervals of 15 seconds, 30 seconds, 1 minute, and 2 minutes. Finally, Fig. 4-12(c) shows the cumulative records of a subject's observing behavior on a multiple schedule consisting of a 3-minute fixed interval and a 40:1 fixed ratio. The large letters R and I indicate that the respective ratio and interval schedules are in effect. It can be seen that when the interval was in effect the subject's observing behavior was characteristic of behavior obtained under fixed-interval schedules. When the fixed-ratio schedule was in effect the behavior was typical of that for fixed-ratio reinforcement.

A great many operant conditioning experiments have been performed with psychotic and hospitalized patients. Some of this research has been directed toward correlating behavioral pathology with the patient's performance (e.g., Lindsley, 1960). There have also been a number of attempts to change major behavior patterns of patients by the use of operant techniques in controlled situations. As an example of this, Bachrach, Erwin, and Mohr (1965) reported the conditioning of a female patient who refused to eat. All medical and psychotherapeutic techniques had been exhausted, except for the sheer necessity of intravenous feeding. At the beginning of the experiment the patient weighed 47 pounds. The experimenters specified the behavior to be conditioned as an increase in the frequency of food intake. Since food obviously did not constitute a reinforcer for this patient, the first task was to discover what reinforcers were effective. It was found that the patient enjoyed television, reading, music, and visits from people. Accordingly, the patient was deprived of these sources of reinforcement. They were made contingent upon eating and weight gain. After a year of the operant conditioning control of the patient's behavior, her initial weight had more than doubled and continued to increase.

It appears that many psychotic states may reflect a long history of general nonreinforcement, whereas others may be indicative of a history of too much reinforcement or of unusual functional relations having been established between behavior and its controlling variables. We will return to the subject of the operant conditioning of behavior in later chapters.

SUMMARY

Some of the principles that have been established by experimental psychologists are relevant to a scientific interpretation of human behavior. This is especially true in the area of *learning.* Autonomic and glandular respondents can come to be elicited by a wide variety of stimuli through being repeatedly associated with stimuli that are naturally adequate to elicit them. Classical conditioning experiments have demonstrated this under controlled laboratory conditions.

Another classification of learning involves the behavior of the whole organism in free-responding situations. This is called *operant* or *instrumental* learning. Operant conditioning consists of strengthening a particular response to the point where it occurs consistently and promptly in the presence of a particular stimulus.

A reinforcer is any variable that strengthens behavior—that increases the likelihood of behavior recurring. Reinforcing events can be positive or negative. A *positive reinforcer* is a variable which strengthens behavior when it is presented. A *negative reinforcer* is a variable which strengthens behavior when it is withdrawn. Negative reinforcers are, of course, noxious or aversive in character.

Stimuli can develop reinforcing properties through conditioning (i.e., when they are frequently associated with stimuli that naturally act as reinforcers). They are called *conditioned reinforcers.* When conditioned reinforcers have been paired with not one, but a variety of primary reinforcers they are called *generalized reinforcers*—they have the capacity to serve as reinforcers for a wide variety of responses. Symbols and tokens such as money, degrees, awards, etc., appear to be generalized reinforcers. Generalized negative reinforcers may consist of quite subtle behavioral cues indicative of detachment, hostility, or anxiety.

Learned responses generalize to similar stimuli. This is called *stimulus generalization*. There is a gradient of stimulus generalization, with responses being strongest to stimuli most resembling the stimulus present during learning. A large number of factors determine whether a response will be made to stimuli that are very dissimilar to the stimulus present during learning. Among the most important of these are the number of stimulus dimensions changed and the general level of arousal of the organism. Human language habits can mediate stimulus generalization in complex ways.

A response to a generalized stimulus can be weakened by always rewarding a response to the conditioned stimulus and simultaneously presenting the generalized stimulus, but never rewarding it. In this way a discrimination is learned. A similar kind of discrimination learning involves presenting a cue, called *discriminative stimulus*, that serves to mark the occasion when certain behavior will be reinforced. Discrimination learning can be further enhanced by also presenting a cue, called an *S-delta*, that marks the occasion when the behavior in question will not be reinforced.

Time intervals between responses and their reinforcement are particularly important. In general, the longer the reinforcement for a behavior is delayed, the less power it has to strengthen that behavior.

It appears, at least under certain circumstances, that the two types of conditioning are not really separate, but that respondent conditioning is embedded within instrumental or operant chains. The expectancy mechanism, the *fractional anticipatory goal response*, has been discussed. Through the combined effects of respondent conditioning and stimulus generalization, anticipatory responses develop as the organism approaches rewarding events or as they come closer to him in time or distance.

When reinforcers are removed and the organism continues to respond *in the same situation* the learned behavior is weakened. This process is called *extinction*. Many factors affect the resistance of behavior to extinction. One of the most important is whether during learning every response is rewarded, as opposed to reward for only some responses. Behavior learned under partial reinforcement is markedly more resistant to extinction.

When reinforcement is intermittent, behavior during both learning and extinction shows characteristics which vary with the way it is programmed—the way the reinforcing contingencies are arranged. Some of the basic schedules of reinforcement have been discussed. Tentative generalizations have also been made about human behavior under analogous contingencies of reinforcement.

REFERENCES

Adelman, H. M., & Maatsch, J. L. Learning and extinction based upon frustration, food reward, and exploratory tendencies. *J. exp. Psychol.*, 1956, **52**, 311-315.

Azrin, N. H. Effects of punishment intensity during variable-interval reinforcement. *J. exp. anal. Behav.*, 1960, **3**, 123–142.

Bachrach, A. J., Erwin, W. J., & Mohr, J. P. The control of eating behavior in an anorexic by operant conditioning techniques. L. P. Ullmann & L. Krasner (Eds.), *Case studies in behavior modification.* New York: Holt, 1965.

Bijou, S. W., & Sturges, P. T. Positive reinforcers for experimental studies with children—consumables and manipulatables. *Child Develpm.*, 1959, **30**, 151–170.

Diven, K. Certain determinants in the conditioning of anxiety reactions. *J. Psychol.*, 1937, **3**, 291–308.

Feldman, R. S. The specificity of the fixated response in the rat. *J. comp. physiol. Psychol.*, 1953, **46**, 487–492.

Ferster, C. B., & Skinner, B. F. *Schedules of reinforcement.* New York: Appleton, 1957.

Fink, J. B., & Patton, R. M. Decrement of a learned drinking response accompanying changes in several stimulus characteristics. *J. comp. physiol. Psychol.*, 1953, **46**, 23–27.

Hilgard, E. R., & Humphreys, L. G. The retention of conditioned discrimination in man. *J. gen. Psychol.*, 1938, **19**, 111–125.

Holland, J. G. Human vigilance. *Science*, 1958, **128**, 61–67.

Hovland, C. I. The generalization of conditioned responses. I. The sensory generalization of conditioned responses with varying frequencies of tone. *J. gen. Psychol.*, 1937, **17**, 125-248. (a)

Hovland, C. I. The generalization of conditioned responses. III. Extinction, spontaneous recovery, and disinhibition of conditioned and of generalized responses. *J. exp. Psychol.*, 1937, **21**, 47–62. (b)

Kendler, H. H. *Basic Psychology.* New York: Appleton, 1963. P. 285.

Kimble, G. A. *Hilgard and Marquis' conditioning and learning.* (2nd ed.) New York: Appleton, 1961.

Lindsley, O. R. Characteristics of the behavior of chronic psychotics as revealed by free-operant conditioning methods. *Dis. nerv. syst.* (monogr. suppl.), 1960, **21**, 66–78.

Long, E. R., Hemmack, J. T., May, F., & Campbell, B. J. Intermittent reinforcement of operant behavior in children. *J. exp. anal. Behav.*, 1958, **1**, 315–339.

Loucks, R. B. The experimental delimitation of neural structures essential for learning: The attempt to condition striped muscle responses with faradization of the sigmoid gyri. *J. Psychol.*, 1936, **1**, 5–44.

Lundin, R. W. Personality: An experimental approach. New York: MacMillan, 1961.

Maier, N. R. F. *Frustration: The study of behavior without a goal.* New York: McGraw-Hill, 1939.

McNamara, H. J., & Wilke, E. L. The effects of irregular learning conditions upon the rate and permanence of learning. *J. comp. physiol. Psychol.*, 1958, **51**, 363–366.

Razran, G. The observable unconscious and the inferable conscious in current Soviet psychophysiology: Interoceptive conditioning, semantic conditioning, and the orienting reflex. *Psychol. Rev.*, 1961, **68**, 81–147.

Razran, G. Extinction, spontaneous recovery, and forgetting. *Amer. J. Psychol.*, 1939, **52**, 100–102.

Roessler, R. L., & Brogden, W. J. Conditioned differentiation of vasoconstriction to subvocal stimuli. *Amer. J. Psychol.*, 1943, **56**, 78–86.

Rosenbaum, G. Stimulus generalization as a function of level of experimentally induced anxiety. *J. exp. Psychol.*, 1953, **45**, 35–43.

Skinner, B. F. Some contributions to an experimental analysis of behavior and to psychology as a whole. *Amer. Psychologist*, 1953, **8**, 69–78.

Solomon, R. L. Kamin, L. J., & Wynne, L. C. Traumatic avoidance learning: The outcomes of several extinction procedures with dogs. *J. abnorm. soc. Psychol.*, 1953, **48**, 291–302.

Spence, K. W. *Behavior theory and conditioning.* New Haven: Yale Univer. Press, 1956.

Stoddard, L. T., Sidman, N., & Brady, J. V. Reinforcement frequency as a factor in the control of normal and psychotic behavior. Paper delivered at Psychonomic Soc. Meetings, Chicago, 1960.

Wendt, G. R. An analytic study of the conditioned knee jerk. *Arch. Psychol., N. Y.*, 1930, **19**, No. 123.

White, S. H. Generalization of an instrumental response with variation in two attributes of the CS. *J. exp. Psychol.*, 1958, **56**, 339–343.

Yerkes, R. M., & Margulis, S. The method of Pavlov in animal psychology. *Psychol. bull.*, 1909, **6**, 257–273.

Chapter 5

Principles of Learning II. Anxiety, Punishment, and Conflict

INTRODUCTION

In the previous chapter many of the essential features of operant and respondent conditioning were discussed. Much remains to be said, about learning situations and their implications for human adjustment. In this chapter the focus of attention will be upon the effects of painful or frightening stimulation, escape and avoidance conditioning, and the topics of punishment and conflict. A basis will thus be provided for an analysis of frustration in the chapter that follows.

THE ACTIVATION SYNDROME

Stimuli which are aversive in some way, such as being painful or frightening, elicit the many respondents that underly the emotions of anger and fear. Learning psychologists refer to these concomitantly activated autonomic and glandular responses as the *activation syndrome*. The physiological aspects of the activation syndrome were described in Chapters 2 and 3 in connection with the autonomic nervous system and the activation of emotional circuits. Under conditions producing fear and anger there are changes in the respiratory rate, sugar is released from the liver, there is an increase in heart rate, and a pattern of vasoconstriction and vasodilation occurs, as do a decrease in stomach contractions, a decrease in gastric secretions, a lowering of the electrical resistance of the skin, pupillary dilation, etc.

A specific emotion cannot be easily defined by enumerating the reflexes involved, because the same respondents of the activation syndrome are implicated in many different emotions. Furthermore, within the context of the activation syndrome each individual shows a specific pattern of responses which is more or less unique to him. Therefore, psychologists have found it more fruitful to conceive of a particular emotion as characterized by predispositions to behave in a certain way. For example, when an aversive stimulus elicits the activation syndrome, there is an increase on future occasions in the frequency of occurrence of behavior that escapes or avoids any cues associated with the aversive stimulus. When the respondents producing a state of anger are elicited, there is an increased likelihood of hostile or aggressive behavior occurring. Correspondingly, other classes of behavior tend to decrease. The individual who is frightened does not "care about eating" even though he may be hungry. The angry or hostile individual is less likely to respond to positive stimulation: he is likely to reject friendly overtures from others, etc.

ACQUIRED EMOTIONALITY (DRIVE)

It is known that there are only a small number of conditions that function as unconditioned stimuli in eliciting the activation syndrome. Among these are pain, sudden stimuli, falling, and restraint. However, through conditioning, it can come to be elicited by an unlimited range of stimuli.

How these emotional (drive) states are acquired can best be illustrated by a classic animal experiment performed by Miller (1948). The apparatus used in this experiment consisted of two compartments, divided by a small door. One compartment was painted white; the other was painted black. The floor of the white compartment consisted of an electric grid, through which the animal could be shocked. This apparatus is illustrated in Fig. 5-1.

Before the experiment began the animals were tested for the presence of emotionality. They were placed in the white compartment and the presence or absence of indexes of fear such as urination, defecation, crouching, the erection of hair, and tenseness was observed. The rats did not show any of these signs of emotionality. The stimulus complex (whiteness, the proprioceptive cues of the box, visual cues of depth and distance, right angles, etc.) did not naturally elicit fear in these experi-

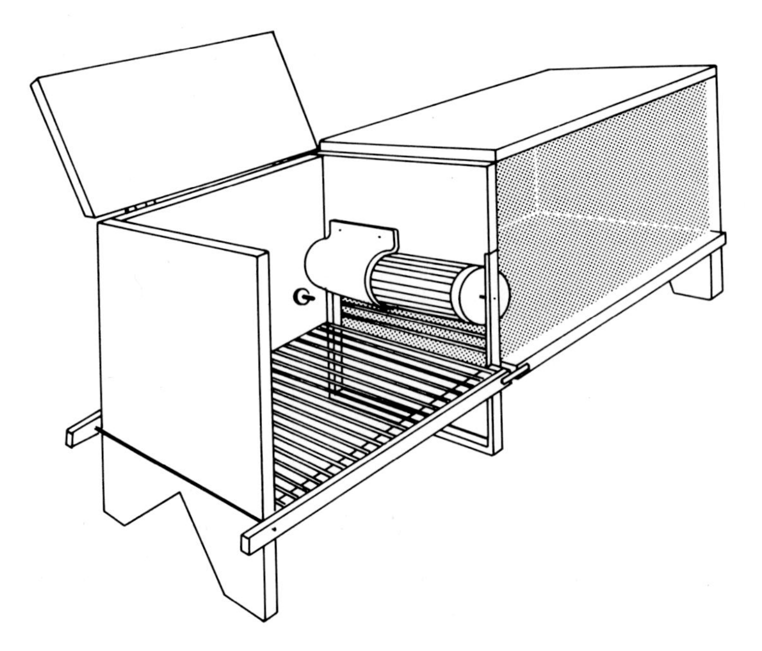

FIG. 5-1. Miller's apparatus for study of the learning of fear. Animals placed in the white compartment were shocked through the grid floor. The black and white striped door dividing the two compartments could be raised (by the animal turning the wheel in later experiments) so that escape into the adjoining black compartment was possible. (After Miller, 1948.)

mental animals. The animals were then given a series of brief electric shocks through the electrified grid which served as the floor of the white compartment. Thereafter, whenever they were placed in the white compartment they showed unmistakable signs of neurotic, fearful behavior. This was true even though shock was never again presented. Fear had become conditioned to the stimulus cues of the box. The stimulus complex now consisted of conditioned aversive stimuli.

Having demonstrated that the fear response could become learned, Miller set out to discover whether the learned fear could mediate new learning. That is, Could the conditioned emotionality in response to the cues of the white box reinforce, or strengthen, new behavior? Miller arranged the apparatus so that the door separating the white from the black compartments could be opened by turning a wheel above the door in the white compartment. The animals were again placed in the white compartment with no further shocks being given. However, they showed many signs of anxiety (conditioned fear). Eventually, by chance, they turned the wheel and escaped into the black box. In just a few trials the animals learned to turn the wheel and escape upon being placed in the white compartment. It is important to note that this learning occurred in the complete absence of any unconditioned stimulus. It was mediated entirely by conditioned aversive stimuli. The basis of the behavior was merely "psychological." The cause of the behavior was the cues of the white box, which had become fear-provoking through conditioning.

Most learned fears and anxieties in humans are based upon different processes; they might involve for example, what others tell the individual, his observations of others and the consequences of their behavior. However, occasionally shocking or aversive unconditioned events occur which provide us with the direct analogue of the learning of fear by Miller's experimental animals. Once a child has been exposed to the dentist's drill, strong fear may become attached to stimuli associated with it. These may include the dentist's office or the dentist himself. This acquired fear may also mediate the learning of new responses, although the situation is more complex than the analogous wheel-turning of Miller's experimental animals. Techniques of diverting mother's attention away from dental appointment dates; learning to give excuses (rationalizations) for not going to the dentist; or simply learning to throw a very loud temper tantrum—all may be included among the new learning mediated by such an acquired fear.

It will be recalled that when a positively reinforced response is learned, removal of the reinforcement results in the more or less rapid extinction of a conditioned response. Behavior learned under aversive stimulation is quite different. Habits based on fear tend to persist in the absence of any primary reinforcement. In the Miller experiment, for example, during extinction the animals continued to respond rapidly for 100 successive trials. Two of the animals were still responding after 600 trials, and the other stopped after making 300 responses. Once fear has become conditioned, organisms make strong and persistent responses to avoid the stimuli associated with the original fear response.

Miller (1951) has demonstrated that both the rate and the persistence of learning based upon conditioned fear depend upon the intensity of the original fear-provoking stimulus. He systematically varied the intensity of shock given to the animals in the white compartment. Under very weak shock the cues of the white compartment did not become conditioned aversive stimuli strong enough to mediate the learning of the wheel-turning response. Strong shock intensities led to a more rapid acquisition of the wheel-turning response than medium shock intensities. Other experiments in which the intensity of the aversive stimulus has been systematically increased have shown that response rate increases correspondingly up to a point, following which there is a gradual decline. Very strong aversive stimuli appear to have a response-depressing effect, which becomes dominant over escape behavior (e.g., Kaplan, 1956).

Intensity is not the only relevant variable. Both the schedule of reinforcement and the number of conditioning trials affect fear conditioning. For example, Kalish (1954) administered shock of the same intensity to different groups of animals using different numbers of fear-conditioning trials. He found that the greater the number of conditioning trials, the faster was the learning of the fear-reducing responses. Kaplan (1956) showed that escape behavior maintained under fixed-ratio schedules has characteristics very similar to behavior maintained under fixed-ratio schedules using positive reinforcement. In this experiment light served as an aversive stimulus. A given number of responses were required to turn the light off. The typical break in responding followed each reinforcement on the fixed-ratio schedule. Furthermore, as the ratio increased, the break became longer before the next run of responses that could eventually lead to escape from the light. Higher rates could also be achieved by gradually increasing the ratios, just as in the case of positive reinforcement.

We have been discussing situations in which the organism is allowed to learn a response in order to escape an aversive stimulus. This is called *instrumental escape conditioning.* In *instrumental avoidance conditioning* the subject is allowed to learn to avoid the aversive (unconditioned) stimulus. The situation is so arranged that if the subject makes a response, or a series of responses, the aversive stimulus does not occur.

An experiment by Sidman (1953) illustrates the procedures used in instrumental avoidance conditioning. Sidman shocked animals every 20 seconds unless they pressed a bar during that interval. By pressing the bar they avoided the shock for 20 seconds. Animals readily learned under these conditions to avoid the occurrence of shock. They developed high rates of responding.

Another common experimental procedure is to use a warning signal such as a buzzer or a light which just precedes the shock. When a warning signal is used the animal's avoidance behavior is a direct response to the signal, which itself becomes a conditioned aversive stimulus. Responding is dependent upon the presentation of the signal. This is quite different from the experiment where no warning signal is used, but in which the response delays the presentation of the aversive stimulus for a given period of time. The rate of responding in those experiments far exceeds that necessary to avoid shock. No clear stimulus is discriminated out of the environment because no specific stimulus regularly appears just before the onset of the unconditioned stimulus. This lack of stimulus cues to provide for clear discrimination appears to result in a more intense level of arousal, and in excessive responding.

Under the conditions of Sidman's avoidance conditioning experiment animals also tend to eliminate all forms of behavior other than bar-pressing. This is because almost every response they make, except bar-pressing, may get punished. This may well be analogous to the punishing contingencies in the life of the rigidly "perfect child." So much of the spontaneous behavior of these unfortunate children may get punished without any discriminable cue preceding the punishment that they are forced to cling to an anxious generalized compliance with parental dictates in order to avoid more severe threat.

Conditioned avoidance responses can be generated at high rates and are extremely resistant to extinction. Furthermore, the conditioned aversive stimulus, the previously neutral cue which serves as a warning signal, may become more aversive than the original (unconditioned) negative reinforcer. Sidman and Boren (1957) devised an experiment in which rats could postpone the shock but still have the warning signal remain, or they could take the shock and simultaneously terminate the warning signal. Under these conditions the animals attempted to terminate the warning signal as soon as possible and take the shock. The warning stimulus appeared to be more aversive than the shock itself. Once they have experienced an aversive event human beings frequently report that the anticipation of it was more aversive than the event itself. Conditioned aversive stimuli can result in suffering and maladaptive avoidance responses which are extremely disproportionate to the realities of the situation.

OPERATIONALLY DEFINED ANXIETY

Learning psychologists have developed a rather simple operational definition of the term anxiety: *Anxiety* is the behavioral consequence of exposure to a neutral stimulus followed by a primary aversive stimulus. It will be recalled that a neutral stimulus paired with an aversive stimulus defines respondent defense conditioning. When sufficient time elapses between the presentation of the neutral stimulus and the presentation of the aversive stimulus for certain behavioral effects to be evidenced, the condition is called anxiety. Synonyms for operationally defined anxiety include the *conditioned fear response* and the *conditioned emotional response* (CER). The conditioned emotional response is sometimes further divided into two types—CER(a) and CER(b). Both of these can be termed anxiety. The CER(a) was described in Chap. 4, p. 104. An animal given a number of conditioning trials with a clicker (neutral stimulus) followed by shock rapidly develops a conditioned emotional response to the clicker. Upon hearing the clicker it becomes immobile, or otherwise plainly shows a fear response.

The CER(b) can be illustrated by an experiment by Estes and Skinner (1941). Animals were conditioned to press a bar for food on a 4-minute fixed-interval schedule. After complete conditioning, a tone was presented for 5 minutes followed by a brief electric shock which was presented through the grid floor of the experimental box. Under these conditions there was a depression in the rate of bar-pressing during the interval between the presentation of the buzzer and the presentation of shock. After successive pairings of these two stimuli the depression of the response rate during the presentation interval became greater. It typically decreased to nearly zero and remained there until the period ended with shock. There was a drastic decrease in the rate of lever-pressing during these periods and there was an increase in crouching, immobility, erection of hair, etc.

The suppression of a positively reinforced response is taken as a sensitive measure of the strength of the conditioned emotional response. It is designated the CER(b). This is one of the most important behavioral effects of the operations defining anxiety. At the human level individuals who are chronically made anxious by conditions in their lives analogous to those producing the CER(b) of experimental animals may similarly not be interested in presumably positively reinforcing events such as those connected with food, sex, and creative work.

Let us explore the effects of the operations defining anxiety further. Suppose a food-deprived monkey receives food after every twelfth pull of a chain. At the same time a lever is present which in the past has been used in avoidance procedure (a response to it delayed shock for 20 seconds). Now suppose that the avoidance contingency is not in effect, but extinction is not yet complete, and that the experimenter introduces a buzzer which is terminated by shock and alternated every 6 minutes with periods of quiet.

A record from one of the sessions of this experiment (Sidman, 1958) is shown in Fig. 5-2. Between points 1 and 2 in Fig. 5-2 the buzzer and shock remained off. At point 2 the buzzer was sounded, continuing to point 3. At 3 a brief shock was presented and the buzzer terminated. Notice that between points 1 and 2, when the conditioned aversive stimulus is absent, the rate of the food-reinforced response is high and the rate of the response with an avoidance history is rather low. A CER(b) rapidly develops, however, under these conditions and between

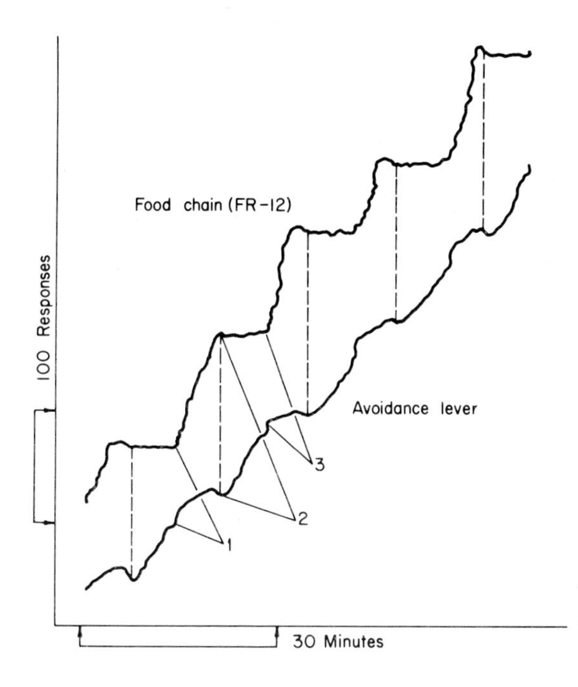

FIG. 5-2. In the experiment illustrated a monkey was taught to pull a chain for food. It had also learned in the past to avoid shock by pushing an avoidance lever. The vertical broken lines indicate the onset of the stimulus warning of shock. Between points 1 and 2 the shock and the warning stimulus remained off. At point 2 the warning stimulus sounded, continuing until point 3. At point 3 a brief shock was presented and the warning stimulus terminated. The warning stimulus suppresses positively reinforced behavior, indicating the development of a conditioned emotional response. (After Sidman, 1958.)

points 2 and 3, the period during which the buzzer is presented, the rate of the food-reinforced response is low. In fact, it reduces to almost zero. On the other hand, during this period the rate of the response having an avoidance history is higher than before the buzzer onset. This experiment shows nicely the suppression of a positively reinforced response which occurs under anxiety. It also shows another very important effect of anxiety: *along with the suppression of positively reinforced behavior there is an increase in behavior having an avoidance history.*

Anxiety is of central importance in human maladjustment and psychopathology. Cameron and Margaret (1951) have suggested that the individual may frequently be subjected to contingencies similar to those which produce anxiety in the laboratory. These involve situations where (1) escape from aversive stimuli is impossible, (2) the individual experiences a loss of support (withdrawal of a positive reinforcer constitutes punishment, or aversive stimulation, but as we will see below, it constitutes anxiety when it is not response-specific), and (3) the individual anticipates punishment. The anxiety produced by the third situation is commonly referred to as "guilt." The operation of anxiety in human adjustment will be discussed in detail in later chapters. For humans primary aversive stimulation operates far less frequently than conditioned aversive stimulation. The latter takes the form of criticism and negative evaluation from others; self-generated images, thoughts, etc. Although the simple animal experiments illustrate the fundamental effects of anxiety, further knowledge of the mechanisms of social learning and imitation and of verbal behavior are essential for an adequate understanding of the operation of anxiety in human adjustment.

THE EFFECTS OF PUNISHMENT

Two operations define punishment: either (1) making an aversive stimulus (conditioned or otherwise) contingent upon a response, or (2) making the withdrawal of a positive reinforcer or a conditioned positive reinforcer contingent upon a response. Punishment is thus defined as a procedure which is the opposite of reinforcement.

Punishment and Anxiety

The concepts of punishment and anxiety are very similar. In the operations defining anxiety, aversive stimulation is inevitably contingent upon some response: whatever behavior happens to be occurring at that time. However, punishment differs from anxiety in that a given response is specified.

Avoidance behavior is also frequently concomitant with punishment. When pressing a lever turns on a shock, pressing the lever is being punished. Under these conditions the animal learns to avoid pressing the lever.

Stimuli accompanying or just preceding a punished response become conditioned aversive stimuli by being paired with the punisher. Conditioned aversive stimuli generated by regularly punished behavior result in a state of anxiety. Positively reinforced behavior tends to decrease, and avoidance behavior tends to increase in frequency. Thus, aside from the effects of punishment *per se*, anxiety and avoidance behavior are always operative to some degree in punishment.

The following experiment illustrates the importance of differentiating between punishment and anxiety: Hunt and Brady (1955) conditioned rats to press a bar for positive reinforcement. The animals were then divided into two groups. In one group, shocks were given during a clicking noise, or at the end of a clicking noise, but they were never made contingent upon lever-pressing. In the second group, the animals received shock only if they pressed the bar in the presence of the clicking noise; thus, this group was being punished.

For both groups a significant depression of the lever response occurred, not only during the presentation of a clicker noise but during periods when the conditioned stimulus was absent as well. However, there were important differences in the behavior of the two groups. When the aversive contingencies were removed the depression in rate of the punished group disappeared more rapidly. The rate of depression during periods when the clicker noise was absent was far less severe in the punished group. In the presence of the clicker noise, conditioned avoidance behavior appeared to be operative for the punished group, whereas response suppression accompanied by an admixture of abortive lever-pressing, and withdrawal and avoidance characterized the other (anxiety) group. Finally, there was generally less indication of emotionality (urination, defecation, the erection of hair, etc.) in the punished group.

This experiment illustrates the general principle that when behavior is inconsistently or indiscriminately punished, in the sense of the punishment not being response-specific, then anxiety, conflict, and incompatible approach and avoidance behavior tend to be generated.

In contrast, when punishment occurs regularly and consistently, and when it is connected to behavioral (response-produced) cues the individual can easily discriminate, then the individual learns to avoid the punished behavior with a minimum of anxiety and conflict. Also, when punishment which has been consistent and discriminable is removed, learning to make the response readily occurs and the individual does so without inordinate anxiety and conflict.

For the dominated child, many prohibitions create a situation in which the child is anxious and timid about self-initiated behavior. The overindulged child, too, is often inconsistently treated so that he learns to be anxious about the world. Punished at one moment, yet coddled the next, he learns neither to inhibit the punished behavior, nor to be spontaneous about initiating new activities. Nor is he able to discriminate those situations in which punishment is likely to occur from those in which it is not.

The individual who has been inconsistently punished has, in effect, been anxiety-conditioned. His behavior is analogous to that of the experimental animals discussed above. Having repeatedly experienced anxiety, he tends to move toward anxiety, or anxiety-producing situations. Punished responses may be abortively made. He also tends not to be able to see that behavior which has been frequently punished may not be punishable under certain circumstances. These individuals have been called *sensitizers*. Sensitization as a personality defense will be discussed at length in Chapter 10.

Other individuals experience consistent and appropriate punishment. They show more appropriate avoidance behavior in situations that tend to elicit punishable responses. Less anxiety appears to be generated for them, and, hence, anxiety does not interfere with their ability to discriminate whether or not punishment will occur in a given situation. These individuals show a variety of behaviors that are quite the opposite of that shown by sensitizers. They are called *repressors*. Repression as a personality defense will also be discussed in detail in Chapter 10.

Punishment and Positive Reinforcement

When conditions are arranged so that the positively reinforced response is punished, and, at the same time, an alternative response is positively reinforced, then punishment may operate very effectively. Consider an experiment by Whiting and Mowrer (1943). In this experiment animals were positively reinforced for taking the shorter of two paths in a simple, single-choice maze. Following this conditioning the animals were trained to reverse their preference. They were divided into three groups: one group was no longer rewarded for choosing the shorter path (extinction group); for another group an impassable barrier was placed in the middle of the short path; and a third group received an electric shock half-way through the short path (punishment). The results, in terms of errors (the number of choices of the short path) are instructive. The mean number of choices of the short path was 230 for the extinction group, 82 for the barrier group, and 6 for the shocked group.

Hence, when a competing response is positively reinforced (The animals were rewarded for taking the long path, competing with their previously learned habit of taking the shorter path) discrimination learning can be greatly enhanced by the appropriate use of punishment. However, as we shall see below, when the individual does not have an opportunity to form a new habit, punishment has little effect in eliminating undesirable behavior. What occurs under these conditions is behavioral suppression which remains operative only so long as the punishing contingencies are in effect.

Research findings on the combined effects of punishment and reward with children parallel these animal findings. Walters and Parke (1967), in a comprehensive review of studies dealing with the punishment of children, concluded that socialization may be considerably enhanced if the punishing agents also provide information about alternative acceptable behavior and consistently reward it when it does occur.

There is also evidence that affectional relationships between punishing agents and their victims increase the likelihood that punishment will be effective. Prior to a punishment experiment in which both the intensity and timing of punishment were varied, Parke and Walters (1967) had the experimenter interact warmly with 40 boys during a play period with constructural materials. Another group of 40 boys were given relatively unattractive materials and the experimenter was merely present in the room without interacting with them. Later punishment was significantly more effective under both the intensity and timing conditions for the group that experienced positive interaction with the experimenter prior to the introduction of punishment.

There is some evidence that negative reinforcement alone may be effective, but only if its termination is immediately contingent upon the elicitation of the desired response. Thus, Aronfreed (1964) has shown experimentally that children learn self-critical responses (and therefore self-control) if punishment is abruptly terminated when self-critical utterances occur.

The Effects of Brief, Mild Punishment

The effects of punishment on positively reinforced behavior also depend upon variables such as the amount of conditioning before the introduction of punishment, the intensity of punishment, the number of responses punished, and the punishment schedule. Skinner (1938) reinforced experimental animals with food on a fixed-interval schedule for pressing the lever. After a steady rate of responding was attained for a long period of time, reinforcement was discontinued. For half of the animals the apparatus was so arranged that each lever-press during the first 10 minutes of the extinction period resulted in a slap on the fore paws by a return kick of the lever. The other half of the animals were not slapped at all. The extinction curve for this latter group provided a control in determining the effects of punishment.

The response rate during the first 10 minutes of extinction was lower for the punished animals than for the unpunished ones. The rate continued to be lower for a while after the punishment contingency was lifted. However, soon the punished group began to show a slightly higher rate of responding, and as extinction proceeded it was found that both groups had made approximately the same total number of responses. Skinner concluded that in the case of a brief period of mild punishment the rate of responding is decreased both during and after punishment, but the rate of responses emitted in complete extinction remains essentially unchanged.

During punishment the stimuli provided by the experimental box, by the animal's behavior, and by the lever, etc., having been paired with an aversive stimulus, become conditioned aversive stimuli. When punishment is discontinued, the rate would be expected to remain low until the stimuli provided by the situation cease to be conditioned aversive stimuli through extinction.

Figure 5-3 shows the results during an extinction experiment by Estes (1944). In this experiment the animals first learned to press the bar for food, then one group was punished by being given an electric shock when they pressed the bar, while the other group remained unpunished. Subsequently, both groups were placed in the apparatus during extinction.

As Fig. 5-3 indicates, the unshocked rats showed a gradual extinction of the conditioned response. The effects of the conditioned aversive stimuli of the experimental situation can be seen for the shocked group during the first 2 hours of the extinction period. These animals made

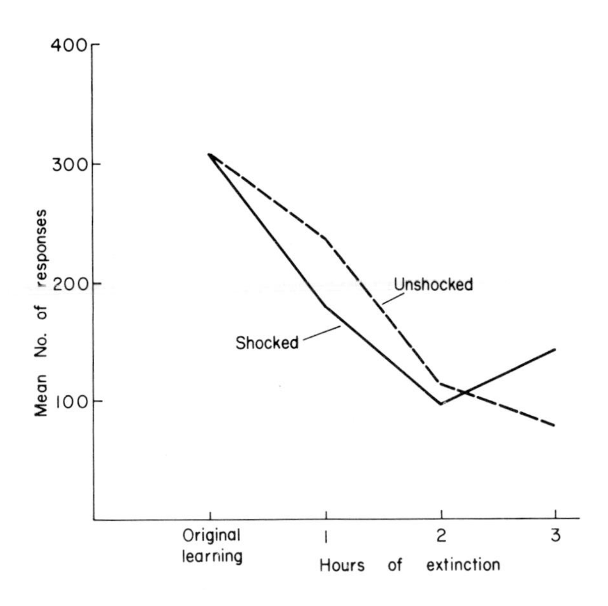

FIG. 5-3. After a group of rats learned to press a bar for food half of them were shocked for bar-pressing. This was followed by extinction of bar-pressing for both groups. During the first 2 hours of extinction the unshocked animals made more bar-pressing responses than the shocked animals. Thereafter there was a marked increase in responding for the shocked group so that the total number of responses made by the two groups during the 3-hour extinction period was approximately equal. Punishment suppressed the bar-pressing response temporarily, but it did not weaken it. (Estes, 1944.)

consistently fewer responses for the first 2 hours of extinction. After approximately 2 hours, however, the response rate increased so that toward the end of extinction they were making far more responses than the unshocked animals, and probably would have continued beyond the point where the experiment was terminated. The total number of responses made by the shocked and unshocked animals during the 3 hours was almost exactly equal. Presumably, the abrupt change in responding after the 2-hour period by the shocked group marks the point at which the conditioned aversive stimuli of the apparatus had become completely extinguished. Punishment suppressed the bar-pressing response temporarily, but it did not weaken it.

In another experiment, reported by Skinner, rats were slapped during the first 10 minutes of extinction. Following the 10-minute period of punishment, each animal was left in the box for 2 hours, but the lever was removed. Two hours later the lever was replaced, and extinction continued with no further punishment. Except for a very slight initial depression in rate, the extinction curve was similar to that of the unpunished animals. When the lever was removed, further responses could not be emitted, and therefore extinction could not be completed. However, in the absence of the lever, punishment was discontinued so that the conditioned aversive stimuli in the experimental box were given the opportunity to extinguish. Conditioned aversive stimuli from the lever itself, and from the behavior of pressing the lever, could not be extinguished during the 2-hour period. This probably caused the slight initial depression in rate.

Taken together, these experiments indicate that brief mild punishment inhibits behavior, and that the inhibition remains for some time after the punishment stops. However, it does not eliminate behavior in any fundamental sense. Although the overt behavior is inhibited, the inclination to behave remains equally as strong as it is without punishment. As soon as the punishing contingency is lifted, the situational cues begin to lose their conditioned aversive properties. After a period of time, during which extinction of the anxiety conditioned to these stimuli occurs, the behavior returns in full strength.

This is common knowledge, although we ordinarily tend to make the mistake of not seeing these facts as part of a body of general principles related to the effect of punishment. Thus, in everyday life it is not uncommon to hear a mother remark, "If I have told that child once, I have told him a hundred times not to do that." In these instances the first "telling" has undoubtedly been accompanied by brief mild punishment; but, after the proportionately brief time required for the extinction of the resulting conditioned aversive stimuli, the behavior has returned in full strength. Presumably the parent again administered brief mild punishment, and the cycle was again repeated; that is, until the parent's own punishing behavior eventually began to weaken. In order for brief mild punishment to be effective at all, it would have to be continuous; even then it would only be effective in suppressing behavior rather than eliminating it.

It will be recalled that one of the major effects of anxiety is an increase in avoidance behavior. Punishment indirectly lowers the rate of a punished response by providing conditions for the reinforcement of incompatible avoidance behavior when incompatible behavior is extinguished, the punished response emerges in full strength. Since the punished response is never emitted while the punishing contingency is in effect, the extinction of the punished behavior does not occur. In these experiments alternative escape responses are not available, and the punishment generates conditioned aversive stimuli. Under these conditions any behavior of "doing something else instead" will be strongly reinforced. Behavioral immobility is a common response to this situation. By holding still, making a movement which resembles the punished one can be avoided.

Human adjustment under similar punishment contingencies involves similar behavioral characteristics. Shy, inhibited people are often actively not behaving. Although they may gain very little in a positive sense, their punishment history has been such that avoiding the execution of behavior constitutes a stronger source of reinforcement for them. When a person is unable to talk about a very aversive experience, psychologists speak of *repression* as a "mechanism of defense." Repression may be interpreted as a displacement of the punished behavior by avoidance behavior. A person may talk about other things, or just not talk, as a means of avoiding talking about aversive situations. Although detailed discussion of the defense mechanisms is not in order until we have explored the nature of verbal operants, it may generally be said that the so-called mechanisms of defense are instrumental avoidance acts. One of the key roles of the psychotherapist, or of a friend, is that of a nonpunishing audience. The more trust another person engenders, the less is defensive behavior necessary.

The Effects of Continuous Punishment

Holland and Skinner (1961) report a series of experiments which illustrate well the effects of continuous punishment. In the first of these experiments a food-deprived pigeon was placed in a standard experimental chamber and pecking responses were reinforced with food on a variable-interval schedule until the rate of responding was stable from one day to the next. Then small electrodes were embedded beneath the pigeon's skin. Brief shocks could be delivered in this manner. In the next part of the experiment every response was punished, and continued to be punished through many daily sessions while the variable-interval schedule remained in effect.

During the 1st hour of the 1st day of punishment, the response rate was considerably lower than it was on the previous day when responses were not punished. By the end of the 3rd day of continuous punishment, however, the rate of responding was nearly the same as the rate of responding during the no-punishment procedure. By the tenth session of punishment the slope of the cumulative record was about the same as the last of the no-punishment sessions. Holland and Skinner concluded that after many thousands of responses have been mildly punished the overall rate of responding is not appreciably lowered.

At the end of the above experiments punishment was discontinued. For the first few sessions the rate of responding was actually higher than before the introduction of punishment. Then, after several sessions without punishment, the rate of responding dropped to the original level and became stable once more. After the rate had stabilized at the original level, each response was followed again by a shock, but this time a moderately strong shock was used instead of the previous mild one.

On the 1st day of the stronger punishment the rate of responding was nearly zero. By the end of the 5th day of the stronger punishment, the overall rate was higher than it had been on the 1st day. The highest overall rate during punishment occurred on the 32nd day of punishment. However, by the 32nd day of punishment the rate had not reached the no-punishment level. From this experiment and other similar ones, it can be stated that generally the stronger the continuous punishment the lower and more stable the rate of responding.

Recall that in the first experiment where mild shock was used, when punishment was discontinued the response rate was actually higher than before punishment. In another experiment a pigeon was reinforced with food under a variable-interval schedule. A moderately intense shock followed each response. Halfway through the session punishment was abruptly discontinued for 10 minutes and then reinstated. The cumulative records from this experiment are shown in Fig. 5-4. Notice the drastic increase in response rate when punishment was discontinued. When every response was again punished the response rate immediately decreased.

Generally, the rate of a response maintained by positive reinforcement may not be decreased by mild continuous punishment, but is decreased by strong continuous punishment. However, this obtains only

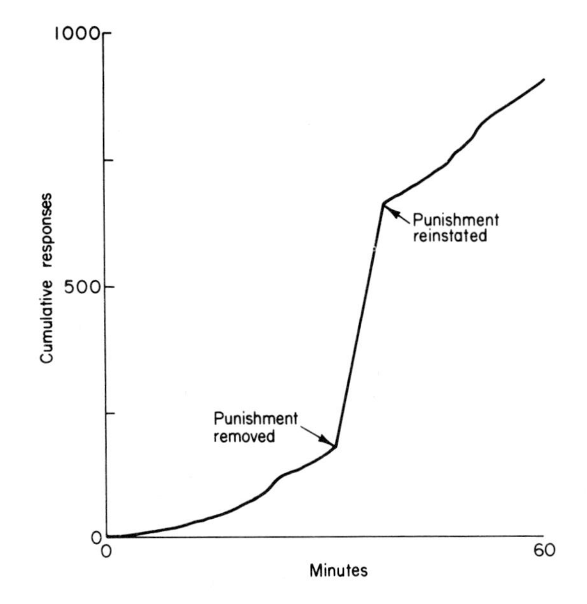

FIG. 5-4. Removal of the punishing contingency. A pigeon was reinforced for pecking a disk, following which every response was shocked. The animal responded at a relatively slow rate despite the shock. When punishment was removed there was a sudden drastic increase in response rate. When punishment was reinstated the rate returned to its previous "punishment" level. (Holland and Skinner, 1961.)

while the punishing contingency remains in effect. Immediately after continuous punishment is discontinued, the response rate increases. In fact, for a while it is typically higher than under no-punishment conditions.

These experiments on continuous punishment indicate a general principle: When punishment is continuous, and not very severe, in time the punished behavior regains its original strength in spite of the fact that it is being punished. Severe punishment, on the other hand, does inhibit behavior. However, the inhibition occurs only so long as the punishing contingency remains in effect. When the punishment is withdrawn, the punished behavior recurs with an even greater frequency than it did before it was punished.

The general principles here appear to have very wide scope. In human culture political, military, or religious suppression may last for centuries, the degree of effectiveness depending upon the power and strength of the punishing agents. When the punishing contingencies are lifted or weakened, however, the punished responses reappear.

Punishment is notoriously ineffective in eliminating criminal behavior. In fact, in countries such as England, where capital punishment has been abolished, the evidence is that the frequency of murder remains essentially unchanged. Capital punishment appears to be no more effective than other methods of controlling murder, insofar as its abolishment does not lead to an increase in murder.

Of course, these topics are extremely complex, and without more detailed analysis they provide only the most superficial analogies to the principles of punishment established in the animal laboratory. Nevertheless, suppose the parents of an adolescent boy were advised to severely punish every evidence of heterosexual responsiveness so that he might be able to control the "animal side of his nature." Assuming that the parents followed instructions, could the reader make any predictions about the boy's sexual behavior during the late-adolescence, young-adult period? Could he say anything about the likelihood of occurrence of sexual behavior on the part of the individual during his overseas tour of duty in the Army in a culture that held permissive attitudes toward the expression of sexuality? After the boy returned home to all the cues of the environment in which the punishment occurred, and married, what is the likelihood of his exhibiting sexual responsiveness and pleasure in marriage?

On the basis of analogy to the experimental laboratory it would be expected that the individual's sexual behavior would be inhibited during later adolescence and early adulthood; that those who might have the opportunity to observe his sexual behavior both at home and overseas would be surprised at the stark contrast; and that, married to the girl back home, he would again suffer from anxieties and inhibitions.

An important condition for such predictions would be that severe punishment be the sole means of controlling the boy's behavior during late adolescence. When positive reinforcement for behavior which is incompatible with the undesired behavior is the primary technique used, quite a different result occurs: the individual *tends to lose his desire to make the responses which the reward was designed to control.* In effect, positive reinforcement, unlike punishment, tends to eliminate, rather than suppress, undesired responses.

Changes in other variables, such as the consistent introduction of a delay between the time the behavior occurred and the time it was punished, would require different predictions. The introduction of a delay would pit immediate positive reinforcement for sexual behavior against the later punishment. Strong immediate reward might easily outcompete the weaker effect of delayed punishment. Under these conditions the behavior might regularly occur in spite of the fact that it was eventually always followed by severe punishment.

The Learning of Apparently Punishing Behavior

Some behavior appears to the objective observer to be masochistic. A maladjusted person may behave in a hostile and aggressive manner toward others, forcing others to reject him; or he may withdraw and seek seclusion when, in fact, what he truly and deeply desires is the company of others. Many different types of neurotic behavior patterns involve the individual's active attempts at self-frustration. Two kinds of laboratory experiments have helped to further our understanding of the mechanisms involved in such apparently punishing behavior.

The first is a well-known experiment by Pavlov (1927) that demonstrated that a painful stimulus can, through conditioning, not only lose its aversive properties but actually become a conditioned stimulus for pleasure. In a respondent conditioning experiment he employed electric shock, cauterization, and pricking of the skin as conditioned stimuli for feeding. When the animals were very hungry, and when conditioning had begun with mildly painful stimuli, eventually even strong aversive stimulation employed as a conditioned stimulus lost its power to elicit the respondents of the activation syndrome. According to Pavlov:

> Subjected to the very closest scrutiny not even the tiniest and most subtle phenomenon usually exhibited by animals under the influence of strong injurious stimuli can be observed in these dogs. No appreciable changes in the pulse or in the respiration occurred in these animals whereas such changes are always most prominent when the noxious stimulus has not been converted into an alimentary conditioned stimulus. [Pavlov, 1927, p. 30]

The second type of experiment is best illustrated by an operant conditioning experiment by Farber (1948) which was specifically designed to attack the problem of apparently punishing behavior. In this experiment animals were trained to go to one arm of a T-maze. The animals underwent 100 trials with food located in that arm of the maze. After the 100 conditioning trials the food was shifted to the other arm. It took the animals an average of 10 trials to change their behavior and traverse the opposite arm of the maze, now rewarded by food. Farber then trained a second group of animals to traverse one arm of the maze for food, but for the last 60 of the 100 practice trials he shocked them after they had passed the choice point and before they reached the food. When reward was shifted to the other side of the maze and shock was discontinued, these animals behaved very differently from the first group. They continued to run to the box with no food for an average of 61 trials. This apparently unrewarded behavior continued for two of the animals for more than 250 extinction trials.

Because of the temporal arrangements involved, the results of this experiment cannot be explained by the same mechanisms as those involved in the experiment by Pavlov mentioned above. That is, for the shocked–fed animals the shock did not function as a conditioned stimulus for food. Rather, the choice point, because of its association with shock, had become a conditioned stimulus for anxiety. It will be recalled that with continued conditioning, conditioned aversive stimuli may become even more aversive than the primary aversive stimulus itself. When shock was discontinued the animals became anxious at the choice point and escaped the aversive cues by making their old response. This response had previously been reinforced from two sources: the positive reinforcement associated with food and the negative reinforcement associated with escape from shock.

The difference between the first and second groups during extinction indicated that escape from an aversive stimulus can be a more powerful reinforcer than food. This anxiety-reduction hypothesis was further tested by employing a third group of rats which were treated exactly like the shocked–fed group, except that after conditioning they were fed on the grid of the choice point in order to eliminate the possibility of that locus being a conditioned stimulus for anxiety. When these animals were tested with the food reward shifted to the other box, their behavior was no different from that of the group of animals that had never received shock.

Delay of Reward and Punishment

Another way to look at the foregoing experiments is in terms of the temporal relationships between punishment and reward. The delay-of-reinforcement principle asserts that the effectiveness of reinforcers depends upon how soon they are presented following the occurrence of the to-be-reinforced behavior.

When punishment and reward occur in temporal sequence, punishment may follow reward, or the reverse. When punishment is followed by reward, either the punishment is effective, or it is not. Anything other than strong, immediately administered punishment would be expected to be less than maximally effective. Thus, if it is weak, or delayed, or both, it might be expected that the rewarding event(s) that follow it would render it ineffective.

This is commonly found to be the case in human maladjustment. Thus, when the rewarding and punishing contingencies in the lives of emotionally disturbed individuals are studied, it is often found that those whose inadequate behavior persists in spite of the fact that it is frequently punished also receive delayed "secondary gains." Extreme emotional withdrawal, for example, is a behavioral tendency that is not only disabling, but also frequently meets with social punishment. Why, then, does it persist?

In the first place, the extremely withdrawn individual is rewarded because the very mechanism of withdrawal which is being punished may provide escape from further punishment. However, severe withdrawal may also lead to secondary gains. He may be rewarded by being almost completely cared for by others—whether inside or outside of an institution. Passive-aggressive wishes may be satisfied by the fact that others are forced to give their attention to him, willingly or not. Satisfaction may be available in the situation for certain other aggressive motives. For example, the obvious fact that he is a failure might serve as a continual source of frustration to achievement-oriented parents, and corresponding pleasure to the individual himself. In short, the later secondary gains may be considerable, and they may easily outweigh the effects of immediate social punishment.

Now consider the opposite case, where punishment for the behavior occurs only after it has been rewarded. In this case the punishment must be strong and immediate in order to offset the effects of the behavior having been rewarded. When the punishment is delayed, the probability is increased that the reward for the behavior will be the more important determiner of what gets learned. In such instances the delay may be long enough so that the cues associated with punishment simply do not get attached to the punished behavior.

Thus, the dog that is punished for chasing automobiles only some time afterwards may never learn to inhibit anticipatory automobile-chasing responses. The person who "inexplicably" punishes him, however, may become a conditioned stimulus for anxiety.

At the human level, too, threat of future punishment may not prevent the individual from engaging in punishable behavior, although fear, guilt, or anxiety may be experienced immediately after his performance of the punishable act. When the delay is very long even this anticipatory fear following the punishable act may not exist. Instead, he may merely learn an antipathy for the punishing agent.

Whenever behavior is both punished and rewarded, corresponding positive and inhibitory tendencies are aroused. Irrespective of which tendency is the stronger, the existence of the opposing tendency must be taken into consideration for a complete interpretation of the individual's behavior. We will now turn to an analysis of how these incompatible response tendencies influence behavior.

CONFLICT

The classic experiment on conflict, which was performed in Pavlov's laboratory by Shenger-Krestovnikova (Pavlov, 1928), involved conditioning a dog to salivate when a circular figure was presented, but not to salivate when the stimulus was an ellipse. After the animal had learned the discrimination it was presented with a sequence of stimuli systematically altered to be more elliptical in shape. When the ratio of the axes was 9.8 the animal's ability to discriminate completely broke down. It became very agitated, and began to struggle, howl, bite the apparatus, and show other signs of emotional disturbance. The animal was also found to be incapable of making simpler discriminations, although it had been capable of doing so during the first part of the experiment. This breakdown in the animal's ability to discriminate, along with the concomitant emotional disturbance, was referred to as "experimental neurosis."

Since the Shenger-Krestovnikova experiment a great variety of laboratory situations have been devised to create a point of equilibrium between two or more incompatible behavioral tendencies. One of the most common of these involves the use of aversive stimuli to condition avoidance behavior that is incompatible with previously learned positive responses. Typically the animal is taught to traverse a straight alley in order to obtain food. After this learning is well established, the animal is subjected to a strong electric shock, or some other aversive stimulus, whenever it attempts to do so. When the animal is unable to leave the field, or otherwise escape from the situation, it tends to become emotionally disturbed. The activation syndrome is elicited, and there is a disorganization of behavior. Prolonged or intense conflict situations produce all the responses of the general adaptation syndrome (see pp. 138).

The resemblance between the symptoms developed by experimental animals in conflict situations and those of emotionally disturbed human beings is sometimes striking. For example, in one series of experiments Masserman (1966) trained animals to manipulate a device that flashed a light, rang a bell, and then deposited a pellet of food in a food box. Following this training he subjected them to an air blast or to a shock when they began to manipulate the device. In from two to seven trials (the animals were given one trial per day) they developed behaviors which were remarkably like those of human neurotic responses. Masserman described their behavior as follows:

> Neurotic animals exhibited a rapid heart, full pulse, catchy breathing, raised blood pressure, sweating, trembling, erection of hair, and other evidences of pervasive physiologic tension. They showed extreme startle reactions to minor stimuli and became "irrationally" fearful not only of physically harmless light or sounds but also of closed spaces, air currents, vibrations, caged mice and food itself. The animals developed gastro-intestinal disorders, recurrent asthma, persistent salivation or diuresis, sexual impotency, epileptiform seizures or muscular rigidities resembling those in human hysteria, or catatonia. Peculiar "compulsions" emerged, such as restless, elliptical pacing or repetitive gestures or manner-

isms. One neurotic dog could never approach his food until he had circled it three times to the left and bowed his head before it. Neurotic animals lost their group dominance and became reactively aggressive under frustration. In other relationships they regressed to excessive dependence or various forms of kittenish helplessness. In short, the animals displayed the same stereotypes of anxiety, phobias, hypersensitivity, regression and psychosomatic dysfunctions observed in human patients. [Masserman, 1966, p. 247]

Similar behavioral abnormalities have been produced in conflict situations involving the discrimination between similar tones, light intensities, rhythms, shapes, etc., even though these stimuli were not conditioned stimuli for reward or punishment. Since such abnormalities can be produced without an aversive stimulus present, it appears that a difficult discrimination is itself aversive. When the subject is forced to remain in the presence of cues eliciting incompatible responses the effect is aversive. Fear, rage, and whatever responses to frustration have been learned, or that the organism is biologically predisposed to make, may be expected to occur.

Restraint appears to be a relevant variable. Cook (1939) punished two groups of animals while they drank. One group was restrained and the other was not. When the CS for water was made almost identical with the CS for punishment, the restrained rats showed the symptoms described above, whereas the unrestrained ones did not. This finding has not been contradicted by later experiments. The majority of them support it.

Types of Conflict

The conflict situations that we have been discussing have involved a positive response or an approach response of some kind as opposed to an inhibitory or an avoidance response. Other possible incompatible response combinations include those involving the simultaneous activation of incompatible avoidance tendencies, the simultaneous activation of incompatible approach tendencies, and situations involving the simultaneous activation of two or more incompatible approach tendencies, each of which as avoidance tendencies associated with it. These four categories of conflict have been called approach–avoidance, avoidance–avoidance, approach–approach, and double approach–avoidance conflicts, respectively.

Approach–Avoidance Conflict. The diagram below schematizes the approach–avoidance conflict situation. In this diagram O represents the organism, S represents the effective stimulus complex, the symbols + and − indicate the positive and negative tendencies, and the arrows show their respective directions.

$$S^{\pm} \quad \overleftarrow{\overrightarrow{\qquad\qquad\qquad}} \quad O$$

Notice that in the approach–avoidance conflict situation the incompatible positive and negative tendencies are associated with the *same stimulus*. The conflicted individual tends to both approach and avoid the stimulus object or event represented by the letter S in the diagram. This might be a generalized stimulus which the animal cannot discriminate as being a positive or a negative stimulus; it might be a food box in which the animal has also received shock; or at the human level it might represent another person toward whom the individual feels ambivalent, and toward whom he exhibits responses having both approach and avoidance characteristics. The essence of the approach–avoidance conflict is that the same object or event arouses incompatible tendencies within the subject to both approach and avoid it. The intensity of the conflict depends upon the degree to which these tendencies are aroused.

Avoidance–Avoidance Conflicts. The incompatible response tendencies involved in avoidance–avoidance conflicts are elicited by two stimulus objects or events. In this type of conflict two avoidance tendencies are elicited by each of two stimuli or stimulus complexes, and the avoidance tendencies that they elicit are mutually exclusive. If the organism avoids one he must, by virtue of that fact, approach the other. Hence, the individual remains trapped between two negative alternatives. This can be schematized as follows:

$$S_1^- \longrightarrow O \longleftarrow S_2^-$$

In this diagram S_1^- represents one aversive stimulus situation, and S_2^- represents the other. The letter O represents the organism which is trapped between the two aversive alternatives.

The concept of response gradients, and the manner in which their relative strengths change with increasing distance from the stimuli that generate them, will be discussed at length below. However, here it might be noted that if the animal were shocked at both ends of a straight alley equally often and with equal intensity, and if S_1^- and S_2^- represented those shock points at either end of the alley, then it would be predicted that the animal would tend to remain in the center of the alley at the point designated O in the above diagram. This is because moving toward either end would increase the fear of being shocked. Attempts to avoid S_1^- would bring the animal closer to the feared stimulus S_2^-.

Many human conflicts also involve two choices, both of which are negative. The individual may see the acquisition of new skills as an uncertain alternative to staying with present work that he finds unrewarding. Loneliness or an increase in dependency needs may constitute an unpleasant alternative to living with an undesirable marital partner, and so forth. In these situations the person tends to remain caught "between the devil and deep blue sea."

Approach–Approach Conflicts. The approach–approach conflict can be schematized as follows:

$$S^{+} \longleftarrow\!\!\!-\!\!\!-\!\!\!-\!\!\!-\; O \;-\!\!\!-\!\!\!-\!\!\!-\!\!\!\longrightarrow S^{+}$$

This conflict involves two positive choices that are mutually exclusive. Whereas in the avoidance–avoidance conflict the organism tends to remain at an equilibrium point somewhere between the two choices, in the approach–approach conflict the balance of forces is unstable and momentary. As soon as the organism begins to move toward one or the other alternative it is disrupted: the strength of the opposing alternative is then less than it was at the point of choice and the organism tends to approach the stimulus it first chooses. A jackass simply does not starve between two equally attractive bales of hay!

The balance between two incompatible positive response tendencies is so unstable that as a conflict situation it probably does not deserve special attention. When conflict is experienced between what appear to be two equally attractive alternatives, invariably both alternatives will be found to involve negative factors as well. This is referred to as a double approach–avoidance conflict.

Double Approach–Avoidance Conflict. The double approach–avoidance situation involves two alternatives, or two incompatible stimuli, both of which possess positive and negative features, and hence elicit both approach and avoidance tendencies. It can be schematized as follows:

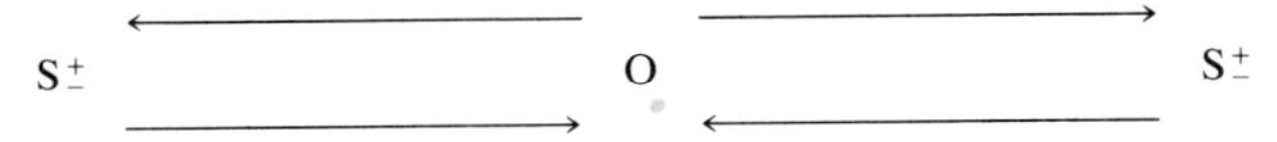

Here the individual must choose between two stimuli, each of which elicits conflicting positive and negative responses. The choice is between two approach–avoidance conflict situations.

When a choice between two positive stimuli or events appears conflictful it is very likely that it actually represents a choice between two alternatives, each of which elicits ambivalent responses from the individual. When Mary has a difficult time deciding whether or not to marry Tom or Harry, our best guess is that in reality she finds neither completely satisfying as a prospective marital partner. In the double approach–avoidance conflict increments may also be added to the already existing inhibitory tendencies insofar as the choice of one means losing the advantages that would have been gained had the other one been chosen.

The Original Gradient Theory

In simple animal learning experiments in which the straight runway is used, the consummatory or goal response becomes strongly conditioned to stimuli near the goal. As learning proceeds it becomes generalized to stimuli further and further back in the runway. Learning psychologists have employed the r_g–s_g mechanism to account for the gradient of approach which results from this generalization (see pp. 113–118). ***Approach gradient*** merely refers to the fact that measures of the approach response show that its strength increases with nearness to the goal.

The variation of the strength of behavioral tendencies with distance from the eliciting stimuli has been fundamental in the theoretical analysis of conflict. The most influential theoretical model has been that of Neil Miller and his associates (Dollard & Miller, 1950). Before discussing this theory in detail, it should be pointed out that recent experimental work (which will be discussed in the next section) has shown that some aspects of the theory must be revised. However, Dollard and Miller's original gradient theory has been very useful in generating research and in the application of the work of the animal laboratory to human behavior. Conflict will undoubtedly continue to be discussed in terms of this theory, even as the theory is being modified.

The Dollard and Miller theoretical model is based upon the following three assumptions: (1) The tendency to approach a goal is stronger the closer the subject is to it. (2) The tendency to avoid a feared stimulus is stronger the closer the subject is to it. (3) The strength of avoidance increases more rapidly with nearness (to the goal) than does that of approach. Miller believed that on the basis of these assumptions the behavioral phenomena associated with conflict could be explained.

Dollard and Miller represented their assumptions graphically as shown in Fig. 5-5, which illustrates an approach–avoidance conflict. The strength of the tendency to approach or to avoid the goal is given on the ordinate and the distance from the feared goal, indicated at the left, is given on the abscissa. The solid line is the approach gradient and the dashed line, the avoidance gradient.

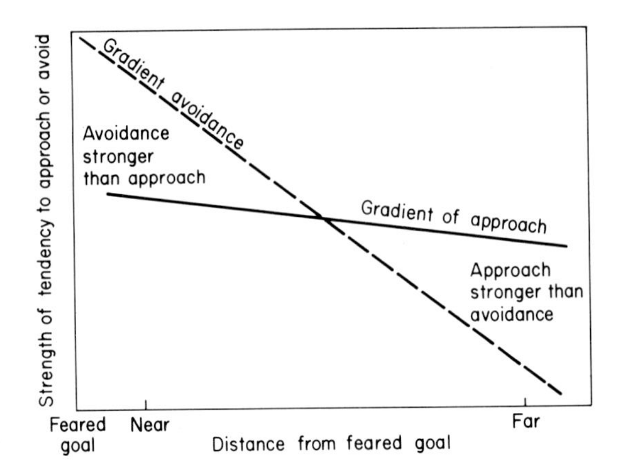

FIG. 5-5. Graphic representation of an approach–avoidance conflict in which it is assumed that the avoidance gradient is steeper than the approach gradient. This assumption predicts that near the feared goal the stronger of the two tendencies is avoidance, whereas far from the feared goal the approach tendency is the stronger. The two gradients cross at the point of behavioral oscillation, and the subject should tend to remain near this distance from the feared goal. (Dollard & Miller, 1950.)

In accordance with the first two assumptions, both the tendency to avoid and the tendency to approach increase in strength as the nearness to the goal increases. However, the avoidance gradient is much steeper, indicating the more rapid increase of the strength of avoidance with nearness to the goal, in accordance with the third assumption.

The assumption that the avoidance gradient is steeper than the approach gradient is the crux of the theory. The subject would be predicted to remain at the point where the gradients cross. This is called the point of *behavioral oscillation*. Responses that carry the individual away from this point in the direction of the feared goal result in a sharp increase in fear, because the avoidance gradient rises rapidly above the approach gradient. The avoidance response is the stronger response in this region.

On the other hand, responses that carry the subject beyond the point of behavioral oscillation in the opposite direction from the feared goal bring him into the region where the approach gradient is higher than the avoidance gradient. The dominant tendency is, therefore, approach. When the subject is in this region he is motivated toward the point of behavioral oscillation. Thus, until some new variable enters the picture to change the relative strength of the approach and/or avoidance gradients, the individual is predicted to remain essentially at the point of behavioral oscillation, and the conflict remains unresolved.

The point of behavioral oscillation indicates the distance from the feared goal at which the individual will remain. It also indicates the intensity of the conflict. How these relationships change as the variables affecting the approach and avoidance gradients change is illustrated in Figs. 5-6 and 5-7. Figure 5-6 shows a single approach gradient with two different avoidance gradients, a weak one and a strong one. According to the theory of Dollard and Miller, as the strength of avoidance motivation increases the entire gradient is raised. Inspection of these gradients reveals how the position of the intersection of the avoidance gradient with the approach gradient—the point of behavioral oscillation—changes systematically with changes in the strength of avoidance. When the avoidance motivation is strong, the point of behavioral oscillation, (point A in Fig. 5-6) is far from the feared goal. Neither of the incompatible approach and avoidance tendencies is strong at this distance from the goal and the intensity of the conflict is minimal.

On the other hand, when the avoidance motivation changes from strong to weak the point of behavioral oscillation (B in Fig. 5-6) moves closer to the feared goal. At this point both of the incompatible tendencies of approach and avoidance are more strongly elicited. Thus, para-

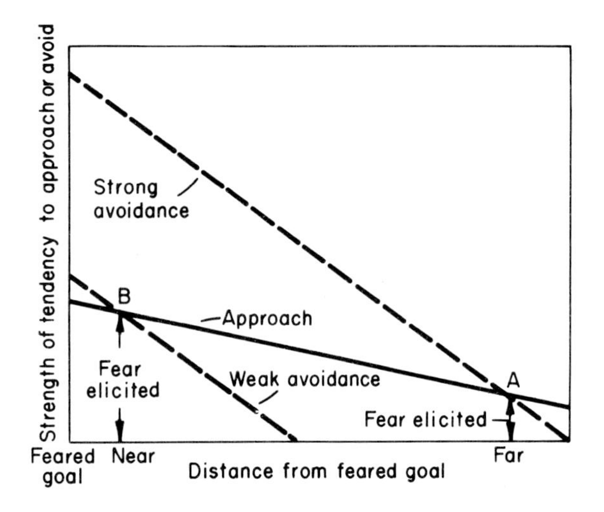

FIG. 5-6. Single approach gradient and two avoidance gradients. When the avoidance gradient is strong the subject tends to remain at the point where the two gradients cross, far from the feared goal. Therefore the intensity of conflict is not very great. Lowering the avoidance gradient (weak avoidance) has a paradoxical effect: the organism approaches nearer the feared goal where the intensity of the conflict is greater. (Dollard & Miller, 1950.)

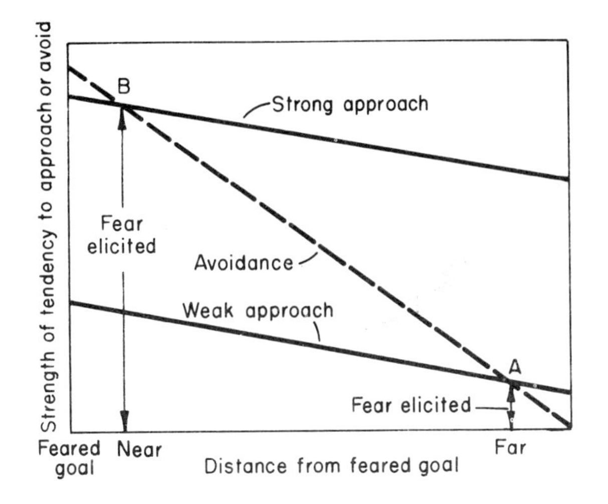

FIG. 5-7. Single avoidance gradient and two approach gradients of different strengths. In the case of the weak approach gradient the point of behavioral oscillation is far from the feared goal, hence the intensity of the conflict is not very great. Strong approach brings the subject closer to the feared goal and the conflict is very intense. (Dollard & Miller, 1950.)

doxically, as the avoidance motivation decreases from strong to weak, the intensity of the conflict increases.

Figure 5-7 illustrates the changes in the point of behavioral oscillation which occur when the strength of the avoidance motivation remains constant and the strength of the approach motive varies. Here, again, only the height and not the relative steepness of the gradient is assumed to change. Inspection of Fig. 5-7 shows that as the approach motivation changes from weak to strong the point of behavioral oscillation moves from A to B. The conflict again becomes more intense because at point B the individual is near the feared goal where the incompatible approach and avoidance tendencies are both stronger.

The gradient model of conflict was, of course, based on laboratory animal studies. However, Dollard and Miller (1950) have drawn a large number of analogies between the behavior of conflicted human psychotherapy patients and the gradient conflict model. In general, they have tended to think of the avoidance gradient as representing fear. Two very common conflicts which clinic patients are found to have are a fear of expressing strong feelings, especially of anger, and a fear of sexual expression. The feared goal is taken to be, let us say, the expression of sexuality. An approach gradient such as the one shown in Fig. 5-5 is supposed to represent the corresponding goal-directed approach behavior. Analogously, the avoidance gradient of Fig. 5-5 is taken as a representation of the incompatible fear response, elicited by cues associated with the feared goal.

Consider an individual who has learned to be afraid of sexual expression, yet has a normal sex drive. If the fear were extreme the fear gradient would be raised so high that the point of behavioral oscillation would be far from the feared goal. The sex–fear conflict for this person would be represented by the approach and strong avoidance gradients of Fig. 5-6. It would be expected that he would show little interest in the opposite sex. He might not marry. Nor would he tend to recognize his own sexual needs. In short, he would tend to remain far from the feared goal and the intensity of the conflict would be minimal.

Suppose, on the other hand, that the fear was less strong: indicated by the weak avoidance gradient in Fig. 5-6. Under these conditions it would be expected that some definite approach responses would be made; he might marry. However, he would be quite incapable of moving beyond a certain point on the approach gradient. For example, he might be frightened to make any but the most pristine advances toward his mate. The point of behavioral oscillation would be characterized by relatively strong incompatible approach and avoidance tendencies.

Unless some variable changed to raise or lower one or the other of the two gradients, the individual would remain essentially at the point where the two gradients crossed. Furthermore, should he find himself at a point near the feared goal it would be predicted that he would hastily retreat to the point of behavioral oscillation. If, for example, he suddenly found himself alone in the presence of a very seductive woman it might be expected that he would become frightened and flee. That is, between the point where the gradients intersect and the feared goal, fear would dominate, because the fear gradient is higher than the approach gradient throughout this region.

On the other hand, if the individual were placed in a situation in which he was far from the feared goal, it would be expected that approach forces would operate, because in this region the approach gradient remains above the avoidance gradient. Continuing with the present example, if our subject was isolated or restricted in some way so that there was little or no contact with members of the opposite sex, it would be expected that the motivation to do so might become apparent. Considerable ingenuity might be shown in that direction on the part of our subject.

Strong conflicts put the individual in an intense state of arousal, or high drive. Dollard and Miller refer to this state of high drive as *misery.*

Seriously conflicted individuals for whom the point of behavioral oscillation occurs near the feared goal appear to experience much more misery than those whose fear is intense and who therefore remain far from the feared goal at a point of behavioral oscillation where incompatible tendencies are elicited in little strength.

Conflict and Personality

Experimental personality research (see pp. 329) has shown that some people have high physiological reactivity to threatening situations and measurable indications of anxiety, but they tend not to recognize that they are anxious or afraid. They tend to defend themselves against the fear by denying the existence of it, and by avoiding threatening situations. These individuals, called repressors, show few outward signs of misery.

The explanation of many of their characteristics might well be that they react to stimulus situations such as those eliciting hostile or sexual responses with strong fear. Their strong avoidance functions as a defense. It keeps the intensity of the conflict at a minimum, because the approach and avoidance gradients intersect far from the feared goal. The mechanism of denial, presumably thus motivated by strong fear, has the effect of short-circuiting threat by divorcing these individuals from stimulus cues which would be frightening. Since this is the case they experience little anxiety or misery. In short, their psychological "life space" appears to be analogous to the psychological facts of conflict depicted by the approach and strong avoidance gradients of Fig. 5-6.

Another group of people show a contrasting pattern of personality defense. Rather than attempt to avoid anxiety, they tend to move toward anxiety and anxiety-producing situations in an effort to cope with them. Their attempts at controlling anxiety in this way do not appear to be successful, however, insofar as they tend to be people who are manifestly anxious. These people are called sensitizers. In terms of a conflict analysis their psychological life space is analogous to the approach and the weak avoidance gradient in Fig. 5-6. Much more will be said about these dimensions of personality in Chapter 10.

Ideational and Temporal Gradients

It may be apparent to the reader that the preceding discussion has stretched the notion of a gradient in terms of distance from the feared goal to include cues of nearness or distance that are not necessarily spatial. Repression, denial, or a short-circuiting of ideational cues are presumed to have the effect, at least at the level of conscious experience, of taking the individual away from the feared goal, just as spatial distance would in a conflict situation where the distance dimension involved only physical distance. Also, presumably ideational cues, such as thoughts about performing an act the consequences of which the individual fears, may bring him "near" the feared goal in a psychological sense. Dollard and Miller have assumed that the consequences of this kind of psychological nearness are analogous to nearness to a feared goal in a spatial sense, and that the same principles apply.

The strength of approach and avoidance tendencies may also vary with temporal nearness to the feared goal. In some situations this is confounded with spatial distance. The animal that is spatially distant from food at the end of the alley is also temporally distant from it simply because it takes time to traverse the alley. However, the cues that elicit the incompatible approach and avoidance tendencies are often in temporal sequence unconfounded with spatial distance. The student becomes tense and anxious as the day of an important examination approaches. The young couple meet, carry on a courtship, decide to get married, set a date for the wedding. If an avoidance gradient exists it may be relatively weak, and hence not affect behavior until very near the final hour: a case of the gradient increasing near the feared goal.

Epstein and Fenz (1962) made a study of novice parachute jumpers. They felt that the need for excitement, adventure, or prestige, pitted against the fear of injury or of death represented an acute approach–avoidance conflict for these inexperienced parachutists. They predicted, and found, that in word association tests an increased number of parachute-related (goal-relevant) responses were produced on the day of the jump compared to 2 weeks prior to it. There was also a predicted increase in the GSR to words related to parachute jumping and an increase in perceptual deficit as the day of the jump approached. On the basis of his research with parachute jumpers, Epstein has also developed a tentative, complex, but noteworthy theory of how the inhibition of anxiety is learned and becomes anticipatory (see Epstein, 1967).

Another experiment, by Maher, Weisstein, and Sylva (1964), has indicated that the occurrence of behavioral oscillation is a function of the amount of the total anticipated time to the event that has elapsed. The temporal conflict apparatus used in this experiment is shown in Fig. 5-8. The finger of the clock moved in an anticlockwise direction from zero to wherever it was set. The timer could be activated by depression of either of two switches, right or left, below the clock face. However, once activated, depression of the other switch reset the clock back to the starting point, from which it continued to move to zero. Given this arrangement the subject could change his mind and make an alternative selection any time before the clock ran out.

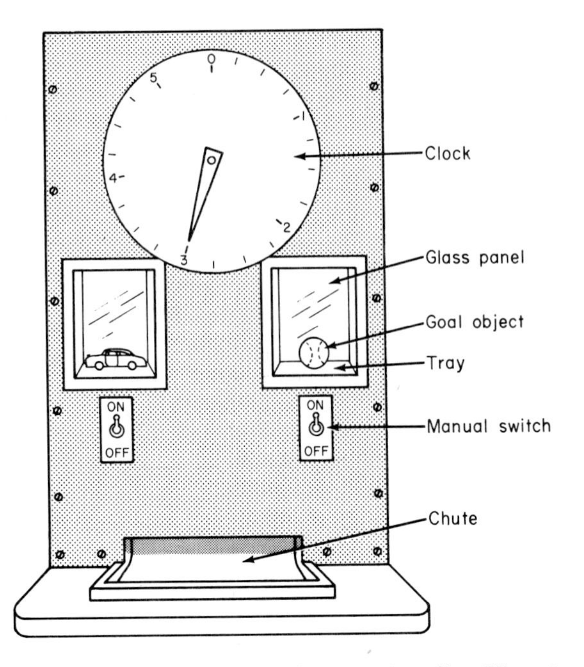

FIG. 5-8. Apparatus used for the study of temporal conflict. When the timer reached zero the object corresponding to the side chosen was delivered. Before the timer reached zero the subject could change sides chosen, thereby resetting the timer. (Maher, Weisstein, and Sylva, 1964.)

Six- to nine-year-old children were put in a series of double approach–avoidance conflicts by having goal objects such as candy or small toys on each side, but one preferred to the other; or by having them on one side and nothing on the other and having the subject guess the right side. The choice of side was indicated by depression of the appropriate lever. The children were allowed to change their mind during the guessing period.

When various time periods during which a subject could make his choice were used it was found that the typical oscillation (change of choice) occurred after 8/30, or approximately one-third, of the total time had elapsed. The maximum number of oscillations was produced when there was a risk of getting no reward. Otherwise, the larger the number of goal objects at stake, the greater the number of oscillations.

The New Gradient Model

The assumptions underlying the gradient model of conflict of Dollard and Miller were largely supported by a classic experiment by Brown

(1948). Brown conditioned two groups of rats to avoid shock: a strong-shock group and a weak-shock group. Two other groups were approach groups. The approach response was a running response to a food box. The strong-approach group was given 48 hours of food deprivation and the weak-approach group was given 1 hour of food deprivation. Test trials involved measuring the strength of pull for these groups by attaching to each animal a small harness that was hooked to a restraining spring. Strength of pull was measured at two points, one near the positive stimulus (or near the negative stimulus in the case of the avoidance group) and one distant from it (or distant from the negative stimulus in the case of the avoidance group). The gradients found in this experiment were the same as those shown in Fig. 5-5.

These data were accepted as definitive for many years. However, as research in the area expanded it became apparent that there were a number of technical problems in the Brown experiment so that the results must be taken as less reliable than had been generally supposed. For example, since each animal pulled twice, it is possible that both fatigue and the extinction of avoidance were confounded with measures of the avoidance gradient (Maher & Nuttall, 1962). Further, in the strong-avoidance condition 16 of the 120 animals failed to pull at all at the "far" position. Hence, statistical evaluation of the results was severely handicapped.

At least two other considerations have interested recent investigators. In the first place, Brown's gradients were established using only two points: "far" and "near." This fact has tended to encourage the assumption of linear gradients. Dollard and Miller were aware that this assumption was a sketchy one. However, since the actual shape of the gradients had not been determined experimentally, they felt that the assumption of linear gradients did little violence to the known facts and it simplified illustration. The weight of evidence now indicates that the gradients are nonlinear.

Secondly, in the Brown experiment different subjects were used to establish the approach and avoidance gradients. Hence the slopes of those gradients cannot be taken to represent the situation which exists within the same individual. Unless, of course, the assumption is made that the approach and avoidance gradients obtained separately using different subjects, can be directly used to predict the approach and avoidance tendencies within an individual.

Studies in which the shape of the approach and avoidance gradients have been determined directly have produced a variety of results. The majority have found S-shaped gradients (e.g., Maher & Nuttall, 1962; Smith, 1960).

Only one study has been reported so far in which the shape of approach and avoidance gradients was determined within the same subjects. However, the results of this study (Rigby, 1954), in conjunction with other findings that support it, have been the basis for a revision in the way psychologists view the gradient model of conflict.

In Rigby's experiment animals were conditioned by a procedure which presented a light as a CS for food and a buzzer as a CS for mild electric shock. The animals were restrained in the presence of the CS, which preceded the delivery of food by a fixed period of time. Later the delivery of food was replaced by mild shock, after preliminary training during which the CS preceded the UCS immediately. That is, the CS was turned on but the animals were restrained from approaching the food tray for various periods of time. Their responses to obtain food or avoid shock were measured by polygraphic recordings of their movements against a restraining stock in a forward or backward direction. Measures were taken at various temporal "distances"—00, 2.5, 5.0, 7.5, and 10 seconds.

The approach and avoidance gradients obtained in this experiment are shown in Fig. 5-9. The slope of the gradients tended to increase as positive or negative reinforcement approached in time. The approach and avoidance gradients were also nearly identical at each point measured. That is, their slopes tended to increase in parallel fashion. Thus, con-

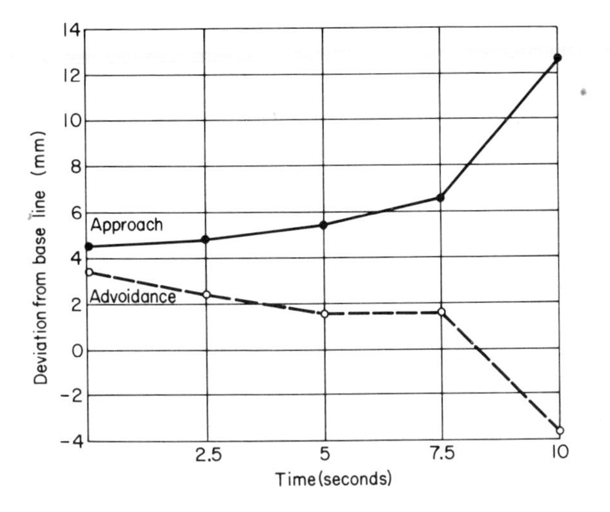

FIG. 5-9. Approach and avoidance gradients determined within the same subjects by Rigby (1954).

trary to the Dollard and Miller assumption, the avoidance gradient did not have a steeper slope.

When both positive and negative CSs were presented together an intensification of conflict-related behavior occurred as the time goal approached. There were also wide individual differences in the kind of behavior exhibited by the animals, ranging from immobility to violent escape responses.

The Dollard and Miller conflict model requires that if the animal is placed distant from the goal in the approach-dominant zone it will approach the goal up to the point of behavioral oscillation. At that point the approach and avoidance gradients intersect. The animals should stop and show vacillation. Accordingly, Maher and Noblin (1963) gave approach and avoidance training to a group of rats and then continued to give trials in a straight alley until each subject reached the trial on which its stopping point was between 33 and 39 inches from the goal. Then two subgroups of animals were given subsequent trials with the start box moved back 12 and 24 inches, respectively, from where it had been in the original training. If the heights of the approach and avoidance gradients are dependent upon distance from the goal as the Dollard and Miller model supposes, then the animals should have come to a stop and oscillated around a point that was approximately the same distance from the goal on the next trials, irrespective of the distance from the goal at which they were started.

This did not happen. The farther away from the goal the animal was when it started, the farther away was its stopping point. Animals came to a stop after they had run a constant distance, irrespective of how far away from the goal their starting point was. This result led Maher and Noblin to conclude that the gradients are not determined by the variable of distance from the feared goal. Instead, there is merely a correlation of distance with cues that actually generate approach and avoidance gradients. They pointed out that during avoidance training running had been followed by punishment. Therefore running had become a CS for fear. Animals tended to stop after having run a constant distance because fear was generated to a considerable degree by the approach response itself. The intensity of fear apparently increases not with nearness to the goal, but with approach responses.

At the human level, the individual who is afraid of the water may become frightened after having made responses preparatory to swimming in it. The student may begin to fear failure after having initiated study for the final examination. A self-effacing person may begin a piece of creative work such as a painting, and as soon as he sees that it is good, do something that makes himself feel uncomfortable and dissatisfied with what he has done. In these instances it is as if the approach responses themselves acted as a CS for anxiety or fear.

Still another experiment which has led to a revision of the Miller *et al.* conflict model was performed by Trapold, Miller, and Coons (1960). In order to determine what actually happens when the approach motivation is increased in an approach–avoidance conflict they constructed a 100-foot runway and taught animals to traverse it for food until they developed stable performance. Then they subjected them to shocks of an increasing intensity until a stringent criterion of avoidance was reached. Following this the approach gradient was systematically raised by increasing the animals' hunger until it was so strong that they approached completely to the goal.

As would be predicted by the theory of Dollard and Miller, in the early trials the animals approached part way to the goal and then stopped. Also, as would be predicted, as the strength of the approach tendency increased relative to the avoidance tendency, the locus of behavior moved progressively toward the goal.

A second feature of the experiment, however, would not have been predicted by the theory. Part of the experimental procedure involved placing the rats in the avoidance-dominant zone near the goal and again in the approach-dominant zone far from the goal. It was expected, of course, that when placed near the goal they would retreat to the place where the gradients crossed, and when placed far from the goal they would approach the point of intersection of the gradients. However, in both instances the animals moved *toward* the goal. Furthermore, this finding proved not to be an isolated instance. Taylor and Maher (1959) also reported that they had observed the response of running toward the goal by animals introduced into the supposed avoidance-dominant zone.

It is relevant that Trapold *et al.* were able to conclude that behavioral oscillation occurs only when the approach and avoidance tendencies are nearly equal in strength. That is, as the two opposing tendencies approach one another in strength, neither tends to dominate. The dominance of one or the other is likely to fluctuate. Speculations of Greenfield (1960) are compatible with these findings. Greenfield pointed out that the latency of responses in conflict situations increases as the alternative response tendencies become more incompatible, and further, that the stronger the incompatible response tendencies are, the longer the response latency. These hypothetical relationships are shown in Fig. 5-10. When the alternative responses are completely incompatible, neither dominates. This results in indecision, which results in increasing response latency. The more intense the incompatible responses, the longer the indecision, and the longer the response latency.

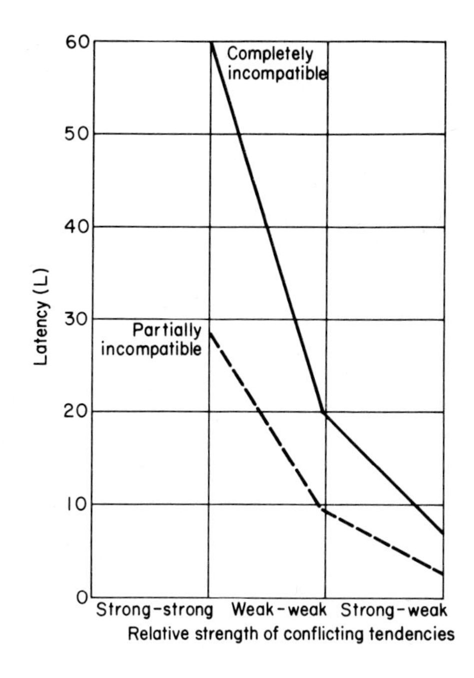

FIG. 5-10. The dependence of response latency (a measure of conflict) upon the relative strength of conflicting tendencies and their degree of incompatibility. (Adapted from Greenfield, 1960.)

The Maher Revision

On the basis of studies such as those described in the preceding section Maher (1964) has formulated a revision of the Dollard and Miller conflict model. Maher's model is presented in Fig. 5-11. He has argued that the gradients representing most competing response tendencies in human behavior as well as those in many animal experiments represent learned responses, and that they are therefore likely to have similar slopes. Since these gradients are essentially parallel, the one that is higher than the other will determine behavior. In Fig. 5-11 the approach gradient is higher than the avoidance gradient. Presumably this is the more common case.

The higher gradient should determine a consistent response: approach or avoidance, depending upon which tendency it refers to. However, this is incompatible with the known fact that behavioral oscillation is characteristic of conflict situations. Maher (1964) resolves this by the following assumption: "The stronger of two competing response tendencies mediates behavior only when its absolute momentary strength ex-

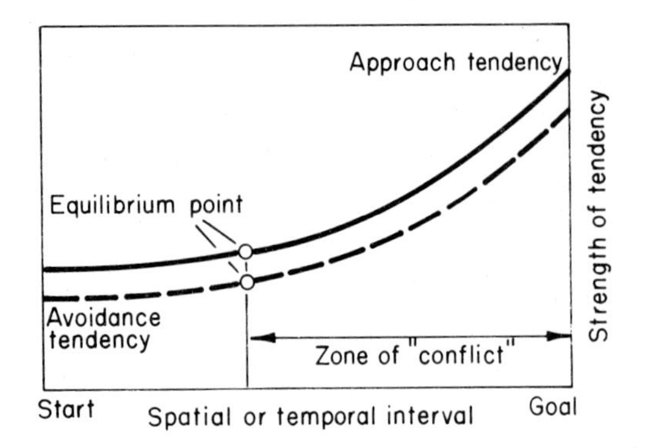

FIG. 5-11. Maher's conflict model. This model assumes parallel approach and avoidance gradients. Nearer the goal both gradients rise in height. Thus the difference between them (in terms of strength) becomes a smaller proportion of their total heights. Therefore beyond the equilibrium point either may dominate. See text for further explanation. (Maher, 1964.)

ceeds that of the weaker by an amount which is a constant fraction of the weaker." In Fig. 5-11 it can be seen that the two gradients, being parallel, maintain a constant difference in height. That is, because they are parallel there is a constant difference between the two gradients at every point—from the start to the goal. The relevant feature of parallel gradients is this: the closer the goal, the higher both gradients rise. This means that the constant difference between them becomes a smaller and smaller proportion of the height of the weaker. Thus, as the goal approaches, the stronger exceeds the weaker by an amount that is a steadily diminishing fraction of the weaker.

Think of it by imagining that at the start, far from the goal, the weaker gradient has a value of 15 units, and the difference between it and the stronger gradient at the same point is 5 units. That is, the height of the stronger exceeds that of the weaker by an amount (5 units) that is a constant fraction of the weaker; namely, $^5/_{15}$ or $^1/_3$. This fraction remains constant as long as the gradients do not rise. As long as this is true, the stronger of the two tendencies will mediate behavior.

But what happens when the gradients rise as the goal approaches? The difference between the two gradients does not change—it remains 5 units. However, the weaker gradient rises higher and higher, paralleling the rise of the stronger gradient. Let us say that near the goal it is no longer 15 units, but 30. Thus, the fraction by which the difference between the height of the stronger and the height of the weaker exceeds the height of the weaker near the goal is much smaller than at the start. It is now $^5/_{30}$ or $^1/_6$, instead of $^1/_3$.

From the point where the gradients begin to rise it can no longer be predicted with certainty that the stronger of the two gradients will determine behavior. This is labeled "zone of conflict" in Fig. 5-11. Behavioral oscillation begins to occur at the beginning of the zone of conflict, labeled "equilibrium point" in Fig. 5-11. Anxiety and anxiety-produced responses would also begin to occur at this point. Since incompatible response tendencies become more closely equal in strength as the goal approaches, either may dominate. Hence, the fact that animals placed in this zone may run toward the goal, rather than away from it, is not surprising.

A final comment—the student may wonder why psychologists would predict oscillation or even the dominance of the weaker response tendency when, because the gradients are parallel, the absolute strength of the weaker never surpasses that of the stronger, nor could it be equal to it. This is no surprise to learning psychologists. The strength of all response tendencies, or habits, fluctuates from moment to moment for the same individual in the same situation. Moment to moment changes in attention, momentary fluctuations in body metabolism, slight variations in the environment—these factors and many more produce variability in behavior even under the controlled, relatively constant conditions of the laboratory. The joint effect of two competing response tendencies, the strength of both depending upon the momentary fluctuation of many variables, allows for even less predictability. Reread Maher's assumption (p. 175–176). He has hypothesized a rule that indicates when our ability to predict the dominance of one tendency over the other will break down. Assuming parallel gradients, the best that can be said is that the probability of the weaker tendency gaining momentary dominance increases with nearness to the goal.

SUMMARY

Laboratory findings relative to certain basic aspects of learning, all of which are of central importance for an eventual scientific interpretation of human behavior, have been discussed in this chapter. The activation syndrome can be conditioned to neutral cues in the environment. The resulting acquired emotionality can have a very powerful effect on behavior. Furthermore, conditioned fear, or anxiety, is extremely resistant to extinction. Conditioned fear can serve as a basis for the learning of new behavior, if the consequences of the new behavior include escape from the cues that have been conditioned to elicit it.

Responses that allow the organism to avoid cues that are associated with fear may also be readily learned. When avoidance learning involves a warning signal of some kind it tends to be adaptive without excessive responding; otherwise the response rate far exceeds that necessary to avoid the feared stimulus and the behavior tends to become rigid insofar as all responses other than the avoidance response tend to become eliminated.

Another kind of conditioned emotional response, or anxiety, has been called the CER(b). This refers to the suppression of positively reinforced behavior that accompanies, and serves as an index of, anxiety conditioning. In general, anxiety has two important effects: behavior with an avoidance history increases and positively reinforced behavior decreases under its influence.

Punishment involves making an aversive stimulus contingent upon behavior or withdrawing a positive stimulus. When behavior is inconsistently or indiscriminately punished anxiety and conflict are generated. Punishment may be highly effective when it is consistent and when an alternative response is positively reinforced. Punishment alone, however, does not eliminate behavior, nor change the potential to behave. It merely has a suppression effect that no longer operates when the punishing contingencies are removed.

Behavior that appears to be self-punishing is invariably found upon analysis to involve powerful sources of positive and/or negative reinforcement. Immediate rewards may offset delayed punishment as a controlling variable. Similarly, it is possible for strong delayed rewards to successfully compete with immediate punishment for the control of behavior.

These and similar situations involve conflict between incompatible response tendencies. The gradient model of conflict of Dollard and Miller has been very fruitful in generating research and extending principles established in the animal laboratory to complex human behavior. Recent experimental results, however, have shown that modification of the original conflict model is required. Maher has formulated a parallel gradient model that appears to be more in line with the facts of conflict behavior and that promises to further our understanding of the principles involved in these complex and important behavioral processes.

REFERENCES

Aronfreed, J. The origins of self-criticism. *Psychol. Rev.* 1964, **71**, 193–218.

Brown, J. S. Gradients of approach and avoidance responses and their relation to motivation. *J. comp. physiol. Psychol.*, 1948, **41**, 450–465.

Cameron, N., & Margaret, A., *Behavior pathology*. Boston: Houghton, 1951.

Cook, S. W. The production of "experimental neurosis" in the white rat. *Psychosom. Med.*, 1939, **1**.

Dollard, J., & Miller, N.E. *Personality and psychotherapy.* New York: McGraw-Hill, 1950.

Epstein, S., & Fenz, W. D. Theory and experiment on the measurement of approach avoidance conflict. *J. abnorm. soc. Psychol.*, 1962, **64**, 97–112.

Epstein, S. Toward a unified theory of anxiety. In B. A. Maher (Ed.), *Progress in experimental personality research.* Vol. 4, New York: Academic Press. 1967.

Estes, W. K. An experimental study of punishment. *Psychol. Monogr.*, 1944, **57**, No. 3 (Whole No. 263).

Estes, W. K., & Skinner, B. F. Some quantitative properties of anxiety. *J. exp. Psychol.*, 1941, **29**, 390–400.

Farber, I. E., Response fixation under anxiety and non-anxiety conditions. *J. exp. Psychol.*, 1948, **38**, 111–131.

Greenfield, N. Summation of conflict states. *Percept. mot. Skills*, 1960, **11**, 103–110.

Holland, J. G., & Skinner, B. F. *The analysis of behavior.* New York: McGraw-Hill, 1961.

Hunt, H. F., & Brady, J. V. Some effects of punishment and intercurrent anxiety on a simple operant. *J. comp. physiol. Psychol.*, 1955, **48**, 305–310.

Kalish, H. I. Strength of fear as a function of the number of acquisition and extinction trials. *J. exp. Psychol.*, 1954, **47**, 1–9.

Kaplan, M. The maintenance of escape under fixed-ratio reinforcements. *J. comp. physiol. Psychol.*, 1956, **49**, 153–157.

Maher, B. A. The application of the approach-avoidance conflict model to social behavior. *J. conflict Resolut.*, 1964, **8**, 287–291.

Maher, B. A., & Noblin, C. D. Conflict equilibrium and distance from the goal: a demonstration of irrelevance. Paper read at Southeast. Psychol. Assoc., Miami Beach, 1963.

Maher, B. A., & Nuttall, R. The effect of repeated measures and prior approach training upon a spatial gradient of avoidance. Paper read at Amer. Psychol. Assoc., St. Louis, 1962.

Maher, B., Weisstein, N., & Sylva, K. The determinants of oscillation points in a temporal decision conflict. *Psychon. Sci.*, 1964, **1**, 13–14.

Masserman, J. H. Experimental neuroses. In *Frontiers of Psychological research: Readings from the Scientific American.* San Francisco: Freeman, 1966. p. 247.

Miller, N. E. Studies of fear as an acquirable drive. I. Fear as motivation and fear-reduction as reinforcement in the learning of new responses. *J. exp. Psychol.*, 1948, **38**, 89–101.

Miller, N. E. Learnable drives and rewards. In S. S. Stevens (Ed.), *Handbook of experimental psychology.* New York: Wiley, 1951. Pp. 435–472.

Parke, R. D., & Walters, R. H. Some variables influencing the effectiveness of punishment for producing response inhibition. *Monogr. Soc. Res. Child Development*, 1967, **32**, (whole No. 109).

Pavlov, I. P. *Conditioned reflexes.* (transl. by G. V. Anrep) London & New York: Oxford Univer. Press, 1927.

Pavlov. I. P. *Lectures on conditioned reflexes.* (transl. by W. H. Gantt) New York: International Publishers, 1928.

Rigby, W. K. Approach and avoidance gradients and conflict behavior in a predominantly temporal situation. *J. comp. physiol. Psychol.*, 1954, **47**, 83–89.

Sidman, M. Avoidance with brief shock and no exteroceptive warning signal. *Science*, 1953, **118**, 157–158.

Sidman, M. *Tactics of scientific research.* New York: Basic Books, 1958.

Sidman, M., & Boren, J. J. The relative aversiveness of warning signal and shock in an avoidance situation. *J. abnorm. soc. Psychol.*, 1957, **55**, 339–344.

Skinner, B. F. *The behavior of organisms: An experimental analysis.* New York: Appleton, 1938.

Smith, N. An empirical determination of an approach gradient. *J. comp. physiol. Psychol.,* 1960, **53**, 63–67.

Taylor, Janet A., & Maher, B. A. Escape and displacement experience as variables in the recovery from approach-avoidance conflict. *J. comp. physiol. Psychol.,* 1959, **52**, 586–590.

Trapold, M. A., Miller, N. E., & Coons, E. E. All-or-none versus progressive approach in an approach-avoidance conflict. *J. comp. physiol. Psychol.,* 1960, **53**, 293–296.

Walters, R. H., & Parke, R. D. The influence of punishment on children's behavior. In B. A. Maher (Ed.) *Progress in experimental personality research,* Vol. 4. New York: Academic Press, 1967.

Whiting, J. W. M., and Mowrer, O. H. Habit progression and regression—a laboratory investigation of some factors relevant to human socialization. *J. comp. Psychol.,* 1943, **36**, 229–253.

Chapter 6

Frustration

INTRODUCTION

As is the case for other terms which we have discussed, frustration has been taken over into scientific psychology from everyday language. Normally when this happens terms tend to become more meaningful and precise, because systematic relations are established between the operations defining them and a variety of other concepts and phenomena. At the same time many of the surplus meanings of such terms are dropped because of their vagueness. Therefore, in a certain sense such terms become less meaningful.

The term "frustration" is applied to several different operations. Unhappily, although these operations are distinctly different from each other, some of them differ only slightly from operations employed as definitions of other concepts. Furthermore, the effects produced by these operations are multiple and the factors influencing them are not always clear. Nevertheless, the phenomena related to frustration, which will be discussed in the present chapter, are of central importance for an understanding of the adjustment process. Further elaboration of them will occur throughout later chapters.

FRUSTRATION OPERATIONALLY DEFINED

Operationally, frustration involves preventing a response from being made. There are three classes of operations that do this: (1) The use of physical or psychological barriers to prevent a response from occurring

—an example of the latter would be learned restraint, e.g., rules and regulations designed to govern behavior. (2) The withholding or the removal of sources of reinforcement—when reinforcement is withheld for a relatively short period of time it is referred to as *frustration by delay.* When reinforcement is withheld indefinitely it is referred to as *extinction.* (3) The elicitation of incompatible responses—in other words, placing the individual in a conflict situation.

The first of these has been called *frustration by thwarting,* the second, *frustration by removal of the maintaining stimuli,* and the third, *frustration by conflict.* The topic of conflict was discussed in some detail in the previous chapter; our immediate concern will be with the operations of removing the maintaining stimuli and with thwarting.

FRUSTRATION BY REMOVAL OF THE MAINTAINING STIMULI

When responses are extinguished after having been learned on a continuous reinforcement schedule (crf), extinction is relatively rapid and the animals show a good deal of emotionality (see pp. 9–10). Laboratory cats for whom food is ordinarily constantly available, for example, when left unfed by an undependable research assistant may "cry all night." That is, extinction for crf animals involves a good deal of emotion. Cumulative records show that animals respond with bursts and depression of activity during the early trials of extinction. Typically the animals not only respond vigorously, but they attempt to bite the experimenter, gnaw at the bar and/or other parts of the apparatus, and show other signs of emotional disturbance. The most frequently reported behavior associated with the early trials of extinction is increased activity and response intensification (e.g., Finch, 1942).

Learning that has been established and maintained on partial reinforcement schedules, on the other hand, tends to result in extinction behavior that lacks this emotional basis. It also results in behavior that is more resistant to extinction.

Amsel's r_f–s_f Mechanism

Amsel (1958) has hypothesized that the removal of reinforcing stimuli produces an emotional disturbance which he refers to as a frustrative response. He designates the frustrative response R_F. The frustrative response involves components of the activation syndrome. The subjective feeling that accompanies it is unpleasant. This response is qualitatively specific insofar as, although it may involve components of unpleasant emotional experiences such as anger and fear, it is specifically associated with that complex of tensions that a person experiences when he says, "I feel frustrated."

The frustrative response, R_F, is sometimes referred to as *the primary frustration response* in order to distinguish it from r_f, its fractional component, which is analogous to the r_g, or fractional anticipatory goal response discussed earlier (see pp. 113–118). We recall that in simple animal learning situations, fractional components of the goal response (R_G) occur in response to stimuli more and more remote from the rewarding event as learning proceeds. This is because stimuli continuously associated with the reward take on conditioned reinforcing properties. When immediately adjacent stimuli thus become conditioned reinforcers after a number of exposures, stimuli adjacent to them but more remote from the primary source of reinforcement also begin to take on conditioned reinforcing properties. The result is that stimuli further removed from the goal begin to elicit anticipatory, fractional components of the goal response. It is assumed that the r_g and its associated stimulus cue, s_g, heighten approach motivation. Presumably this is the explanation for an approach *gradient.* Coming home for dinner John walks down the street, but runs up the front steps.

Amsel postulates an r_f–s_f mechanism, the operation of which is analogous to the r_g–s_g mechanism. Nonreinforcement is assumed to arouse R_F. As nonreinforcement continues, r_f, like r_g, is assumed to become anticipatory, and to occur earlier and earlier in the chain of responses that lead to the place where the R_F is elicited.

Like r_g, the r_f is assumed to be a cue-producing response. Its associated stimulus cue is designated s_f. These anticipatory cues of frustration are something with which the subject must cope. Just as the r_g–s_g mechanism is thought to function as a motivational mechanism, r_f–s_f is assumed to have motivational properties, but frustration is assumed to be inherently aversive. Hence, many responses incompatible with the ongoing behavior are elicited. They include attempts to avoid or escape the frustrating situation, aggressive responses, etc.

When learning involves partial reinforcement, nonreinforced trials of simple animal learning situations can be thought of as being themselves extinction trials. The only difference between them and actual extinction is that during extinction the subject is exposed continuously to nonreinforcement, without reinforcement ever occurring again. Since the nonreinforced trials that occur during learning produce an R_F which becomes anticipatory, avoidance tendencies which interfere with learning may be elicited. This accounts for the fact that under partial reinforcement, performance tends to be poorer *on the early trials* than it does under continuous reinforcement. However, as learning proceeds under partial reinforcement it would be expected that performance would improve as it usually does with frequency of reinforced occasions. Since cues of varying remoteness from the goal evoke both anticipatory r_gs and anticipatory r_fs, and since the positively reinforcing stimulus events associated with the goal eventually gain control over the animal's behavior, it follows that the animal learns to perform in the presence of frustration-produced cues. In other words, on later trials s_f becomes more and more strongly attached to approach tendencies.

The fact that the frustrative response is conditionable in this manner accounts for the effects of partial reinforcement during extinction. Since partially reinforced animals have associated the frustration stimuli with making the response, it would be predicted that they would continue to respond when the primary source of reinforcement is removed. It would also be predicted that pronounced emotionality would be absent during extinction when reinforcement during learning has been partial. Continuously reinforced animals on the first extinction trials ought to be faced with a full-blown R_F with strongly aversive properties. For partially reinforced animals, on the other hand, components of R_F have been associated with "feeling good," with anticipatory goal responses.

These basic principles are highly generalizable. The individual's reaction to frustration when it is defined as removal of the maintaining stimuli is undoubtedly related to the amount and kind of rewards and punishments (nonreinforcement) the frustrated behavior has received in the past. One psychologist (Skinner, 1948) has even seriously suggested that much of the frustration, anxiety, and unhappiness of adults could be ameliorated if children were deliberately trained in frustration tolerance. This could be accomplished by arranging for frustration to occur in small doses. In one scene from his utopian novel, lollipops are tied around the children's necks but they are not allowed to eat them! As farfetched as this may seem, it has, at least in principle, much in common with the child-rearing practices of cultures where the Puritan ethic has dominated, although the object there appears to have been less the amelioration of human unhappiness than creating adults who would show persistence in the face of hardships.

FRUSTRATION BY THWARTING

In frustration by thwarting the maintaining stimuli are present, but the organism is prevented from getting reinforcement. Interference with conditioned behavior in almost any manner can be considered thwarting. This is probably the most common type of frustration-producing situation. At the human level, too, sometimes the obstacles are physical, as when an individual suddenly finds that he has lost the key to his house and the doors and windows are locked. Anyone who has been in such a circumstance is familiar with the feeling of annoyance that is produced.

More commonly for people, frustration involves the thwarting of learned expectancies. These may involve ideas people come to have about themselves, about others, about what is "right" and what is "wrong," etc. They are most often learned at the verbal and symbolic levels without actual primary sources of reinforcement ever having been involved. Unobtainable standards may be set by parents. When the child is thwarted by being unable to live up to parental expectations considerable frustration is the inevitable result. Similarly, unrealistic images and expectations may be created by the individual's peer group or by the mass media of communication, advertising, the entertainment world, popular art and literature, etc. The individual's evaluation of himself in reference to the ideal images and components of the popular myths of the culture may result in frustration because he does not have the personal characteristics, physical or otherwise, that would allow him to match or at least approximate the ideal.

Sometimes, because of their peculiar reinforcement histories, people behave toward others in ways that are arbitrary and unfair. The so-called authoritarian personality is apparently not uncommon in our culture. Such individuals, although they behave in a dependent and submissive manner to people of higher status than their own, enjoy the submission of, and show a lack of concern for, others over whom they have authority. These individuals actively produce anxiety, threat, and frustration for others. Students are familiar with the occasional college professor who never gives an A grade, and whose major purpose in life appears to be to overwhelm, degrade, and intimidate the student. Professors are familiar with administrators whose only concern is the number of publications of their faculty, with no consideration for quality as a major virtue, and who thereby arrange conditions for the degradation of people. Constructive work and productive scholarship may not be possible under such circumstances. Unfortunately, people in authority who show such a disrespect for others tend to cause a spiral effect. Others learn that in order to be reinforced they must also assume such values (or lack of values). Thus, a vicious circle may be created and the effects may be widespread.

Aside from frustration produced by physical barriers, social and legal regulations, and by the behavior of other people the individual's own physical or psychological limitations may be sources of frustration. Physical handicaps, debilitating illness, the color of one's skin, lack of physical attractiveness, and the like may constitute severe sources of thwarting. The source of frustration may also be perceived to be the self. The alcoholic may fully realize the ultimate aversive consequences of his behavior and the immediate suffering which he imposes upon his family; yet he may be unable to do anything about his compulsion to drink. He may have been taught by the culture that he is responsible for his actions, and that what he is doing is sinful or bad. Considerable guilt and self-hatred may be generated by what are perceived to be self-imposed frustrations. Similarly, individuals who have been exposed to early rejection and deprivation, or otherwise anxiety-conditioned, often fail when they attempt the major undertakings of life; working responsibly and independently in their occupation; love and marriage; relating responsibly to others, etc. Feelings of inferiority, guilt, and self-contempt may be generated by the individual's awareness that his failures have been self-imposed, thus adding further increments of frustration.

A list of thwarting circumstances in life situations could be extended indefinitely. The reader will probably be readily able to supply examples of his own. A more detailed analysis of the dynamics of human behavior under thwarting will be reserved for later in the text.

GENERAL CONSEQUENCES OF FRUSTRATION

The most general consequence of frustration is an intensification of behavior. Many learning psychologists reason that the state resulting from frustration functions as an irrelevant drive to heighten the total level of arousal. This heightened arousal increases the intensity of any behavior evoked in the situation. Although one of the most primitive reactions to frustration appears to be anger and aggression, the actual behavior which occurs under frustration is a combined result of many sources of reinforcement. If withdrawal has received particular reinforcement in the individual's past, frustration will be likely to intensify the withdrawal reaction. Actively nonresponding and withdrawal are not infrequent responses to frustration for people in our culture because overt displays of anger and aggression are among the most commonly punished kinds of behavior during socialization. Sometimes direct aggression does occur as a response to frustration but more frequently it is displaced onto people and objects other than the original frustrating agent.

Thus, frustration may produce many effects on behavior that depend upon the sources of reinforcement operative in the situation and in the organism's learning history. Generally, reactions to frustration are negative in character. The behavior is less adaptive, less constructive, more primitive, and less flexible than it is under nonfrustrating conditions.

With these general reactions to frustration in mind, let us turn to an analysis of more detailed aspects of frustration-produced behavior.

FRUSTRATION AND REGRESSION

In addition to withdrawal, aggression, and agitated behavior, the sources of reinforcement operative in the frustrating situation may combine to produce a phenomenon known as *regression.* Regression is the reverting to earlier, previously learned behavior. There are many experiments demonstrating a regressive effect as a function of frustration. A simple yet instructive one was performed by Hull (1934). Hull first conditioned animals to traverse a 20-foot runway in order to receive food. After the behavior was well established, he lengthened the alley to 40 feet. In the new 40-foot alley the animals ran halfway and stopped. In a relatively short time however, they learned to traverse the entire 40-foot alley for food. The experiment was continued until this habit was well established. Then frustration was introduced. Food was removed from the goal box. Under these conditions the animals tended to stop at about the point where they were originally fed. That is, the previously learned and abandoned habit reappeared.

Regression and Inefficient Performance

Behavior that is inefficient and weak in strength is most often a response to frustration by extinction. It is also frequently regressive in character. This can be most clearly demonstrated by careful observation of behavior during learning and extinction in simple animal learning situations. When an animal is learning to press a bar in a Skinner box the early trials are characterized by inefficient and often effortful behavior. At least for the first few trials animals typically reproduce almost exactly the behavioral sequence that occurred just prior to their receiving reinforcement for the first time. If the animal circled and approached the bar from the left on the initial trial, or if his initial response was made while standing on his hind legs and holding his head and neck high in a manner of stretching then these same peculiarities of behavior would tend to be repeated.

Often several responses which are actually irrelevant to the animal's receiving reinforcement occur in combination or in sequence. Gradually, as learning progresses, these responses drop out and are replaced by others. Eventually all irrelevant responses drop out. The animal's responses become very efficient and well organized. If the behavior has been learned on a schedule of reinforcement that yields a gradual decline in response rate and response strength during extinction, then as the original response becomes weaker under nonreinforcement some of these earlier irrelevant responses reappear. In the case of human behavior it is as if the individual were saying to himself "since nothing else seems to work, maybe I'll have some success if I do things the way I used to." Actually, the responses which recur are not likely to be well integrated. Also, the individual is usually unaware of his regressive behavior.

Constructiveness of Behavior

Under frustration-produced stress behavior is less constructive than it would normally be. A famous study by Barker, Dembo, and Lewin (1941) illustrates this. They gave 30 children 30 minutes of unrestricted play with a set of toys. The following day the children were presented with the same toys together with a much more attractive set of toys. When the children had become involved in the play situation, the more attractive toys were removed to another part of the room and a wire mesh screen was locked into place between them and the children. They were then allowed to play with the original set of toys for 30 minutes. Their behavior during this latter period was rated for "constructivity" by two observers through a one-way window.

During the frustration period they spent a considerable amount of time attempting to reach the inaccessible toys or trying to leave the experimental situation. Compared to the prefrustration period only about half as much time was spent in what the authors called "primary play"; that is, play in which the child's entire attention was directed toward the toys. Furthermore, the constructiveness of the primary play which did occur declined significantly.

Some children were frustrated very much by the removal of the attractive toys, whereas other children were hardly frustrated at all. When the children were divided into strongly and weakly frustrated groups in terms of the amount of barrier and escape behavior exhibited it was found that there was a significant decline in constructiveness of play during frustration for the strongly frustrated groups, but not for the weakly frustrated group.

Probably the best explanation of the lack of constructiveness of behavior following frustration, as evidenced in the Barker, Dembo, and Lewin study is that frustration is likely to produce competing responses which interfere with ongoing activity. Thus, Child and Waterhouse (1952) have argued that "frustration of one activity will produce lowered quality of performance in the second activity to the extent that it leads to the making of responses that are incompatible with or interfere with the responses of the second activity." Child and Waterhouse postulated that the many intefering responses that occurred under frustration in the Barker, Dembo, and Lewin experiment indicated that a good deal of interference was occurring covertly as well, and that these competing responses were responsible for the decline in constructiveness of play. They were able to show very high and significant correlations between constructiveness of play and time spent in other activities. For example, the correlation between constructiveness of play and the time spent in barrier and escape behavior during the frustration period was −.72.

Primitive Regression

Under severe frustration not only does the individual perform the behavior being frustrated inefficiently and show a lack of constructiveness in his other activities, but with repeated nonreinforcement or punishment he may revert to even earlier learned patterns of behavior. If these earlier patterns are not reinforced regression may continue to attachments that were earliest, more primitive, and most satisfying. This is especially likely to happen when earlier behavioral patterns have received prolonged intermittent reinforcement and when the reinforcement for current behavior patterns is insufficient to maintain the behavior in strength.

There are occasionally dramatic cases of regression. Wallin (1949) describes the case of a woman who, following the death of her husband, began to adopt her daughter's adolescent circle of friends and to act and dress very much as if she were one of their peers. Shortly, her behavior showed even more marked regression until she was dressing in short dresses and pigtails and behaving like a 6- or 7-year-old child. Presumably, because of continued nonreinforcement at each level, she eventually regressed to a completely childish and infantile stage. She was institutionalized and remained completely infantile and helpless until her death some years later.

More commonly, individuals regress to strong emotional habits learned earlier in their lives and the regressed behavior is maintained by multiple sources of reinforcement operative in their life situations. Further regression does not occur unless the individual is exposed to added sources of frustration and stress. Severe regressions of this sort have been described by Grinker and Spiegel (1963) in their analysis of reactions to severe stress occurring in combat situations:

> Since the dependence caused by regression cannot be gratified, the resulting frustrations produce conflict between the patient's needs and reality. The result is a neurosis, dominated by depression due to the feeling of not being loved and by aggressiveness as a reaction to the depriving environment. This creates new anxiety, because of fear of retaliation, and further augments regression . . . .

> Psychosomatic disorders may be the only means by which regressive tendencies are permitted expression. Others take flight into marriage or alcoholism; some turn their needs around and become solicitous over parents and family. A few react to their needs by overcompensating aggressiveness, even volunteering to return overseas. By this means, anxiety is avoided and their needs are repressed or satisfied only by displacement. [1963, p. 359]

The latter reactions which Grinker and Spiegel mention constitute mechanisms of defense against anxiety. If anxiety is sufficiently reduced, further regressions may be held in check. These mechanisms will be discussed at length in later chapters.

Milder Forms of Regression

More common than these dramatic forms of regression are behavioral episodes in which both the frustration and the ensuing regressive behavior are circumscribed. Thus, the young wife frustrated in her married life may return home to mother. The child, fearing loss of love upon the birth of younger siblings, may revert to bed-wetting. Whether or not regression occurs depends upon how strongly reinforced, or how tenuous, the current behavior is and upon the rewarding or punishing contingencies associated with the older habits.

Variables Affecting Regression

Variables affecting regression can readily be illustrated by an experiment by Whiting and Mowrer (1943). The apparatus used in this experiment is shown in Fig 6-1. Animals were trained first to take the path S–G. In the second part of the experiment the habit S–G was extinguished in three different ways for different groups of animals: nonreward, a physical barrier placed at point C, and shock at point O. Each of these three groups of animals was further subdivided into three groups. One group was taught the habit S–A_s–G, another the habit S–A_i–G, and

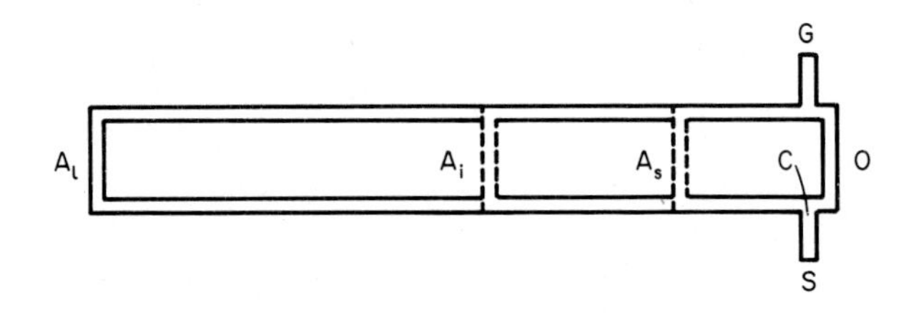

FIG. 6-1. Floor plan of the apparatus used in the Whiting-Mowrer experiment. Groups of animals were taught the habit S–G; this habit was extinguished by nonreward, a barrier at C, or shock at O; and the new habits S–A_s–G, S–A_i–G, or S–A_l–G were taught. Then all pathways were open for the third part of the experiment. (Whiting & Mowrer, 1943.)

a third the habit S–A_l–G. In the final part of the experiment all pathways were open for 75 trials.

The results of this experiment that have special bearing on our understanding of regression include the following: (1) The progression from the original habit, S–G, was more difficult the longer the pathway the animal was required to traverse in the learning of the second habit. (2) Regression to the original habit during the free-choice condition was fastest for group S–A_l–G (the habit acquired with the greatest difficulty), and slowest for group S–A_s–G. (3) The groups that progressed to the new habit under shock conditions showed the least regression when all the pathways were opened (in this latter case two strong sources of reinforcement had been operative: food and escape from shock). This experiment demonstrates that habit regression is a function of the strength and quality of the sources of reinforcement operative in the situation. Difficult and weakly learned behavior will be the first to be abandoned. Earlier behavior that had been nonreinforced will be returned to more rapidly than earlier behavior that had been punished.

Adjustive Regression

Regression is not always maladjustive. The student under the stressful and competitive atmosphere of the university may spend a weekend with his family and friends and receive encouragement and assurance which allow him to return and function with less anxiety and fewer task-interfering responses. The businessman or professional person may relax and allow himself to be childish at times.

Whether or not this type of regressive behavior becomes fixated depends upon the source and quality of reinforcement in the regressive situation. If the only source of reinforcement in the regressive situation is negative reinforcement (anxiety reduction), the regressive behavior may remain fixated. If positive sources of reinforcement are operative as well, some immunity to frustration may be gained.

FRUSTRATION AND FIXATION

A very early series of experiments by Hamilton (1916) clearly demonstrated that behavior becomes stereotyped, fixated, and repetitive under certain conditions of thwarting. Hamilton's apparatus, which was designed for both humans and animal subjects, contained four exit doors from which a subject could escape. The correct (unlocked) door was randomly varied with the restriction that the same door was never unlocked on successive trials. In the presence of strong aversive stimulation subjects developed stereotyped responding. Typically, they would show persistent attempts to escape through the same locked door on subsequent trials. A similar experiment by Patrick (1934) yielded essentially the same results.

For a more careful examination of the nature of behavioral fixation let us turn to an experiment by De Valois (1954). This experiment demonstrates a general relationship between the intensity of stimulating conditions and behavioral variability. Although the experiment itself is very simple it brings the conditions producing behavioral rigidity into clear relief. De Valois employed a maze, the plan of which is shown in Fig. 6-2. Four groups of animals were used. They were run through the maze after 6-hour or 22-hour water deprivation, and under weak or strong shock. During the first 24 trials the rat could turn either to the left or to the right at each choice point. The center alley was kept closed. The number of changes on successive trials at each choice point was measured. After trial 13 each of the four groups of rats was further subdivided, with one subgroup continuing under the original conditions, and the other switching to a new motivation (e.g., from shock to water deprivation). From trials 25 to 36 the center alley was opened.

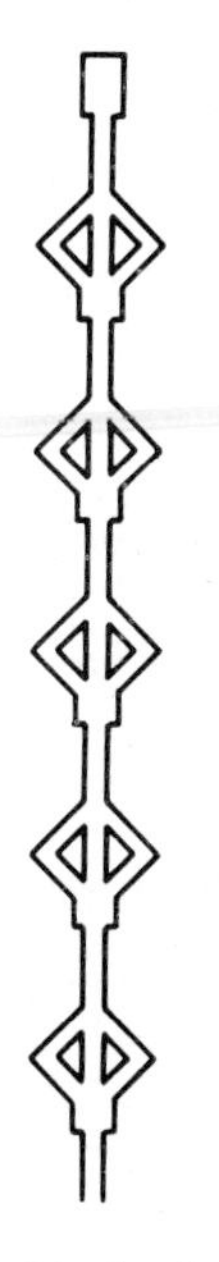

FIG. 6-2. A section of the maze used in the de Valois experiment. In the first experiments the center lane was closed and animals were run through under strong or weak motivation. Then the center lane was opened. Weakly motivated animals took the short path more often than strongly motivated animals. (De Valois, 1954.)

The results were as follows: during the first 24 trials De Valois found that the strong motivation groups (strong shock, 22-hour thirst) showed less behavioral variability (as measured by the number of changes on successive trials at each choice point) than the weak motivation group did. Furthermore, animals shifted from weak to strong motivation tended to show less variability than animals maintained on constant motivation or shifted to weak motivation. Increasing the intensity of arousal decreases the variability of behavior.

The results following trial 13 are particularly interesting. With the center alley opened the animals could take the direct, shortest path to the goal. What actually happened is that the animals run under constant strong shock did not enter the center alley at all! Animals shifted from strong to weak shock on trial 13 made very few entrances to the center alley. Generally, animals run under strong motivational conditions showed a much smaller tendency to take the shorter path to the goal.

The importance of this experiment is that it demonstrates that responses learned under conditions of intense arousal not only are more rigidly performed, but also tend to show greater resistance to change. Therefore, it is apparent that behavioral rigidity and fixation may occur in situations that do not involve any of the operations defining frustration. In this particular experiment, merely the introduction of strong thirst or strong shock and a crf schedule were sufficient for producing fixated behavior.

It is important to distinguish between the behavior resulting from the reinforcing contingencies described above and the stereotyping of behavior which can occur under frustration. The most extensive work on the relation between frustration and fixation has been carried on by Maier (1949). Maier's experimental procedures involved placing rats on a stand and forcing them (usually by means of an air blast) to jump to one of two stimulus cards. This apparatus is shown in Fig. 6-3.

When animals jumped to a "correct" card the force of the jump easily knocked the card over so that they actually jumped through a window in front of which the card had been placed, landing on a platform containing food. When they jumped to an "incorrect" card, the window behind the card was latched so that the animal would bump his nose and fall several feet into a net below.

Under these conditions animals can readily be trained either to jump to a card marked with a particular symbol (e.g., a square), and not to a card marked by a different symbol (e.g., a triangle); or to jump always to one side (right or left), irrespective of which symbol card appears on that side. Thus, animals readily learn symbol responses or position responses.

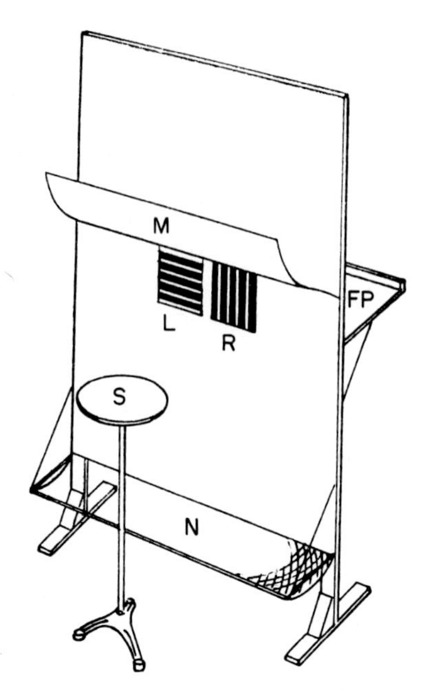

FIG. 6-3. The Lashley jumping apparatus used in Maier's experiments. The animal is placed on the jumping stand (S) and forced, by an air blast or shock, to jump to the stimulus windows (L) and (R). The "correct" window is unlocked. If the animal jumps to it the force of his jump causes it to open and he lands on the food platform (FP). If he jumps to the "incorrect" window, which is locked, he bumps his nose and falls several feet below into the net (N). M is a protective shield preventing him from jumping high. (Lashley, 1930.)

Maier introduced frustration by making the problem insoluble. The cards were designated as correct or incorrect in a random sequence so that no response could be learned which would always escape punishment. Under these conditions the animals soon refused to jump. Therefore, an air blast or a strong shock was administered in order to force the animal to respond. With the added air-blast or shock animals sometimes attempted to make escape responses: they would try to jump directly into the net or out of the experimental situation. These responses were highly resistant to change. Maier called them *stereotype responses*. Escape responses actually were infrequent. More commonly, the resulting behavior involved one or the other of two kinds of highly stereotyped responses: position-stereotype response, or symbol-stereotype responses. In the former, the animals developed a fixated position habit.

They always jumped to the same side, whether that side was rewarded or punished. In the latter, the animal always jumped to the same card, irrespective of punishment or reward.

Following training on the insoluble problem, the problem was then made soluble by changing back to the original conditions. Before the introduction of the insoluble problem animals had little difficulty in learning the correct response. After the experience of frustration in the insoluble problem situation, however, the stereotype response learned in the insoluble problem situation persisted. When the animals were unable to change their behavior after 200 trials Maier called these stereotype responses *fixations*. In one survey of the results of such experiments Maier concluded that about 75 percent of the animals with position stereotypes were unable to change their responses when the problem was made soluble.

Several lines of evidence indicate that the fixations produced in these experiments are compulsive in nature. It can be imagined that if the animal were human he might indicate that he "knew his behavior was self-punishing and irrational, but that he couldn't help himself." In fact, that is exactly what human neurotics express concerning their compulsive, self-punishing behavior.

Remaining within the context of this particular experiment for the present, however, there is ample evidence relative to the compulsive nature of the animal's fixations: (1) An analysis of the animals' response latencies in the insoluble problem indicates no preference for either window, whereas when the problem is again made soluble the response latencies are significantly greater to the negative card than to the positive card. This indicates that the animals can discriminate between the two cards, even though their actual response remains fixated. (2) Fixated rats show more avoidance behavior to negative cards than to positive cards while they are exercising their fixation, again showing that a discrimination has actually been learned, even though the behavior has not been brought under stimulus control. (3) Those that do change their fixated behavior adopt the required responses either immediately or very rapidly, indicating again that the correct response has been learned, but that the animal has been prevented from exercising it. (4) A particularly interesting finding (Ellen, 1956) is that animals can be made to break their fixations, provided that the new response required does not interfere with the expression of the fixated response. Thus, animals with a left position fixation will jump to a positive window if that window is a third window which has been added to the left of the other two. (It will be seen later that human neurotics behave analogously: adaptive behavior may be learned and within the individual's repertoire, however it is only under certain favorable conditions that he is able to exercise it.)

The results of these experiments cannot be explained by any single principle of learning because the reinforcing contingencies are quite complex. In this sense they are analogous to complex human behavioral situations. Maier himself felt that these strong fixations give the animal a way of responding to an insoluble problem. Without the tension relief provided by the fixated response, such a situation would remain highly stressful. Evidence from elsewhere (e.g., Wolpe, 1958) indicates that the mere fact of making a response, of "doing something," may be incompatible with anxiety. However, undoubtedly the mere fact of responding, only partially contributes as a source of reinforcement in these situations.

When shock is very intense an avoidance response may persist for many hundreds of trials (see pp. 139–143). If the animals in the Maier-type situation were not forced to jump by noxious stimulation the fixated behavior would probably not occur.

Still another factor is that during the insoluble problem the animals are placed under a 50–50% random schedule of reinforcement, whereas the soluble problem schedule involves 100 percent reinforcement and 0 percent punishment for the correct response. The combination of escape from noxious stimulation and punishment coupled with partial positive reinforcement would be expected to produce intense and persistent responding.

The number of trials, or the number of exposures, is generally the most important variable in any learning situation. The operation of this variable under very intense conditions of stimulation such as occur in the Maier experiments is likely to be obscured. However, the evidence indicates that the degree of fixation or stereotyping is a function of the number of exposures the animals receive. Increasing the amount of time spent on the insoluble problem increases the time necessary for new responses to be learned for both animals which fixate and those which merely show stereotyped responses.

Fixations may be highly specific. Maier found that animals which had developed a right position fixation in the insoluble problem situation were as easily able to learn a left position response in a maze situation as were nonfixated animals. Within the stimulus complex in which the animal is frustrated, however, increased generalization would be expected because of the great amount of anxiety generated.

Severe frustration in complex human situations typically produces behavior that is associated with widely generalized cues, rather than being under the control of situation-specific stimuli. Since punishment and anxiety are an inherent part of many frustrating situations, positively reinforced behavior is frequently suppressed and the person may become generally inhibited. Not only may he be uninterested in positive reinforcement, but the effects of anxiety and frustration become internalized. He may not be able to express himself in a wide variety of situations because of the generalized fear that any type of self-expressive behavior will be punished. These people suffer from a great deal of unrelieved tension which we normally call misery. An important prerequisite to constructive behavioral change may be that they express themselves in the presence of nonpunishing audience. This allows for the extinction of anxiety to occur.

In connection with our present discussion of the development of severely fixated behavior it should be noted that it is not an infrequent occurrence for human neurotics to be so severely inhibited that self-expressive behavior simply does not occur in the presence of a nonpunishing audience. In that case the extinction of anxiety is impossible and psychotherapy, which places heavy reliance on such a process, must, of necessity, be relatively limited as a technique for producing behavioral change.

Altering Fixated Behavior

It is instructive to note the techniques that experimental psychologists have used to change the strong fixations of animals in Maier-type experiments. Wilcoxon (1951) found that he could break fixations by forcing the animal to jump repeatedly to a locked negative card in the fixated window, while an open window appeared on the nonfixated side. The importance of the combined, appropriate use of reward and punishment is again underscored. Even severely fixated habits may be changed when the punishment is consistent and intensified and an alternative response is amply rewarded.

Maier employed a technique which he called *guidance*. The experimenter placed his arm or some other obstacle in such a manner that the fixated response was impossible; the only response the animal could make was the correct one. The experimenter literally guided the subject to reinforcement. This technique of guidance is a rather interesting one and may have important implications for human behavioral change. It is as if the experimenter, an external agent, assumed the protective role of a good parent or a good teacher. Guidance appears to protect the animal from fixating when it is later put in an insoluble problem situation. Animals with a history of previous fixations which have been broken by guidance show fewer fixations than animals which have not received it (e.g., Maier & Ellen, 1952; Maier & Klee, 1943).

At the human level the analogue of guidance would be protective nurturance. The key role of nurturance in modifying frustration-instigated behavior has to do with the reduction of anxiety and of the emotional components of the frustrative response. As the individual is released from the suppressive and inhibitory effects of these conditioned emotional responses the likelihood of behavior with a positive reinforcement history occurring spontaneously is greatly increased. Fortunately such behavior rarely includes increased dependency. For most individuals in our culture dependency responses do not have a history of positive reinforcement.

Because of the previous learning of many individuals in our culture the evaluation of their personality by others may produce anxiety, frustration, and be experienced as threatening. When one person punishes another there is always the danger of directly depriving the punished individual of self-respect. The deleterious consequences of such conditioning can readily be shown experimentally. For example, Beier (1949) measured several aspects of students' abilities, particularly abstract reasoning. He then administered a projective personality assessment test (Rorschach) to half the group of students. All of the students were then retested as to their abilities. Beier found that the group which had been given an evaluation of their personality showed a decrease in flexibility, a significant decrease in their ability to carry on abstract reasoning, and that they became more rigid, anxious, and disorganized in their thinking as compared to the control group. Altering anxiety and frustration-produced behavior involves the arrangement of conditions which are the opposite of those which dissolve the individual's self-confidence.

Finally, it should be noted that a distinction must be made between positively motivated and frustration-instigated behavior. The same behavior may be maintained by positive reinforcement or by frustration. A boy may steal a watch because to him it is a wonderfully important and attractive object. Treatment in this case would involve giving the boy a watch or helping him to save up for one. On the other hand, the theft may be frustration-instigated. The boy may steal in order to gain attention from significant people in his life who have neglected him. It is very common for large department stores to have a major problem with women stealing. Frequently this is a frustration-instigated response. What they really want is attention and love which they do not receive directly from their husbands and families.

When the behavior is frustration-instigated, positive reinforcement is likely to have little effect. Thus, giving the boy a watch or providing the housewife with a charge account will at best only change their behavior

Stealing to gain symbolic affection lacking in the person's life. Petty thievery may be symptomatic of a thwarted need for affection. (Photograph by Tor Eigland.)

temporarily. Of course, if the behavior is correctly diagnosed as being frustration-produced, then positive reinforcement aimed at removing the source of frustration is effective. Because of the compulsive nature of frustration-instigated responses, punishment is also unlikely to be effective in changing the behavior. In fact, punishment may produce further conflict which increases the individual's frustration, and may even result in an increased likelihood of occurrence of the very response being punished.

FRUSTRATION AND AGGRESSION

A number of years ago a group of Yale psychologists formulated the frustration-aggression hypothesis (Dollard, Miller, Doob, Mowrer, & Sears, 1944). Although this hypothesis had several correlates, the core of it involved two major assumptions: first, that aggression always follows frustration, and second, that the occurrence of aggressive behavior always presupposes frustration.

On the basis of accumulated research findings it now appears unlikely that either of these two assumptions is defensible. Whether or not aggression occurs in response to frustration depends upon the history of reward and punishment that aggressive behavior has received, as well as upon the history of reward for alternative behavior. Furthermore, it now appears that aggression is not an innate response to frustration.

Buss (1961) has defined aggression as, "a response that delivers noxious stimuli to another organism." For people, this means the delivery of noxious stimuli in an interpersonal context. He also implies that frustration is not the major antecedent of aggression. Rather, Buss states that aggression is the major antecedent of aggression.

Consider the distinctions between the operations defining aggression, punishment, and anxiety. They are very slight indeed. The delivery of noxious stimulation to another organism constitutes punishment, if it is response-contingent. On the other hand, it constitutes anxiety if it is not response-contingent. Aggression is distinguished from these two operations in two ways: (1) The noxious stimulus in the case of aggression is always delivered in an interpersonal context. (2) Punishment is frequently delivered in an interpersonal context also; however, as a part of the socialization process children are taught to label the delivery of a noxious stimulus by one individual to another "punishment" under some conditions and "aggression" under others. In many human situations frustration and aggression are also confounded. The delivery of a noxious stimulus in an interpersonal context may constitute a severe source of thwarting. Furthermore, the individual may be taught to interpret rules and regulations, the behavior of other people, and other sources of thwarting as a personal attack. Since frustration and aggression are at times confounded it is best to give some attention to the aggressive response before further investigating the relationship between these two concepts.

Noxious Stimulation and Aggression

It was noted briefly above that when an individual is the victim of aggression, counteraggression is likely to occur. Buss (1961) has indicated that when neither flight nor fight has been previously learned, and when the attacker and the victim are equal in status, a curvilinear relationship exists between the tendency to aggress and the intensity of attack. That is, although a weak attack evokes little tendency to aggression, as the strength of the attack increases the likelihood of aggression also increases. However, at some point along the intensity-of-attack dimension, attack may become so strong that it evokes flight or other nonaggressive responses.

Annoyers

Direct attack that can injure an individual constitutes one class of noxious stimulation. There is also a very large class of noxious stimuli which are generally referred to as *annoyers.* Included in this class are simple sensory stimuli which function as irritants: high-pitched or loud noises, intense light, pungent odors, etc. Annoyers also include stimuli that the individual has learned to find annoying: physical characteristics (e.g., extremes of ugliness or of beauty); the mannerisms and behavior of others (e.g., effeminacy in men, drunkeness, or garrulousness); and appearance (e.g., dress, or grooming).

Aggression as a response to most objects and behavior classed as annoyers is more likely to remain at subthreshold levels. These stimuli are likely to create a predisposition to aggression rather than the overt response itself. The situational variables associated with them are generally more likely to lead to avoidance and escape than to attack.

The Mode of the Aggressive Response

Direct physical assault is generally strongly prohibited in our culture, although physical aggression is somewhat differentially reinforced among different social classes and subcultures. In strictly legal terms the punishment for stealing a man's wallet may be far more severe than for putting out his eye. Be that as it may, when direct aggression is inhibited, aggression may be expressed verbally, indirectly, or passively. Words can be highly effective in injuring others. Words signifying rejection—verbal or behavioral cues indicating avoidance, disgust, dislike, hatred, etc.—may threaten the individual's concept of himself. Since such attacks are not directed at specific behavior they are likely to increase anxiety. Under certain conditions enough stress may be created at the symbolic and verbal level to result in personality disorganization.

When various factors in the situation inhibit the expression of aggression, noxious stimulation may be delivered indirectly. Instead of the direct physical or verbal attack, circuitous routes may be taken: malicious gossip, indirect criticism, deceit, or the arrangement of conditions to thwart the individual without his being aware of it.

Indirect aggression may also involve frustrating others by means of a passive resistance. At the social level the self-conflagration of Buddhist priests, or the refusal to leave voter registration lines by American Negroes exemplify passive forms of aggression. In interpersonal relations there are many forms of passive aggressiveness that one individual may use against another. This is particularly true in child–parent or subordinate–superior relationships. Generally such behavior involves the aggressive individual's frustrating the other individual so that it is difficult to determine whether the frustration is or is not beyond the control of the frustrator.

Hostility and Aggression

Hostility is "an implicit verbal response involving negative feelings (ill-will) and negative evaluation of people and events" (Buss, 1961). This implies that it may not involve overt aggressive behavior, and that it is mediated by symbolic and language conditioning. Hostility is a conditioned anger response. Implicit responses of perceiving, categorizing, and evaluating stimuli (observing–labeling responses) become hostile responses when they are associated with anger reactions. People, situations, and events that cue off anger become labeled with negative implicit verbal responses at the same time that the anger response occurs. Hence, even though the anger subsides, negative evaluations (hate, resentment, jealousy, suspiciousness, etc.) remain. The persistence of such responses is a function of the mechanisms of verbal conditioning and mediation. These mechanisms will be much more fully explicated in Chapter 7.

Although hostility may cue off aggression, it does not necessarily do so. The victim may be completely unaware of the fact that implicit negative-evaluative responses are occurring in the hostile individual. In fact, the victim and aggressor may be closely associated with one another for long periods of time without the former being aware of any hostility. When conditions change, perhaps suddenly, so that overt aggression becomes permissible, it may appear inexplicable to the victim. Consider the interaction between Jewish and German neighbors prior to and following the onset of the National Socialist program. In many human relationships, as between husband and wife, parent and child, undercurrents of hostility and resentment may remain completely unnoticed. Their final overt expression may be a complete source of bafflement to the person who is the object of such responses. Attraction (implicit positively conditioned evaluations) may function in a very similar manner.

Aggression, Punishment, and Displacement

The punishment of aggression may be effective in inhibiting the aggressive response. It has been found that the amount of aggression is inversely related to the amount of punishment anticipated for it (e.g., Doob & Sears, 1939; McClelland & Apicella, 1945). Simply punishing aggression is generally a poor technique, however, because the inhibition of the aggressive response remains only as long as the punishing contingencies are in effect. Furthermore, punishment is not always strong enough to inhibit aggression.

There is considerable evidence for the generalization of aggression and also for its displacement onto other objects or people when its direct expression is blocked. For example, Miller (1948) reported a boy who had been punished for biting and pinching his foster parents, and who became a problem child in school because he would bite, scratch, and pinch his schoolmates. Perhaps the most shopworn of all illustrations of the displacement of aggression is that of the man who, afraid to counter-aggress against his boss resorts to beating his wife and kicking his dog. Occasionally people even verbalize these dynamics to the guilty individual. The reader has undoubtedly at some time or another heard someone say "Look, I know that so-and-so hurt you, but why take it out on me?" The phenomenon is genuine enough. However, its interpretation in terms of stimulus generalization is hardly adequate. Actually many variables combine to determine the subsequent aggressive response.

The target of aggression and the strength of the displaced response is determined by such variables as the amount of reinforcement the individual has had in the past for attacking that, or similar individuals; the amount of discrimination training the individual has had; and the discriminative stimuli present in the situation for the occurrence of alternative, competing responses. Anger lowers the threshold for all aggressive responses. Although the inhibited response might be that of striking the instigator, verbal aggression or indirect and passive forms of aggression may occur either toward him or toward displaced targets. The mobilization of hostility is particularly likely to lead to extensive generalization.

In one of Hemingway's war novels one deserting soldier shoots another. When asked by a baffled third soldier the reason for his conduct his reply is, "I always wanted to kill a sergeant."

Catharsis

Catharsis is a broad term denoting the expression of an emotional impulse. Generally the free expression of an emotional impulse is thought to constitute a "release mechanism." Catharsis seems to change the strength of the emotional impulse and decrease the likelihood of the occurrence of strong behavior related to it. In the case of anger there is a sharply decreased likelihood of occurrence of further aggression following the direct and immediate expression of aggression.

The principle of catharsis was originally expressed by the group at Yale in the following hypothesis:

> The expression of any act of aggression is a catharsis that reduces the instigation to all other acts of aggression. From this and the principle of displacement it follows that, with level of original frustration held constant, there should be an inverse relationship between the expression of various forms of aggressions. [Dollard *et al.*, 1939, pp. 53–54]

Now, however, we know that the occurrence of such a cathartic effect depends upon the presence or absence of anger. When anger is not present, either encouraging the subject to be aggressive (providing reinforcement for aggression), or reducing his inhibitions about aggressive action (removing punishment for aggression) will *increase* aggressive behavior. That is, in the absence of anger the expression of aggression may have the exact opposite effect from catharsis. On the other hand, when anger is present, catharsis temporarily decreases aggression because of the temporary lowering in the level of physiological tension.

Feshbach (1961) hypothesized that watching a prizefight film would give angry individuals vicarious catharsis. He found that an angry group who watched a prizefight film, and therefore had the opportunity for vicarious catharsis, did show less aggressive word associations than an angry group which watched a neutral film. This substantiated the fact that the opportunity for an emotional release decreases the likelihood of further expression. On the other hand, a nonangry group who were given the opportunity to watch the prizefight film gave significantly more aggressive associations than the nonangry groups which watched the neutral film, indicating that when anger is not present the encouragement of aggression increases, rather than decreases, subsequent aggressiveness. Many other studies have substantiated these results.

Since inhibited emotions are likely to disrupt the performance of skilled behavior, the presence or absence of catharsis may be an important factor in allowing for skilled performances. Mention need be made here of only one study which is typical of those demonstrating the role of catharsis in subsequent performance. Worchel (1957) angered three groups of subjects in a classroom situation by insulting them. One of the groups was then allowed to gripe in a group fashion about the procedure; another was allowed only to talk about neutral topics; and the third was required to remain silent. All groups had previously been administered a digit symbol task. When the subjects were administered the same task again following these experimental manipulations, the catharsis group scored significantly higher than did the control groups.

Frustration, Aggression-Anxiety, and Reinforcement

A special feature of extinction when learning has occurred under intermittent reinforcement is the absence of pronounced emotionality. If frustration has been introduced in small doses subsequent frustration does not represent a drastic change. What appears to be frustration tolerance may be the result of either of two processes: the actual absence of emotionality due to a history of reinforcement, or the inhibition of emotion due to anticipated punishment.

Block and Martin (1955) obtained ratings for children for control of their emotional reactions in a number of different play situations. When extreme overcontrollers and extreme undercontrollers were frustrated by having attractive toys which they had been allowed to play with placed behind a wire mesh screen, the behavior of the undercontrollers was marked by aggressive attacks against the screen and a lack of constructive play with less desirable toys. The overcontrollers were able to play constructively with less desirable toys and showed little barrier-attack behavior. The overcontrollers may be individuals who have learned to suppress emotionality for fear of punishment, or their reinforcement history may be such that emotionality simply does not occur in certain frustrating circumstances. The fact that the majority of overcontrollers in this study exhibited alternative constructive behavior under frustration would appear to indicate that their lack of aggressive behavior was more likely due to the fact that frustration did not cue off emotionality for these individuals.

Several studies indicate the operation of aggression-anxiety in response to frustration. Thus, Doob and Sears (1939) presented subjects with a number of hypothetical social situations involving frustration and had them indicate their choice of sentences describing aggressive, non-aggressive, or substitute responses to the situation. Part of the study involved asking the subjects which item would have been the most satisfying for them, assuming it were a real-life situation, and which item would have resulted in the most trouble (i.e., punishment) if that response had actually been made. The subjects indicated that aggression was the most satisfying response. It was also the response for which the most punishment was anticipated; the greater the amount of overt aggression indicated, the greater was the amount of punishment anticipated for such behavior.

Hokanson (1961) made a comparison between groups of high-hostility and low-hostility subjects using various physiological indexes of anxiety first, in a normal state and then in a situation involving frustration. There was no initial difference between the indexes of the two groups. The high-hostility group, however, had a significantly higher level of anxiety under frustration than the low-hostility group did. It was assumed that the high-hostility subjects had frequently expressed hostility and frequently been punished and that they therefore tended to develop more anxiety in situations which were likely to cue off aggression.

The Strength of Frustration and Aggression

The strength of frustration is assumed to be determined by the strength of the response tendency being blocked, the degree of blocking which occurs, and the number of frustration sequences. Ordinarily when we think of the strength of frustration we refer to the amount of emotionality experienced by the individual. This varies with the strength of the response tendency being frustrated and the degree of interference with it. On the other hand, the individual may experience little emotion under conditions which are ostensibly frustrating. He may even find frustration pleasant. This, as we have seen, depends upon relationships among the reinforcing contingencies in the situation which determine the occurrence of masochism. Masochistic behavior will not be discussed at length here. Suffice it to say that when behavior has been strongly, positively reinforced, and then punishment or frustration is introduced just prior to reinforcement, the punishing or frustrating conditions may themselves take on positive reinforcement value.

Generally, aggression does appear to vary directly with the strength of instigation to the frustrated response. Sears and Sears (1940) frustrated infants by withdrawing the bottle before hunger was satisfied. The strength of the instigation to aggression was inferred by these investigators from measures of the latency (in seconds) to crying. The bottle was withdrawn after $^1/_2$, $2^1/_2$, or $4^1/_2$ ounces of milk had been consumed. The latency of crying was found to be 5.0, 9.9, and 11.5 seconds for these respective groups.

Doob and Sears (1939) had college men rate the strength of various instigations to aggression. They also reported their typical response when these instigations occurred. The results indicated that the higher the rating of the instigation, the greater was the likelihood that aggression would occur as the response in the situation. Allison and Hunt (1959) carried out a similar study involving better controls, with essentially the same findings. A positive relationship between "frustration-response incidence" (including aggression) when an animal was placed in a frustrating situation and the number of hours of food deprivation was also reported by Finch (1942).

It has also been confirmed by a number of studies that the strength of the instigation to aggression varies directly with the degree of interference with the frustrated response. Graham, Charwat, Honing, and Weltz (1951) had 50 incomplete sentences scaled for frustration value. Interestingly enough, the incomplete sentences which were scaled for frustration were themselves representative of aggressive acts (e.g., "He hit me so I . . ."). The sentence completion of the subjects (adolescents) was then rated for aggressiveness. The relationship between the strength of external instigation and the strength of aggressive response was essentially linear for these data. This relationship (originally hypothesized by the Yale group) was further supported in a study by McClelland and Apicella (1945). To frustrate their subjects they used verbal derogation and induced failure in a motor task. With increases in the experimenter's hostility there was a significant shift in the number of angry verbal responses exhibited by the subjects.

The Arbitrariness of Frustration

There is considerable evidence that the frequency and intensity of aggression in response to frustration depends upon whether or not the frustration is interpreted by the victim as being arbitrary. It is an everyday observation that when people are unjustifiably frustrated they are more likely to respond with aggression than when the imposition of frustration can reasonably be justified in some way, or when it appears to be beyond the control of the instigator. This has also been demonstrated in the laboratory a number of times. For example, Pastore (1952) readministered the items of the Doob and Sears study referred to above in terms of the arbitrary vs. the nonarbitrary nature of the situations presented. He believed that nonarbitrary situations would tend to result in less aggressive responses, and tested it by devising parallel arbitrary and nonarbitrary situations. The subjects' responses to the questionnaire confirmed his prediction.

Lee (1955) found that students had significantly more aggressive verbalizations when the teacher of a small class acted in opposition to their expressed desires if he used as the rationale that "he wanted to," than when he used "the good of the students" as a rationale for his behavior. Others (e.g., Allison & Hunt, 1959; Cohen, 1955) have substantiated these results in questionaire type studies.

Anticipated Rewards for Aggression

Although there is almost no experimental evidence available, it seems likely, both on the basis of theoretical considerations and on the basis of everyday experience, that another factor which determines whether or not aggression will occur in response to frustration is the instrumental value of the aggressive response. The only study available, to the present writer's knowledge, is an informal reanalysis by Buss (1961) of the frustrating situations presented in the original investigation of Doob and Sears. Quite clearly, those situations in which there was a likelihood that an aggressive response would remove the source of frustration elicited the highest percentage of aggressive responses from the subjects. Consider, for example, the item "A guest of yours whom you know only slightly persisted in telling you how to drive and in general pestered you with back-seat criticism." Seventy-two percent of the responses to this item were aggressive in nature. Presumably, a little verbal or instrumental aggression expressed in such a situation would be rewarded by reducing the likelihood of further similar frustration. In contrast, consider another item: "In an informal gathering of people with whom you are not well acquainted, you found that you couldn't solve what everyone else considered a very simple parlor trick." The frequency of aggression reported for this item was only 9 percent. In this case presumably aggression would be more likely to be punished, or at least, have little reward.

FRUSTRATION AND REINFORCEMENT PROGRAMMING

The relationship between the frustration response and the programming of reinforcing contingencies is an especially important one for human adjustment. As an example, consider the development of emotional dependency and the correlated behavior of seeking approval, help, praise, and the company of others. The incidence of dependency behavior in children is positively related to the quality and intensity of parental nurturance. Normally, the very young child receives an overwhelming amount of help, care, attention, and affection from the mother. She feeds him when he is hungry, changes him when he is soiled, tucks him into bed, warms him, soothes him when he is in pain, and comforts him when he is afraid. Because of the fact that she is continually associated with tension relief and pleasantness, her very presence takes on conditioned reinforcing properties for the very young child. Thus, the mother and later the father, become very powerful agents in the subsequent control of the child's behavior both because of their capability of supplying primary reinforcement and because of their generalized reinforcement value.

If the parents have frequently and indiscriminately rewarded the child (indiscriminate in the sense of smothering the child with attention and affection irrespective of any demands on the part of the child), then the child may learn to lack social interest. The parents may even elicit slightly aversive associations. The reinforcing value of the parents and their subsequent effectiveness may thereby be considerably reduced.

In contrast, when the nurturant behavior on the part of the parents is appropriate to the child's needs, attention has strong reinforcement value for the child. Then the withdrawal of nurturance or punishment for dependency behavior may be highly frustrating. How the child behaves under these conditions depends upon a number of variables: the amount and quality of previous reinforcement for dependency; the intensity and quality of the frustration; the previously learned responses to frustration; the amount of discrimination training the child has received relative to this specific type of frustration; and the availability of rewards for alternative behavior.

Psychologists have found the incidence of dependency behavior and the amount of anxiety concerning dependency both to be positively related to the severity with which dependency has been socialized (e.g., Sears, Whiting, Nowlis, & Sears, 1953; Whiting & Child, 1953). Figure 6-4 shows the relationship between the amount of frustration or punishment and the amount of dependency engendered by it, as hypothesized by Sears *et al.* (1953). Notice that a curvilinear relationship is postulated. As frustration or punishment increases, the amount of dependency behavior is hypothesized to increase. Under intense frustration or punishment, however, dependency behavior is expected to decrease.

In Sear's study maternal punishment and lack of nurturance was found to increase the incidence of dependency behavior in preschool boys. In the same study, the incidence of dependency behavior for preschool girls *decreased* as a function of maternal punishment and lack of nurturance. Sears and his co-workers argued that what appeared to be objectively the same amount of punishment for boys and girls actually represented more severe punishment for girls because the girls have a greater identification with their mothers so that punishment and frustration from the mother added further increments of self-punishment.

Experimental evidence generally supports the hypothetical relationship of Fig. 6-4. The data are especially consistent in showing that when

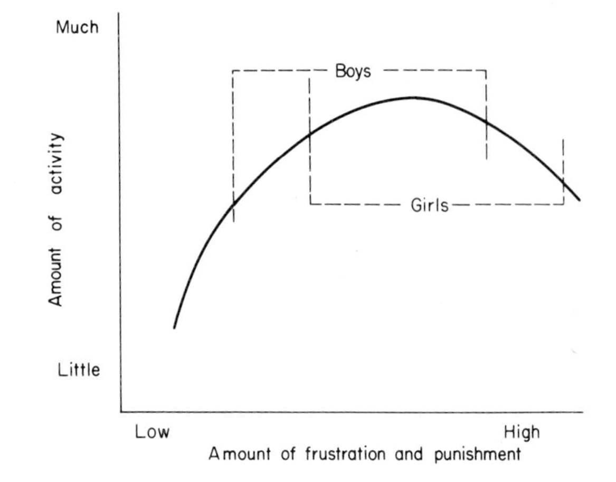

FIG. 6-4. Schematic representation of the amount of dependency behavior as a function of the amount of frustration or punishment. (Sears *et al.*, 1953.)

frustration is intermittent, dependency behavior becomes intensified and more presistent.

The frustration of dependency behavior has been shown both to increase response persistence and to facilitate social influence. Typical findings are those of Zigler (1961) in a study that employed two groups of institutionalized children. One group had been deprived of social contact and reinforcement prior to institutionalization. Another group came from homes where the amount of nurturance was more nearly normal. When tested for their persistence in a game, the more-deprived children were found to persist longer in the game than the less-deprived children.

Hartup (1958) hypothesized that the frustration of dependency behavior would intensify responding and promote faster learning. After first interacting nurturantly with two groups of children, he withdrew attention, affection, and approval from one group. The children in this group learned the experimental tasks more rapidly than did the group of children who experienced consistent nurturance. Other studies have supported this general finding.

Gewirtz and Baer (1958) required children to learn a simple two-choice discrimination task. The experimenter gave the children verbal approval for responses designated as correct. It was found that when the children were tested after a 20-minute period of isolation their responses were more readily modified than when they were brought to the experiment directly from the classroom.

Thus, frustration can be desirable in that moderate amounts serve to maintain activity and persistence. Furthermore, the frustration of dependency behavior may serve to facilitate social influence. We could equally as well have discussed the relationship between frustration and other socially relevant categories of behavior. More detailed aspects of the role of frustration in the shaping of socially relevant behavioral patterns will be discussed in Chapter 8.

Frustration is not always desirable. Since it produces an increase in general emotionality, the behavior which occurs under frustrating conditions, especially when the frustration becomes intense, is likely to be less integrated, less adaptive, and less flexible than behavior under nonfrustrating conditions. Moreover, since frustration is aversive, anxiety and consequent response suppression and avoidance behavior is likely to occur. Under severe frustration whatever behavior occurs in the situation is likely to become extremely rigid and fixated.

SUMMARY

Frustration is defined in terms of operations that result in removing sources of reinforcement, in thwarting, or in producing conflict. In this chapter the first two of these operations have been discussed, along with the many effects of frustration.

Removal of reinforcement may be temporary or permanent. The latter is extinction. The removal of reinforcement results in an aversive reaction called the *frustrative response.* Under the appropriate conditions components of it can be associated with positively reinforced behavior, as in intermittent reinforcement programming, resulting in increased persistence and less emotional disturbance. Thwarting involves preventing the individual from obtaining reward, rather than removing it. It may involve physical barriers, social and legal restrictions, the behavior of other people, or the individual's own physical or psychological limitations.

There are a great many possibilities for the learning of various reactions to frustration. In general frustration produces an increase in behavioral arousal and an intensification of goal-directed behavior. Regression, fixation, and aggression are also commonly associated with frustration. Regression may involve a return to well-defined earlier learned behavior patterns or it can be reflected in a general apathy, inefficiency, and lack of organization. It can be reflected in a return to very early forms of behavior and/or attachments. Temporary mild regression can also be constructive. It may allow temporary security and hence increase the individual's ability to cope with situational demands.

Behavior may become rigid, stereotyped, fixated, and repetitive under severe frustration. However, frustration is not the only factor which produces behavioral rigidity. Any strong motivation can have the same effect. The extinction of fixated behavior may be all but impossible. Special techniques of guidance may be necessary for altering it significantly. The various sources of reward operative in the situation must be considered. The same behavior may be maintained by positive reinforcement or by frustration. Modifying it in one case requires very different techniques from what it does in the other.

Aggression is not a necessary response to frustration, as has often been supposed. Whether or not it occurs depends upon many variables including the strength of the frustration, whether or not the frustration is experienced as arbitrary, and the potential rewards for aggression. Hostility may be learned, but the aggression related to it may be inhibited so that when the situation changes in such a way that aggression is less punishable, it may come as a surprise to the victim. Frustration-produced aggression may be displaced from the original instigator to other objects or people. Displacement depends upon many variables in the situation, including characteristics of the target other than mere similarity to the original instigator.

Many socially relevant behaviors (dependency, achievement, aggression, etc.) are influenced by frustration. Generally, as frustration for the behavior increases, so does the persistence of the behavior. This is true up to a point. When the frustration becomes too strong, the behavior decreases. If frustration is imposed intentionally it must be carefully done because of the many undesirable consequences which may accompany it.

REFERENCES

Allison, J., & Hunt, D. E. Social desirability and the expression of aggression under varying degrees of frustration. *J. consult. Psychol.*, 1959, **23**, 528–532.

Amsel, A. The role of frustrative non-reward in non-continuous reward situations. *Psychol. Bull.*, 1958, **55**, 102–119.

Barker, R. C., Dembo, T., & Lewin, K. Frustration and regression: An experiment with young children. Studies in topological and vector psychology. II. *Univer. Iowa Studies Child Welfare*, 1941, **18**, No. 1.

Beier, E. G. The effect of induced anxiety on some aspects of intellectual functioning. Ph.D. thesis, Columbia Univer., 1949.

Block, J., & Martin, B. C. Predicting the behavior of children under frustration. *J. abnorm. soc. Psychol.*, 1955, **51**, 271–285.

Buss, A. H. *The psychology of aggression.* New York: Wiley, 1961.

Child, I. L., & Waterhouse, I. K. Frustration and the quality of performance. I. A critique of the Barker, Dembo, and Lewin experiment. *Psychol. Rev.*, 1952, **59**, 351–362.

Cohen, A. R. Social norms, arbitrariness of frustration, and status of the agent of frustration in the frustration-aggression hypothesis. *J. abnorm. soc. Psychol.,* 1955, **51**, 222–226.

De Valois, R. L. The relation of different levels and kinds of motivation to variability of behavior. *J. exp. Psychol.,* 1954, **47**, 392–398.

Dollard, J., Miller, N. E., Doob, L. W., Mowrer, O. H., & Sears, R. R. *Frustration and aggression.* London: Kegan Paul, Trench, Trubner & Co., Ltd., 1944.

Dollard, J., Doob, L. W., Miller, W. E., Mowrer, O. H., Sears, R. R., Ford, C. S., Houland, C. I., & Sollenberger, R. I. *Frustration and aggression.* New Haven: Yale Univer. Press, 1939.

Doob, L. W., & Sears, R. R. Factors determining substitute behavior and the overt expression of aggression. *J. abnorm. soc. Psychol.,* 1939, **34**, 292–313.

Ellen, P. The compulsive nature of abnormal fixation. *J. comp. physiol. Psychol.,* 1956, **49**, 309–317.

Feshbach, S. The stimulating versus cathartic effects of a vicarious aggressive activity. *J. abnorm, soc. Psychol.,* 1961, **63**, 381–385.

Finch, G. Chimpanzee frustration responses. *Psychosom. Med.,* 1942, **4**, 233–251.

Gewirtz, J. L., & Baer, D. M. The effects of brief social deprivation on behaviors for a social reinforcer. *J. abnorm. soc. Psychol.,* 1958, **56**, 49–56.

Graham, P. K., Charwat, W. A., Honing, A. S., & Weltz, P. C. Aggression as a function of the attack and the attacker. *J. abnorm. soc. Psychol.,* 1951, **46**, 512–520.

Grinker, R. R., & Spiegel, J. P. *Men under stress.* New York: McGraw-Hill, 1963.

Hamilton, G. V. A study of perseverence reaction in primates and rodents. *Behav. Monogr.,* 1916, **3**, No. 13.

Hartup, W. W. Nurtrance and nurtrance-withdrawal in relation to the dependency behavior of preschool children. *Child Develpm.,* 1958, **29**, 191–201.

Hokanson, J. E. The effects of frustration and anxiety on overt aggression. *J. abnorm. soc. Psychol.,* 1961, **62**, 346–351. (a)

Hokanson, J. E. Vascular and psychogalvanic effects of experimentally aroused anger. *J. Pers.,* 1961, **29**, 30–39. (b)

Hull, C. L. The rat's speed-of-locomotion gradient in the approach to food. *J. comp. Psychol.,* 1934, **17**, 393–422.

Lashley, K. S. The mechanism of vision. I. A method for rapid analysis of pattern vision in the rat. *J. genet. Psychol.,* 1930, **37**, 453–460.

Maier, N. R. F. *Frustration: The study of behavior without a goal.* New York: McGraw-Hill, 1949.

Maier, N. R. F., & Ellen, P. Studies of abnormal behavior in the rat. XXIII. The prophylactic effects of "guidance" in reducing rigid behavior. *J. abnorm. soc. Psychol.,* 1952, **47**, 109–116.

Maier, N. R. F., & Klee, J. E. Studies of abnormal behavior in the rat. XII. The pattern punishment and its relation to abnormal fixations. *J. exp. Psychol.,* 1943, **32**, 377–398.

McClelland, D. C., & Apicella, F. S. A functional classification of verbal reactions to experimentally induced failure. *J. abnorm. soc. Psychol.,* 1945, **40**, 376–390.

Miller, N. W. Theory and experiment relating psychoanalytic displacement to stimulus-response generalization. *J. abnorm. soc. Psychol.,* 1948, **43**, 155–178.

Pastore, N. The role of arbitrariness in the frustration aggression hypothesis. *J. abnorm. soc. Psychol.,* 1952, **47**, 728–732.

Patrick, J. B. Studies in rational behavior and emotional excitement. II. The effect of emotional excitement on rational behavior in human subjects. *J. comp. Psychol.,* 1934, **18**, 153–195.

Sears, R. R., & Sears, Pauline S. Minor studies of aggression. V. Strength of frustration-reaction as a function of strength of drive. *J. Psychol.,* 1940, **9**, 297–300.

Sears, R. R., Whiting, J. W. M., Nowlis, V., & Sears, Pauline S. Some child-rearing antecendents of aggression and dependency in young children. *Genet. Psychol. Monogr.*, 1953, **47**, 135–234.

Skinner, B. F. *Walden two.* New York: Macmillan, 1948.

Wallin, J. E. W. *Personality maladjustments and mental hygiene.* (2nd ed.) New York: McGraw-Hill, 1949.

Whiting, J. W. M., & Child, I. L. Child training and personality. New Haven: Yale Univer. Press, 1953.

Whiting, J. W. M., & Mowrer, G. H. Habit progression and regression—a laboratory study of some factors relevant to human socialization. *J. comp. Psychol.*, 1943, **36**, 229–253.

Wilcoxon, H. C. 1951. *Abnormal fixations and learning.* Ph.D. dissertation, Yale University, New Haven, Conn.

Wolpe, J. *Psychotherapy by reciprocal inhibition.* Stanford Univer. Press, 1958.

Worchel, P. Catharsis and relief of hostility. *J. abnorm. soc. Psychol.*, 1957, **55**, 238–243.

Zigler, E. *J. abnorm. soc. Psychol.*, 1961, **62**, 413–421.

Chapter 7

Verbal Behavior and Thought Processes

INTRODUCTION

Human adjustment is enormously complicated by the use of language. Language forms the basis for vicarious experience; for thinking about past and future events; for solving problems; for forming discriminations and generalizations that would otherwise be impossible; and for understanding the world around us. In short, it is crucial in the adjustment of the individual to himself and to his world. The older philosophies held that ideas and thoughts were innate, and that language merely served to express an individual's "thoughts," "mind," or "soul." The overwhelming evidence now is that words are human habits, subject to the same laws as nonverbal habits. Psychologists talk about verbal "behavior" and study the origins and influences that control it.

There are two broad aspects to the study of verbal behavior. The first has to do with how emitted speech is learned, the conditions which come to control verbal utterances, and the various ways that the individual's verbal behavior affects other people. Just as in the case of the learning of other habits, the learning of verbal behavior is brought about and maintained by special sources of reward, or reinforcement. As we shall see, all verbal behavior is both shaped and eventually controlled by an audience.

A second broad aspect in the study of verbal behavior involves the analysis of how mere sounds produced by the mechanisms of speech come to have the rich and varied meanings which we associate with language. When someone says, "There is a magnificent Christmas tree in the living room," the individual may have vivid images of a colorfully decorated, handsome Christmas tree; whereas the same emitted speech in the presence of a non-English-speaking person would have scarcely any meaning. Words serve as symbols. They represent various aspects of the individual's life experiences. For a complete understanding of the topic of word meaning it will be necessary for us to discuss, first, the process by which words come to represent various experiences for the individual; and second, how the individual learns to respond to verbal symbols in a manner analogous to the way he responds to the actual experiences and events represented by them.

A complete presentation of the topic of verbal behavior and verbal learning is beyond the scope of this book. Certain aspects of verbal behavior and verbal learning are much more important than others for a consideration of human adjustment. We will not be concerned to any great extent with verbal rote learning; the development of facility in grammar or in the manipulation of mathematical symbols; the principles of learning, forgetting, and memory storage, or many other related topics. On the other hand, the nature of the variables that come to control speech; the conditioning and generalization of word meaning; the role played by verbal behavior in the development of attitudes and motives; and topics such as verbal influences on learned drives and repression will be of very great interest.

Because of the close interrelationships among the topics discussed in this chapter, the separation of them must be relatively artificial. First, emitted words will be analyzed as human habits. The types of verbal habits and the relevant variables controlling them will be outlined. A discussion of the development of word meaning and the generalization of meaning will follow, even though, as such learning occurs naturally, the development of emitted speech (verbal operants) and of word meaning are inextricably confounded. Finally, these topics will be somewhat integrated in discussions of the discriminative function of speech, the learning of generalizations and discriminations and of repression.

EARLY SPEECH DEVELOPMENT

Although the very first sounds made by the infant seem to be rather limited forms of crying and squealing, it has been found upon careful analysis that within the first few months of life human infants typically emit almost all of the speech sounds employed in the various languages of human societies. These include German umlauts and gutturals, French trills, and the broad midwestern American "A."

It has also been demonstrated that the amount of infant vocalization can readily be conditioned at an early age. For example, Rheingold, Gewirtz, and Ross (1959) performed two experiments in which different groups of infants of an average age of 3 months were reinforced by a female adult when they made discrete, voiced sounds. The method of reward, or reinforcement, was as follows: the female experimenters smiled broadly, made "tsk" sounds, and lightly squeezed the infants' abdomen. These three responses were performed simultaneously. They required about a second of time. Before the introduction of reinforcement, a base-line vocalization rate was established. The experimenter leaned over the crib with her face about 15 inches away from the infant. She kept her face expressionless while another experimenter tallied vocalizations. In this manner a base-line rate of vocalization was established during 2 consecutive days. On the following 2 days reinforcement was administered, and on the next 2 days extinction procedures were carried out. During these latter extinction days, the behavior of the experimenters was exactly the same as during the first 2 days when the base line was established.

The results of this experiment are presented in Fig. 7-1, where the mean number of vocalizations for each experimental group during consecutive experimental days can be seen. Notice the drastic change in the mean number of vocalizations during conditioning in comparison to the base line, and that the mean number of vocalizations on the 2nd day of

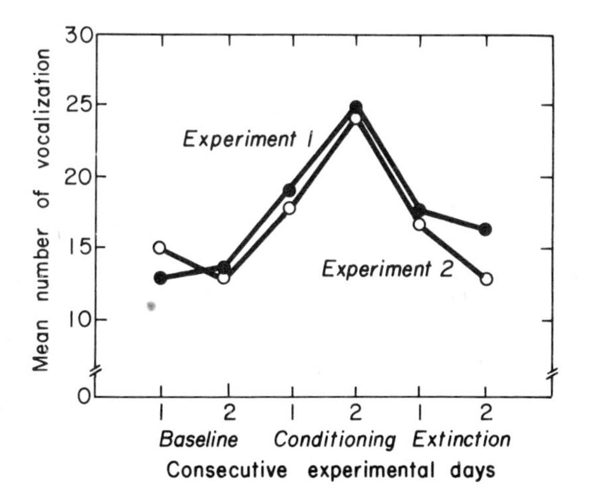

FIG. 7-1. Mean number of vocalizations of 3-month-old infants on consecutive experimental days. (Reingold *et al.*, 1959.)

conditioning was higher than it was on the 1st conditioning day. The effects of extinction may also be clearly seen in Fig. 7-1. With the removal of the reinforcing stimulus the mean number of vocalizations tended to return to the base-line level.

Most parents are concerned that their children develop language. Typically, many hours of parental attention are devoted to training and rewarding the infant's vocalizations. In the American middle class, where conformity, intelligence, and achievement potential are valued, parents may be particularly anxious that their children learn to talk early, just as later their concern centers upon the child's development in school, vocabulary building, etc. The continual, laborious process of differentiating and shaping the child's speech can be readily observed by anyone who normally has the opportunity to do so.

After they have been trained to produce syllables that are more or less like those of adult speech young children of 5 to 7 months typically begin to babble. They repetitively emit speech sounds and fragments of phrases in the absence of any apparent, or at any rate, of any appropriate, stimuli. The seeming lack of reinforcement is only apparent. Miller and Dollard (1941) have described in some detail the manner in which sounds produced by the child may function as conditioned reinforcers:

> Since the mother talks to the child while administering rewards such as food, the sound of the human voice should acquire secondary reward value. Because the child's voice produces sounds similar to that of his mother's, some of this acquired reward value generalizes to it . . . From this hypothesis it may be deduced that children talked to while being fed and otherwise cared for should exhibit more iterative and imitative babbling than children not talked to while being rewarded. [1941, p. 277]

Hence, the sounds which occur frequently in the speech of the parent gain some measure of conditioned reinforcing value. They may serve as a powerful source of reinforcement for imitative babbling.

Staats and Staats (1963) have pointed out a further process in the conditioning of the child's verbal imitative behavior. The parents deliberately reward the child for matching their own speech responses and do not reward, or may even punish, sounds which only very poorly match those of the parents. For example, parents often institute a deliberate training procedure requesting over and over again that the child say "Daddy." At first only crude approximations or even fragments are reinforced, as for example when their child says, "Kaka" or "Da." If a very different sound is emitted, however, the request is repeated and all the reinforcing stimuli—signs of approval, smiles, special attention, etc.—are delayed. Eventually reinforcement is withheld until the child makes better and better proximations, such as "Da," and later "Daddy."

Skinner (1957) refers to this type of verbal conditioned response as an *echoic response*. He originated this term as one of the basic categories of verbal operants (emitted speech). (For others see the following section.) The development of an echoic repertoire is an essential part of early language learning. The child is ordinarily taught both to echo large units such as words, phrases, and sentences, and to repeat small sound patterns such as th, and ā. The phonetic elements may be specifically taught directly, or they may be learned as a matter of course on the basis of many complex echoic variations involved in larger units. In any event, the child eventually becomes capable of echoing novel patterns.

BASIC VERBAL OPERANTS

The subject of verbal learning covers a vast area of research within psychology. Of the many psychologists who have made contributions to our understanding of verbal behavior, perhaps the most original and important interpretations have been those of B. F. Skinner. In his book entitled *Verbal behavior* (1957) Skinner presents an ingenious and detailed analysis of the subject. Although many aspects of verbal behavior are beyond the scope of this book, a discussion of certain of its features will enhance the student's understanding of the complex role of verbal behavior in human adjustment. In the following sections on the intraverbal operant and the other basic verbal operants—the tact, mand, and the autoclitic—at least a few of the detailed aspects of verbal behavior outlined by Skinner will be paraphrased and occasionally Skinner's own minor, but especially apt, illustrations will be employed.

The Intraverbal Operant

Once the child has learned to imitate adult speech sounds, words, phrases, and even larger units of speech (e.g., sentences and announcements) those various units become interconnected through contiguous reinforcement. Each word may be thought of as being a specific discriminative stimulus for the next word. In some instances the associative chains which develop from contiguous usage involve relatively strong verbal habits where words come to control relatively singular verbal responses. Thus, for most people in our culture the phrase "red, white, and . . ." would be responded to with the word "blue." How many times has the stem "red, white, and . . ." served as a verbal discriminative stimulus, marking the occasion upon which the response "blue" has been reinforced?

Language contains very many such strongly reinforced associations. Associational chains such as "Peter Piper picked a peck of . . ."; or "the bigger they come . . ." have been conditioned to elicit the relatively compelling thoughts, "pickled peppers" and "the harder they fall." Verbal behavior based on such associative linkages is generally referred to as *intraverbal behavior.*

Although intraverbal associations represent only one aspect of verbal behavior, the evidence indicates that intraverbal associations account for a much greater proportion of the total verbal behavior of individuals than had heretofore been suspected. Skinner (1957) stated that most "small talk" is intraverbal in nature, and that even serious conversation is not always clearly free of it.

The paradigm for intraverbal associations is given by the serial and paired-associate methods of experimental psychology. In serial learning the subject is required to learn a series of words or nonsense syllables. After the list has been presented the subject is required, upon being shown one syllable or word, to say the syllable or word which will come next in the series. The series is repeated until the subject is able to correctly anticipate each member of the chain. When conditioning is complete the first word of the series will elicit the entire chain, as a function of contiguous associations having been formed during learning. The reinforcement operations in such learning may be quite subtle, although generally the conditioned reinforcer of being right after each correct association probably is the strongest of them. Of course this type of laboratory conditioning involves a somewhat artificial situation in which many of the relevant variables can be well controlled, but it is not unlike the development of contiguous associations in everyday life. The child is taught "Columbus discovered America in 1492," or "Three times three is nine." Television, the movies, and conversations with other people may condition a multiplicity of phrases, sentences, or even longer articulations. The professor who has delivered the same lectures for a number of years might easily be able to talk for 40 or 50 minutes on the basis of well-conditioned intraverbal associations.

Paired-associate learning is similar to serial learning, except that the subject is required to associate pairs of words; that is, to say one word or syllable upon seeing another word or syllable. Thus, JEX may be presented as the stimulus member of the pair followed by QEZ which is presented a short time afterward. QEZ would then be referred to as a *response member* of the pair. The subject is required to say QEZ upon seeing JEX, before QEZ appears. Typically subjects are required to learn a list of such pairs under controlled conditions. In a variation of this procedure subjects may be required to associate two or more response members with one stimulus member. Thus, JEX may be paired both with QEZ and with XEB. After conditioning JEX will then cue off both responses. Depending upon the number of trials and the standard variables of conditioning, one response may be dominant over the other, having been more strongly conditioned.

Intraverbal operants under ordinary life circumstances may similarly involve habit-family hierarchies of varying associative strengths. Many studies of the associational hierarchies of men and women have been done. In one type of study large groups of people are asked to respond to a stimulus word with the first word they think of. Kent and Rosanoff used such a procedure as early as 1910. Of course, relatively wide individual differences were found, but in spite of individual variations certain associations were quite common. For example, of 1000 subjects, 65 percent gave "light" in response to the stimulus word "lamp." About 40 percent of the subjects responded to the word "table" with one of three words. The highest single response to "table" was "chair," given by about 20 percent of the subjects.

Associations to common words in the culture are apparently quite stable over time. In a word association test given to students at the University of Minnesota in 1927 and in 1952 (Jenkins & Russell, 1960), the most common response given to 71 out of the 100 stimulus words did not change. It has also been found that the distribution of responses may vary with cultural background, age, intelligence, and that members of the same group have many associations in common which differ from those which members of other groups have in common. Generally, individuals who have shared similar experiences are more likely to have similar verbal habits. Members of the same profession, for example, show a communality of verbal habits, which differ from those of other professions. Similarly, the verbal habits of members of the same family, particularly those of mother and daughter and of father and son, show a great deal of communality. Table I, taken from a study by Jung (1918), illustrates the similarity in response between a mother and daughter.

Normally intraverbal associations are extremely complex. If we consider a verbal stimulus of only a single word, that word elicits an associational chain composed primarily of other single word units. Thus, the word "knife" may serve as a cue for the response "fork"; "fork," in turn, may elicit "road." However, fork in this example is only the strongest intraverbal to the word knife. Other responses such as cut, kitchen, and bayonet may also exist in some strength, forming a habit-family of words. Moreover, a habit-family hierarchy of associates of varying strength also occurs to the word "road." As we shall see shortly, these associational networks may function implicitly, that is, outside of

TABLE I

RESPONSES OF MOTHER AND DAUGHTER TO SAME STIMULUS WORDS

Stimulus Word	*Mother*	*Daughter*
angel	innocent	innocent
naughty	bad boy	bad boy
stalk	leek's stalk	stalks for soup
dance	couple	man and lady
lake	much water	great
threaten	father	father
lamp	burns bright	gives light
rich	king	king
new	dress	dress
tooth	biting	pains
take care	industrious pupil	pupil
pencil	long	black
law	God's command	Moses
love	child	father & mother

the individual's focal awareness. In our simple example, the individual may think the word road or respond "road" overtly, without being aware of the original controlling stimulus—the verbal stimulus "knife."

Both implicit intraverbal associations and explicit intraverbal associations (those that the individual is fully aware of) are sometimes referred to as *cue-producing responses*. According to Dollard and Miller (1950) a cue-producing response is one "whose main function is to produce a cue that is part of the stimulus pattern leading to another response." The response "knife" in the above example is a cue-producing response. It produces a cue that is part of the stimulus pattern leading to another response, namely, "fork." Many clinical investigators have underscored the importance of such response-produced verbal cues (e.g., Cameron, 1956; Dollard & Miller, 1950; Ellis, 1963). Dollard and Miller (1950) have made important statements concerning the way that intraverbal associations may be deliberately generated in order to supply cues which elicit emotional and other behavior patterns. One of several examples given by Dollard and Miller will clarify this process for the reader. This example comes from *Victory over fear* (Dollard, 1942). The individual in the example is a novelist who has been working on a story based on his own life, although this fact was not generally known.

He had a regular occupation and was writing in addition to carrying on his daily work. His hope of being rewarded for carrying on the work of writing was very nearly balanced by his anticipation of failure and by the effort required to do the writing. It was, therefore, always touch and go as to whether he would get his weekly chapter done.

He had set aside a Sunday for finishing up one chapter and sat down at his desk bright and early. When he sat down he had a sort of feeling that he would not be very productive that day, but he resolutely opposed this notion. It proved, however, to be all too true. He writhed and wrestled for most of the day, attempting to get his plot straightened out and his characters to behave as he felt they ought. Late in the afternoon he decided to stop work on the novel and to start work on why he was blocked. He began writing to himself about as follows:
1. Can't seem to finish this chapter—am I just lazy, loosing my grip?
2. Have been blocked before. It went away—but hate to lose time today.
3. Book won't be any good anyway—I'm just wasting my time—guess I'll quit—no, finish.
4. That last chapter was good, anyway.
5. Perhaps I need someone to tell me it's all right, to go ahead—wish I had a contract for the book.
6. No use, for the moment, anyway.

So the author turned from his desk and gave it up as a bad job. He had tried to find out the source of his blocking by self-study and had failed. But he had a surprise coming. He went out to play with the children before dinner and, as it were, took the weight off his mind. As he was walking back to the house he had a revealing and releasing thought. Only the day before he had been talking to an important friend about his novel. The friend had not been encouraging, felt it would be difficult to get a publisher, thought it might be too obvious that the plot represented the author's life. The writer had not taken any particular notice of these remarks at the time, although they were naturally unwelcome, and he and his friend had gone on to other topics of conversation. But the discouragement had registered even though the author was not aware of it. The fear of exposing himself was added to his other difficulties in writing. It provided just that little weight on the side of not writing the book which was needed to stop him. As soon as he realized that this was the discouraging element which had held him up, he was able to see that his friend's opinion was not entirely decisive. After all, the friend had not read the book and did not know the details of the project. Even if one friend did not like the idea, another might. Perhaps the friend was envious, without being aware of it himself. Certainly the author was resentful and discouraged and he had been stifling both feelings. These realistic and encouraging thoughts were enough to shift the balance in favor of creative work. The author felt released, as if from some witchcraft that had been perpetrated on him, and continued his writing. Without the preparation of the period of self-study he could hardly have realized the significance of the casual thought that struck him while he was walking with his children....[pp. 43, 44]

A subclass of these intraverbal, or cue-producing, responses may eventually prove to play a key role in human adjustment, namely, those intraverbals which serve as cue-producing responses for the arousal of certain aspects of autonomic nervous system functioning and of emotional circuits. These will be discussed at the end of the chapter in connection with irrational ideas which cause and sustain emotional disturbances, and in connection with repression (see pp. 245–249). At this point we may merely note the impossible impact of implicit intraverbal linkages such as "if (significant) people don't love me, that is *terrible* and *catastrophic*," or "unless I am (thoroughly) adequate (in every respect), I am *worthless*," etc. (see Ellis, 1963). Undoubtedly thoughts such as these may be implicit and totally unconscious. They may be elicited by environmental stimuli or other verbal associations and evoke sustained emotional responses and symptomatic behavior.

It is of special significance that the individual need not be aware, and, in fact, typically is not aware of his own implicit responses; not to mention the fact that the variables that determine those responses or the way in which those responses determine other important aspects of his behavior may be completely outside of the individual's field of awareness. In our discussion above we noted that the individual may respond to the word "knife" by saying the word "road." At first glance it would appear that there is no connection between the stimulus word and the subsequent response. However, the response is not free or undetermined. Actually it is mediated by the implicit cue-producing response "fork." Experimentalists have shown that these implicit verbal chains can facilitate learning.

Russell and Storms (1955) identified such verbal chains by an examination of word association data. For example, they found the word "stem" (B) elicited the response "flower" (C) which, in turn, elicited the response "smell (D). Some associative strength would therefore be expected to exist between "stem" and "smell" (B and D), mediated by the intervening response, "flower" (C). A number of such chains of B–C–D word associates were selected. In the experiment proper a nonsense syllable (A) was conditioned to elicit the B response. In the second part of the experiment the A syllables were then paired with the D words. The hypothesis was that this A–D learning would be facilitated by the earlier A–B learning. To test this hypothesis the A–D learning was compared with the learning of controlled paired associates involving A paired with a neutral word, X. The neutral word was chosen so as to control for other variables which might influence the learning, such as the amount of meaningfulness or familiarity. The results indicated that the implicit verbal responses (C links) mediated associations and facilitated learning. It was also found that the subjects were not aware of the mediating link. That is, the mediating intraverbals were, in fact, implicit.

A diagrammatic analysis of the mediating links in both the experimental and control conditions is given in Fig. 7-2, where H_2 represents the new association to be learned and H_1 represents the previously conditioned association. In the case of the experimental situation in the upper part of Fig. 7-2, H_2, the new habit, can be seen to be mediated by the implicit verbal chain B–C–D, established in the learning of H_1. On the other hand, in the control condition, although H_1 also cues off the same implicit verbal chain, it does not involve an association which could serve to facilitate the learning of H_2. With future research more refined

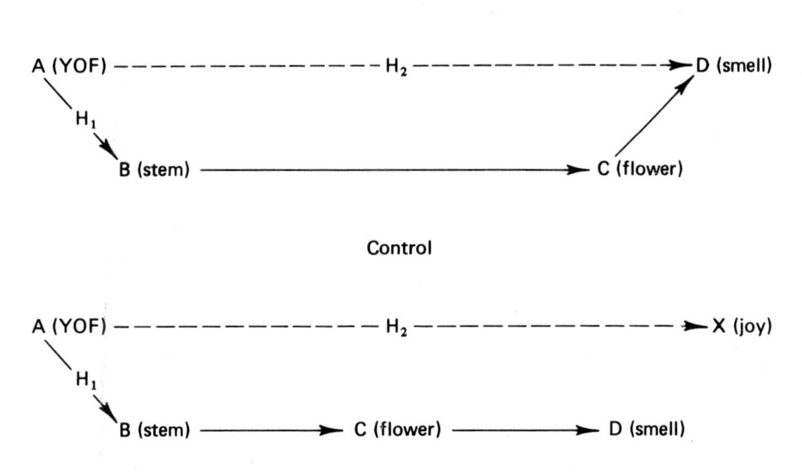

FIG. 7-2. Effect of an implicit verbal response in aiding learning. See text for explanation. (After Russel & Storms, 1955.)

methodological approaches will undoubtedly allow for increasingly sophisticated techniques for studying intraverbal word, sentence, and thought patterns.

The Tact

When a verbal operant comes under the control of a discriminated environmental stimulus the operant is referred to as a *tact*. Tacting occurs when a verbal response is reinforced in the presence of a particular stimulus. The child is reinforced for emitting the response "Daddy" in the presence of his father. At first the child may respond to all males, to relatives, to door salesmen, etc., with the response "Daddy." Each time such stimulus generalization occurs the child is likely to receive a new lesson: "That is *not* your daddy. There is daddy." Eventually the father becomes the exclusive stimulus for the verbal response. According to Skinner a tact may be defined as:

> a verbal operant in which a response of a given form is evoked (or at least strengthened) by a particular object or event or property of an object or event. We account for the strength by showing that in the presence of the object or event a response of that form is characteristically reinforced in a given verbal community. [1957, pp. 81–82]

Notice that some other person always mediates the learning of a tact. After the tact has been learned, other people reward it only on certain occasions. Only objects which are unusual or which occur in unusual surroundings are important to the listener, and therefore the listener only rewards the individual for tacting these objects. Children who tact everything they see are very annoying. Such indiscriminate tacting goes unrewarded or may be mildly punished.

Situational or place variables also influence tacts. Through differential reinforcement verbal behavior generally comes to be influenced by specific environments. Thus, a cocktail party sets up a discriminative function. It becomes and S^D to "pull" certain kinds of verbal behavior (see p. 119). That is, it provides the occasion for reinforcement for certain tacts, intraverbals, and other operants, and the occasion for nonreinforcement (S-Delta) for other verbal responses and the forms which they may take. Similarly, recreational, business, or family environments come to control different aspects of the individual's verbal repertoire. Certain aspects of the individual's verbal repertoire simply do not occur, except under particular stimulus conditions.

Self-Tacts. Idiosyncratic tacts may arise because of the lack of explicitness and the obscurity of the reinforced properties of the original learning situation. Imagine the properties controlling such tacts as: "I am (you are) stupid," "I was (you were) mean," "I am (you are) frustrated," "I am (you are) angry," "I love you with all my heart."

Self-tacts may occur as extensions of tacts the individual has heard other people use or they may be directly reinforced through the mediation of another person. In the former case, the child may, for example, be told "Mother loves Daddy," "Mother loves you," "God is love," "If you're bad no one will love you." In this way the child forms a concept which we ordinarily think of as the meaning of the word "love."

Self-tacts which are reinforced more directly by other persons may come to be controlled by subtle, heterogeneous, or distorted stimulus properties. The child may be taught an extensive repertoire of tacts descriptive of his own behavior. Thus, the child may be told: "You behaved like a big boy at the dentist. You are very *brave*," or "You continually refuse to obey. You are a very *wilful* child." In this manner the child learns that when he, say, deliberately hurts baby sister he is *mean*. Eventually he learns to tact himself as someone who is mean, unkind, stupid, intelligent, considerate, brave, etc.

The self-tacts learned may be grossly inaccurate. Perfectionistic parents may see most of the child's accomplishments as not good enough, when, in fact, the child may be well accomplished or even superior. Positive self-tacts may similarly represent distortions. Children, and adults too, tend eventually to form generalized conceptions of themselves. Particularistic self-tacts tend to become subordinate to more fundamental, generalized self-concepts. The child who has frequently been scolded for being "bad," "willful" or "stupid" may develop a concept of himself as a fundamentally worthless human being. Criticism, perfectionism, unnecessary taboos, rejection, and many other parental practices can develop in the growing child a self-concept of basic inferiority. Since the pain and anxiety produced by these severe negative self-evaluations is intolerable, unrealistic, idealized self-concepts may develop as a defense against anxiety. He may think of himself as "better" than he really is. A variety of generalized *other-concepts* may similarly develop. The child may learn to expect others to be rejecting, hostile, protective, etc.

These self-concepts and other-concepts are frequently implicit. They are rarely within the individual's focal awareness. Nevertheless, they may determine major aspects of his behavior indirectly and they may also determine the experiences he is likely to have. A person whose basic assumption is that he is "worthless" may behave in a variety of ways which do not allow his self-concept to be openly challenged. A person who thinks of himself as inadequate may avoid competitive situations. His self-concept may never be realistically tested. The development of self-concepts and other-concepts will be discussed in great detail throughout later chapters. We will see in Chapter 12 that one of the major aspects of psychotherapy involves techniques that encourage the individual to test reality and otherwise challenge the distorted implicit self-concept and other-concepts which produce his misery and unhappiness.

The individual's ability to tact his own emotions is frequently an important determiner of his subsequent behavior. Commonly, people are able to recognize how they feel, but under certain circumstances it "never occurs to them." Again, under certain circumstances otherwise intelligent people may not "stop and think." A very common instance of this lack of ability has been noted by Dollard and Miller (1950). They state that although neurotic individuals often describe *behavior* that is obviously motivated by fear, they are frequently unable to label (tact) their private internal responses as "fear" or "anxiety." Often the closest such patients can come to describing their own fear is the statement that they "feel they are going insane."

Dollard and Miller have pointed out that society often neglects to tact objects, events, or behavior related to cultural taboos. Thus, tacts dealing with the varied aspects of sexuality are not generally made available in the culture, or again, the individual is not ordinarily permitted to think that different social classes exist. In the latter case a person suffering from neurotic conflict centering around social-mobility strivings may remain unaware of his own feelings, unable to understand himself and his behavior; whereas in the former the individual may be unable to identify, or may mislabel his own sexual feelings, or he may be unable to think intelligently about certain aspects of sexuality.

The Mand

Children are taught not only to tact internal and external stimuli but to make verbal demands and requests in response to their needs, feelings, and wishes. Given patterns of behavior on the part of the child signal to the adult certain areas of discomfort. The child may be asked, "Are you thirsty?" or "Do you want to make potty?" The child may echo the words "thirsty" or "potty." He eventually learns that the appropriate verbal response will elicit specific reinforcing consequences. The young child's response "thirsty" may readily come to mean "I want you to bring me some milk right now!" Verbal operants such as "Please pass the salt," "Wait!," and "Shh!" are referred to as *mands*. Skinner (1957) describes the mand as:

> a verbal operant in which the response is reinforced by a characteristic consequence and is therefore under the functional control of relevant conditions of deprivation or aversive stimulation. . . . A mand is characterized by the unique relationship between the form of the response and the reinforcement characteristically received in a given verbal community. It is sometimes convenient to refer to this relation by saying that a mand "specifies" its reinforcement. [Skinner, 1957, pp. 35–36]

The primary variables controlling the mand involve, first, the intensity of aversive stimulation or of deprivation, and, second, the extent to which the listener (or someone like him) has reinforced the mand in the past (or refused to). Variables affecting the probability of the listener's behavior are also important. They include (1) the effectiveness of rewards and punishments supplied by the speaker; (2) the authority, prestige of the speaker; (3) the intonation, loudness, and amount of repetition of the mand; (4) for some listeners—whether or not a softened form is used (e.g., since the mand is precarious, softened forms may be used which, in part, conceal the character of the mand: "Water!" vs. "I'm thirsty!," "Would you mind getting me a drink?," or "Get me a drink of water and I will tell you all about it," etc.). Many of these variables also determine the degree of social influence and the dominance relationships among individuals.

Generally the probability of emission of the mand is greatest when the stimulus complex closely resembles past reinforced conditions. However, just as in the case of the tact, this operant may be extended. Any feature of the situation which is similar to conditions under which the operant was reinforced in the past may make some contribution. Thus, some measure of stimulus control may be exerted by untrained animals, infants, characters on the movie or television screen, etc. The baseball may be manded to stop inside the baseline before going foul, or the fraternity boy may mand an automobile "Come on, Baby" while passing another on the highway. In more extreme cases a wounded soldier may cry out for "Mama."

Other classifications of the extended mand includes superstitious mands and magical mands. In the former case the mand is reinforced by accidental contingencies, e.g., the audience at a horse race manding their favorite horse to run faster, because he has won when they have done so in the past. In the magical mand, which may occur as a literary technique and which also is a common response to stress, the speaker expresses the reinforcement he *wishes* for, e.g., "Let there be light!" "Go jump in the lake!" or the wounded soldier's "God help us!"

It should also be noted that many types of nonverbal mands exist, mediated by the behavior, gestures, and postures of the individual. In a trivial case, the individual may continue to drink from an empty glass, which may both make a magical mand upon the glass and serve as a cue to anyone present as the occasion upon which the offer of a refill will be reinforced. Again, behavioral and postural mands may occur, signaling for example, "Don't try to stop me," "I need your help, affection," or "You must do better than that."

The Autoclitic

In the following passage Skinner outlines the use of the term *autoclitic:*

> When we ask "Did you see it, or did someone else tell you?", we are asking for more information about controlling relations. We are essentially asking, "Was your response a tact or an echoic or intraverbal response to the behavior of someone else?" Because controlling relations are so important, well-developed verbal environments encourage speakers to emit collateral responses describing them. These responses are in a sense similar to other tacts descriptive of the speaker's behavior (at the moment or at some other time) or even of the verbal behavior of someone else, but the immediate effect upon the listener in modifying his reaction to the behavior they accompany establishes a distinctive pattern. We shall refer to such responses, when associated with other verbal behavior effective upon some listener at the same time, as "descriptive autoclitic." [Skinner, 1957, pp. 314–315]

It is understandable that the student may find this passage of Skinner's difficult to interpret. That is because the technical definition of the autoclitic must be highly abstract in order to include the many specific instances of its occurrence and because of the great variety of verbal responses which can be classified as autoclitic. Actually autoclitics are verbal habits which are very familiar to the student and which are easily recognizable. They are things people are taught to say to help others have more complete information. When a child says, "Grown people smoke cigarettes," we correct him, "No, Johnny, *some* grown-ups smoke cigarettes." The child learns to make the collateral responses "some," "a few," "all" when referring to various things. When the child says "Bill is a bad boy" we ask him how he knows that. Did someone tell him Bill is a bad boy, or did he see Bill doing something bad? Eventually the child learns to add collateral statements such as, "I saw, " "I think," "X told me," etc.

Generally, the autoclitic operant refers to an individual's being taught to tact his own behavior and the reasons for it. There are many classifications of autoclitics: descriptive, qualifying, quantifying, relational, manipulative, etc. Autoclitics describe the kinds of things that are the basis for what the individual says. They are identified by such phrases as "I see" (I observe), "I recall," and "I am told." Another group tacts the strength: "I assure you," "I hesitate to say," "I think," "I believe," etc. Still others describe the relationship between what the person says and other verbal or nonverbal behavior: "I agree," "I confess," "I wish," etc. Another group indicates the internal feelings of the speaker: "I hate to say," "I must tell you,". . . . Qualifying autoclitics are used to modify the intensity or the direction of the listener's behavior: "Sort of," "Maybe," "It is so!" . . . . Quantifying autoclitics are: "a," "the," "all," "some," "almost," etc. Relational autoclitics include many aspects of grammatical manipulation such as the possessive case. Finally, there are a number of manipulative autoclitics such as "but" (exclude something or make an exception) and "and" (add something).

For our purpose the autoclitic need not be greatly elaborated upon. The autoclitic is important because major interest focuses upon how the speaker may acquire verbal behavior descriptive of his own behavior; how he may describe the responses he makes; the state or strength of such responses and the controlling relations. The person who makes a lot of collateral descriptive and qualifying statements may be considered especially intelligent and gifted. In fact, people must be *taught* to use autoclitics. When a person uses such phrases as "I want," "I believe," "I think," they are often taken as an explanation for his behavior. In reality these are autoclitics which he has learned. They may have nothing to do with the true explanation of his behavior. The responses which come under the heading of the autoclitic are of the type which in the past have been especially thought to indicate that behavior and adjustment is determined by an inner agent or inner cause. It may, in fact, take many years before this kind of thinking relative to human behavior is altered in the culture at large or until scientific thinking relative to human behavior comes to significantly change cultural practices.

WORD MEANING

In the act of communication one person can arouse particular meaning in the mind of another person. If the other person has the appropriate verbal conditioning history he can "understand" or, as we shall see, acquire novel symbolic repertoires. In contrast, words spoken, let us say, in Greek to an exclusively English-speaking person do not communicate, and call forth no understanding. The process by which words come to elicit meaning other than the sensory stimulation of sound patterns involves well-established principles of conditioning.

It will be recalled that respondent, or classical, conditioning is brought about by the repeated pairing of two stimuli, one of which naturally elicits the response. A stimulus that is neutral in the sense that it does not naturally elicit the response to be conditioned is paired with an adequate stimulus, one which does elicit the response. When the neutral stimulus (CS) is repeatedly paired with the adequate stimulus (UCS), it acquires the power to elicit the response which, prior to the repeated paired presentation, could only be elicited by the adequate, unconditioned stimulus.

Many psychologists consider the manner in which the meaning of a word becomes established to be analogous to respondent conditioning (e.g., Bousfield, 1961; Osgood, 1953; Staats & Staats, 1963). One of the most comprehensive statements of this process has been presented by Staats and Staats:

> [W]hen a mother shows her child an object, such as a ball, and says BALL, this may be regarded as a conditioning trial in the classical conditioning of a meaning response to a verbal stimulus. The ball elicits a sensory response, part of which may be conditioned to the verbal stimulus, becoming its meaning. . . . This sensory response can be considered to be only a portion of the total sensory response elicited by the object itself, not a precise duplicate, and this has probably led to the descriptive label image . . . . [Staats and Staats, 1963, p. 143]

The verbal stimulus (the word "ball") is considered to be analogous to the neutral, or conditioned, stimulus. The actual object (ball) is thought of as being analogous to the unconditioned stimulus. In Staats and Staats' words:

> [M]any stimulus objects may be considered UCSs in terms of the sensory responses they elicit. For example, a visual stimulus may be considered to elicit a "seeing" response in the individual in the sense that seeing the object is a response. An auditory stimulus may be considered to elicit "hearing" response, a tactual stimulus to elicit "feeling" response, and so on . . . .[p. 143]

Now, notice that the final conditioned meaning is not the same as the sensory responses elicited by the actual object (UCS). It is only a portion of the total sensory response, sometimes described as an image. Osgood (1953, p. 697) describes this by saying, ". . . a minimal but distinctive portion of the total behavior . . . comes to be elicited by another pattern of stimulation . . . ."

Let us summarize briefly in diagramatic form what has been said so far. Before conditioning, two stimuli, one a verbal stimulus and the other an object or event, are paired with each other. The case where the verbal stimulus is the word "ball," paired with the object, ball, can be shown as follows:

CS (word "ball")
UCS (object ball) ——————————→ R ("seeing")

After a number of such pairings the word comes to elicit a representational response. In the case of an object such as a ball, that representational response is often the visual image of a ball. The conditioned meaning response following the above pairing can be simply represented as follows:

CS (word "ball") — — — — — — — — — — — — → CR (image "ball")

The sensory responses of feeling, hearing, and seeing are not the only unconditioned responses that may serve as basis for the conditioning of word meaning. Autonomic and emotional responses may serve the same function. Suppose, for example, punishment is systematically paired with the word "bad." Punishment, and a variety of other aversive stimuli which may be paired with the word bad (soiled clothes, spoiled food, and other frustrating circumstances), ordinarily result in the elicitation of certain autonomic responses such as a change in heart rate, blood pressure, breathing rate, and electrical skin resistance (GSR). "Minimal but distinctive portions" of autonomic response patterns would be expected to become conditioned to the previously neutral word "bad." Of course, along with the slight autonomic responses the word "bad" might be conditioned to elicit a variety of images of events and objects which were present at the time that the pairing of the word with stimuli having aversive characteristics occurred. These conditioned responses would then constitute the meaning of the word "bad" for that individual.

It ought to be noted that the precise nature of that portion of the total response to objects and events which is conditionable is very difficult for researchers to get at in any precise, direct way. We have referred to conditioned meaning responses as images, and as minimal components of total reactions. A more general, and sometimes better, term would be to call them "representational" responses. Whatever its exact nature, the representational response is important not only as a conditioned meaning response, but also because it can mediate other meaning responses: a phenomenon known as semantic generalization.

SEMANTIC GENERALIZATION

Suppose the nonsense syllable DZL is paired with an object, let us say a green light. After conditioning, DZL would come to elicit a meaning response, which would be the representational response to the green light. In the case of a green light this would most likely be a visual image of a green light. This conditioned meaning response can be represented schematically as follows:

CS (DZL) ——————→ CR (green image)

Suppose the syllable DZL is then paired with an electric shock. After this conditioning DZL would elicit two successive meaning responses: one associated with a green light and the other associated with shock. The meaning response of DZL associated with shock would involve conditionable motor and autonomic aversive sensory components, as well as visual ones. The result of the two conditionings can be represented schematically as follows:

CS (DZL) ——————→ CR (green image) ——————→ CR (shock)

Under these conditions it would be found that the green light, which had never been paired with shock itself, would come to elicit the conditioned meaning responses of shock. This is because some of the conditionable sensory components of the object (green light) are identical with those to which the nonsense syllable DZL was conditioned. After DZL was conditioned to arouse these sensory components (primarily the visual image to "greenness") it was paired with shock so that those same components then became associated with shock. Schematically:

UCS (green light) ——————→ R (green image) ——————→ CR (shock)

Thus, the object, never itself paired with shock, would come to elicit an autonomic fear response, mediated by the representational sensory response to the object. This is called *semantic generalization.*

Semantic generalization could be made to occur from the object to the nonsense syllable just as well. If the nonsense syllable were conditioned to the green light and then shock occurred whenever the green light was presented, the nonsense syllable would arouse the anxiety associated with shock, even though it had never been paired directly with shock.

The effects of semantic generalization can readily be observed in life situations. If a person has been punished by another, mention of the *name* of the punishing person may be enough to arouse anxiety or anger. On the other hand, if the person's name has been associated with aversive stimulation then the *sight* of the person may arouse escape, avoidance, anxiety, or anger or whatever other responses were associated with the aversive stimulation.

Semantic generalization may also occur from one word to another. In an experiment by Phillips (1958) five Turkish words were paired with five shades of gray. The five shades of gray were the Munsell color values N1/, N3/, N5/, N7/, and N9/. Value N1/ was the darkest. It was sometimes referred to as "black" by the subjects of the experiment. Value N9/ was the lightest. Subjects sometimes referred to N9/ as "white." The five values of gray constituted a series of equal sense distances.

The five words used in the experiment were not reported directly by Phillips, but they were seven-letter, Turkish words. Suppose the words were "nansoma," "biwojni," "afowrbu," "jandara," and "enanwai." After these words were paired with the Munsell values they would have the following meanings:

CS_1 *NANSOMA* ——————→ N1/
CS_2 *BIWOJNI* ——————→ N3/
CS_3 *AFWORBU* ——————→ N5/
CS_4 *JANDARA* ——————→ N7/
CS_5 *ENANWAL* ——————→ N9/

In the second part of the experiment the Turkish word which had been conditioned to elicit N1/ was presented, along with a tone of sufficient sensitivity to elicit a GSR. This conditioning was continued until the Turkish word which had been paired with N1/ came to elicit the GSR even though the tone was absent. Suppose the word paired with value N1/ was "nansoma." Then the pairing

CS_1 *NANSOMA*
UCS (tone) ——————→ GSR

resulted in

CS_1 *NANSOMA* ——→ N1/ ——→ GSR

Only the Turkish word previously paired with N1/, the darkest shade of gray, was paired with sound, and thus conditioned to elicit a GSR. However, to the extent that the other Turkish words had acquired similar meaning (to the extent that the shade of gray with which they had been paired was similar to the value N1/) they would be expected to elicit a GSR. Thus, the greatest semantic generalization would be expected to occur to the word paired with value N3/ because N3/ was the shade closest to N1/ in the series. Value N5/ would be expected to elicit a GSR of lesser magnitude than N3/ but of greater magnitude than N7/, etc. In other words, a gradient of semantic generalization would be expected.

The experimental results generally substantiated these expectations. The results are shown in Fig. 7-3, where it can be seen that the conditioned meaning response had generalized as had been predicted with the exception of a rise in the generalized GSR at the extreme value N9/. Phillips argued, and rightly so, that because there was a strong tendency for the subjects to consider "white," diametrically opposed to N1/ which was considered "black," other highly overlearned verbal habits confounded the experiment at this point.

THE HIGHER ORDER CONDITIONING OF MEANING

By the kind of conditioning we have been describing the pairing of words with one another can result in new conditioned meaning re-

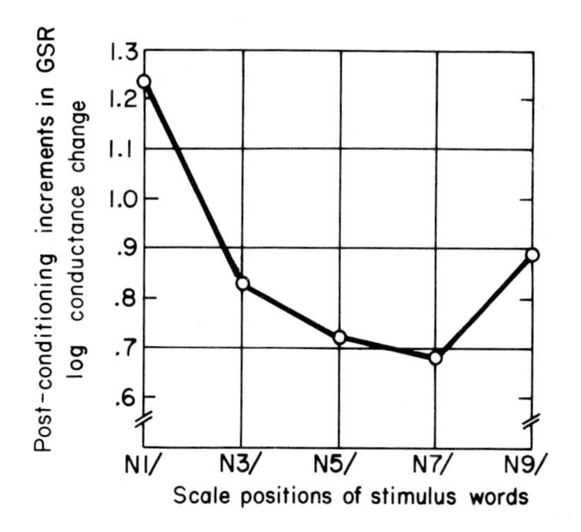

FIG. 7-3. Generalization gradient, as measured by GSR, of the response conditioned to the word which had been associated with N/1. (Phillips, 1958.)

sponses. Mowrer (1954) in a classic paper used the example "Tom is a thief." In this case two meaning responses have already been conditioned: the meaning of the word Tom and the meaning of the word thief. Now , by pairing the two words, the meaning response "Tom" comes to evoke the meaning response "thief." By semantic generalization discussed above, Tom himself will evoke in some measure the meaning "thief." Not only that, but the word "thief" by previous conditioning will tend to evoke a family of other associated verbal responses which are negative-evaluative in nature, e. g., "dishonest," "untrustworthy," "criminal." These intraverbal elaborations have, of course, *their* conditioned meaning responses.

Staats and Staats (1963) have pointed out that this cluster of negative-evaluative meaning will ordinarily come to elicit other responses toward Tom so that the individual will tend to avoid Tom, may rank him low in order of preference among his friends, and show other negative judgment responses toward Tom, perhaps even aggressive responses, etc.

A great deal of learning under ordinary circumstances is on such a purely verbal basis. In fact, words which have no actual reference in reality may become meaningful by being paired with other words which already have conditioned meanings. Thus, the word "ghost" may be paired with a large number of words which themselves evoke representational meaning responses tinged with feelings of fright, dread, unknown expectations, etc.

THE PUNISHMENT OF VERBAL OPERANTS

As in the case of all punishment, the punishment of verbal behavior does not simply reverse the effect of reinforcing consequences; nor is there any evidence that it ultimately reduces a tendency to respond. Rather, it converts the behavior and the circumstances under which it occurs into conditioned aversive stimuli. Reinforcement is automatically provided for behavior which is incompatible with the punished response.

Either the presentation of aversive or conditioned aversive stimuli or the withdrawal of positive reinforcement may be used to punish verbal behavior. The punishing stimuli may themselves be verbal such as the conditioned aversive verbal stimuli, "Bad!" "Wrong!" "No!" or the withdrawal of verbal forms of approval and affection. In the latter case mere silence when the occasion calls for speech may be punishing.

Punished verbal behavior may be emitted slowly and hesitantly. It may be muttered or whispered. It may become covert. If the punishment is severe, repression may occur so that the punished operant does not even occur covertly, at least within the individual's awareness. A simple, but excellent experiment by Dollard and Miller (1950) illustrates very nicely how the effects of punishment of a verbal operant may generalize to covert thoughts. In this experiment subjects were required to pronounce the letter "T" and the number "4" as they were presented. The "T" was always followed by shock, whereas the "4" never preceded shock. Eventually subjects established a discrimination between the two stimuli which could readily be measured in terms of the psychogalvanic skin response. The GSR (employed as index of fear) was very small when the "4" was presented and large when the "T" was presented.

The crucial test of generalization occurred in the second part of the experiment. Only a series of dots was presented to the subjects, who were instructed to think of the letter "T" to the first dot, "4" to the second dot, and so on alternately. The results showed that a markedly strong GSR occurred to the thought "T," while the thought "4" elicited only a minor GSR response. Miller concluded that the involuntary conditioned galvanic responses, which the subject did not know he was making, generalized from the cues produced by pronouncing the words aloud to those produced by thinking them.

The effects of punishment not only show generalization to similar responses, but they generalize to similar circumstances as well. The child punished by his parents for tacting certain experiences and feelings may not be able to "express his thoughts" freely. He may become verbally and emotionally inhibited.

Certain punishing conditions strengthen verbal behavior. Verbal operants may be strong because they reduce conditioned aversive stimulation. In a trivial case, a weak audience variable may be apparent during a telephone conversation. Since pauses or extended periods of silence may be mildly punishing (the withdrawal of positive reinforcement), the speaker may talk to avoid the effects of these conditioned aversive stimuli. Similarly, compulsive talkers may avoid the insecurity associated either with silence or with punishing verbal behavior on the part of others by maintaining a constant barrage of conversation. The defense against anxiety called a "rationalization" tacts the behavior of the speaker so that it is made least subject to punishment.

GENERALIZED REINFORCEMENT

The reader may recall from introductory psychology that verbal operants may be deliberately strengthened by generalized reinforcement. Many experiments have shown that reinforced verbal responses are repeated. Typically in these experiments particular categories of verbal responses such as plural nouns, first or second person pronouns, topics or subject matter areas (opinion statements, animals, books, etc.) are preselected for reinforcement.

In one of the original experiments in the conditioning of verbal operants by generalized reinforcement, Greenspoon (1950) instructed subjects to say a series of individual words. For one group of subjects the experimenter listened to the subject's verbal productions, but remained silent. In the experimental group, however, he smiled, nodded his head, and said "mum-hmm" whenever the subjects emitted plural nouns. As a result their production of plural nouns increased in frequency in comparison to the control group. During the extinction phase of the experiment for the experimental group the experimenter remained silent, just as he had for the control group. During this period the frequency of plural nouns emitted by the experimental group declined and eventually matched that of the control group.

Another variety of operant verbal conditioning experiments, instituted by Verplanck (1955), has been replicated many, many times. In this experiment students are asked to carry on informal conversation with their friends or relatives, who are unaware that they are participating in a psychological experiment. Under these informal circumstances the student is instructed to keep a record of the number of opinion statements the subject makes during the first 10 minutes of the conversation without reinforcing the subject. This can be done in several ways, such as making a small tear on a napkin for each opinion statement, bending a straw, or doodling.

The student is instructed to agree with any statements of opinion emitted by the unsuspecting subject during the second 10 minutes. That is, any statements beginning with "I think...," "I believe...," "It seems to me...," etc., are to be reinforced by the student's nodding, smiling, or making reinforcing statements as "You're right," "I agree," etc.

The subject's operant level for opinion statements is established during the first 10 minutes. Reinforcement is then begun and continued during the following 10-minute period. During the third 10-minute period the student is instructed to extinguish the subject, that is, to stop reinforcement for opinion statements. The effect of generalized reinforcement is indicated by comparing the second 10-minute with the first 10-minute period, and the effects of extinction by comparing the third period with the second period. Students frequently report that the third period is too short for complete extinction to occur. Rather, once the subject has been turned on, he is likely to continue opinion statements for some time before their frequency is reduced to the operant level obtained in the first period.

This type of naturalistic experiment carried on with relatives, friends, and acquaintances takes advantage of the fact that rapport may be readily established. The student's approval acts as a discriminative stimulus for opinion statements, rather than as reinforcement for specific opinions.

> We may say that verbal behavior tends to be emitted if it describes a condition which is or would be reinforcing to the speaker. Distortion of stimulus control through such effects is widely tolerated in some verbal communities and sharply suppressed in others. In *A Passage to India*, E. M. Forster has described many instances of wishful verbal behavior acceptable within the speaker's community. Dr. Aziz, in showing his English guest through some rather undistinguished caves, was "pretty sure they should come on some interesting old carvings soon," but only meant that he wished there were some carvings. In another instance when asked, "Are you married?," he replied, "Yes, indeed, do come and see my wife"—for he felt it more artistic to have his wife alive for a moment, though she had been dead for some time. [Skinner, 1957, pp. 165–166]

Certain components of neurotic reaction patterns involve reinforcement for implicit, covert, and overt verbalizations mediated by the individual acting as his own audience (see pp. 247–248).

THE ELICITATION OF EMOTIONAL BEHAVIOR

Although certain words such as "good" or "bad" can be used as conditioned reinforcers, operant reinforcement generally has little effect in modifying emotional behavior. *In order to arouse an individual emotionally the appropriate stimulus must be presented*; and, in the case of verbal stimuli, the words which cue off emotional responses depend upon the particular conditioning history (often idiosyncratic) of the listener.

That words may come, through conditioning, to elicit emotional reactions can readily be demonstrated experimentally. For example, Dodge (1955) had subjects read stories in which fictitious tribes of people were described, such as the "Melbu" tribe. Meaningful words were then systematically paired with the nonsense name, e.g., the phrase "friendly Melbu" was frequently presented. The results indicated that the positive evaluative meaning elicited by the word "friendly" became conditioned to "Melbu." Moreover, the strength of the conditioning was directly related to the frequency with which the nonsense name and the meaningful word had been paired.

Under natural circumstances words may readily take on conditioned emotional meaning. Parents may readily imply punishment or reward by their general attitude. Consider the following excerpt taken from a therapy session of a 30-year-old female patient, designated Mrs. A:

With my mother I could never discuss anybody's being pregnant. She wouldn't say they were pregnant. She'd say, "Oh, she is so big"—and she would say it as though she were hitting me. It felt to me like she was hitting me. The way my mother would talk about pregnancy was sickening. It was something wrong—can I use the word—it was something dirty. Once when a girl had a miscarriage, she talked about it—I didn't want to listen to it! [Dollard & Miller, 1950, pp. 204–205]

Intraverbal Chains and Emotion

It has been noted that most conversation involves primarily intraverbal operants and tacts. Often the omissions and associations which occur in spontaneous speech are significant in that they point to revealing aspects of the individual's emotional experience and emotional conflicts. It is as if the individual were able to say more by way of implicit verbal associations than his ostensive speech implies. This is especially true in the clinic when the individual is asked to associate freely without restriction as to subject matter or form and without self-editing. However, it is also true in everyday conversation even though such effects may be highly disguised. The following excerpt is taken from the report of a case of impotency by Goldstein and Palmer (1963). At least one of the factors conditioning the patient's impotency was a generalized fear of being punished by an adult authority.

He added that he felt too many people placed too much emphasis on sex, and he was firmly of the opinion that sex should not be a major factor in marital adjustment. "Too many people let their animal nature take them over!"

This remark led Lewis to explain that his father had always sought to teach the boys "self-control" in everything. He said that his father always wanted the boys to try everything but always preached moderation. He added that this was something very hard to teach children. He felt that he himself had learned his lesson quite early, but that his brother Jim always seemed in conflict with his father over it. He said that he didn't want anyone to get the wrong impression, that Jim was "a very wonderful guy," but his brother had always felt cheated by life and always wanted something more than was allowed him. Even as a child Jim had always been the one to demand a second helping at the dinner table or to ask for something extra when there was an outing or a good time to be had. Lewis admitted that he himself would also have liked to have asked for things, but "I knew better." In contrasting himself further with his brother, Lewis remarked that "Jim was always shooting off his mouth about something while I was the quiet one." Jim was the one who always asserted his rights and rebelled against his parents while Lewis was much more compliant. Consequently there was increasing dissension between Jim and his father which during the boys' adolescence rose to a pitch of dissension. Once Lewis and his mother actually had to intervene between Jim and the father when their anger reached the boiling point of fisticuffs. Lewis could not remember ever seeing his father so angry. In addition, Jim was often in scrapes in school or in some other kind of trouble or mischief. Asked if he and Jim quarreled much, Lewis smiled and remarked that he couldn't remember ever being angry at his brother, but then Jim was always bigger and quicker so that there was no use picking a fight with him. "Oh I could get roused up all right, but I always had to back down, 'cause father wouldn't stand for us beating on one another."

At this point Mr. C paused and apologized for getting off the topic, saying that he had meant to talk about his sexual history rather than about his father and brother. [Goldstein & Palmer, 1963, pp. 99–100]

At this point the clinical interviewer might well hypothesize that Lewis C's sexual problem and his reported early experiences with his father and brother might not be emotionally unrelated.

Slips of speech may also reveal supplementary sources of strength. The following example was reported by Freud:

A wealthy but not very generous host invited his friends for an evening dance. Everything went well until about 11:30 p.m., when there was an intermission, presumably for supper. To the great disappointment of most of the guests, there was no supper; instead, they were regaled with thin sandwiches and lemonade. As it was close to Election Day, the conversation centered on the different candidates; and as the discussion grew warmer, one of the guests, an ardent admirer of the Progressive Party candidate, remarked to the host: "You may say what you please about Teddy, but there is one thing—he can always be relied upon, he always gives you a *square meal,*" wishing to say *square deal.* The assembled guests burst into a roar of laughter, to the great embarrassment of the speaker and the host, who fully understood each other. [See Skinner, 1957, pp. 303–304]

Eisegesis

A verbal stimulus may have an effect on another individual which is entirely independent of the actual variables that caused the statement to be made. Thus, an individual may overhear someone saying, "I think he (she) is beautiful," and take it as a reference to himself because similar self-tacting is currently strong for other reasons. Paranoid personalities are especially prone to this type of misinterpretation.

Fragmentary eisegesis may occur in the process of mishearing. Skinner (1957) gives the example of a young man who had done more than his share of dancing with a middle-aged chaperone at a ball. When, in the middle of a dance the chaperone exclaimed, "I'm too danced out!," and led the young man off the floor he replied, "I wouldn't say you were stout at all!"

The neurotic individual who manifests a good deal of self-blame and self-rumination may suspect many casual remarks of others to have been spoken with double intent.

Psychic Driving

It has been found in the clinic that the simple procedure of repetitively playing back a client's tape-recorded statements may have widespread effects. The typical procedure here is selection of a key statement having to do with an important aspect of the patient's problem. The statement may be his own verbal cues, or it may be cues verbalized by others, but based on the effective identification of the client's problems. It is best if the statement is a relatively short one. The part of the tape carrying the selected materials is cut out, made into a loop, and driven continuously. Cameron (1956) sees the driven material as "a verbalization of a part of a community of action tendencies." That is, the verbalization functions as a cue which sets a cluster of action tendencies relative to the particular material in motion. It appears that as the repetition continues these action tendencies (behavioral and emotional) become more intensified. The particular material used produces responses relative to it. Thus, if the material is hostile in nature the client may show increasing signs of hostility, or if the material is depressive in nature the client may become more depressed. Apparently the effects of this technique, referred to as *psychic driving*, are not necessarily of short-term duration; there may be continuing effects. Clients may return continually to the driven material, and show a good deal of rumination and reorganization relative to it. A wide variety of reactions may occur to the psychic driving technique. The development of insight relative to the material; rejection of the material and attempts to escape the situation; rejection and later acceptance; and the development of various defenses against the material are all common reactions.

In order that the reader may more fully appreciate the effects which verbal stimuli may have, two examples reported by Cameron will be given. The first involves a 50-year-old woman who suffered from feelings of inadequacy and considerable ambivalence toward her husband, apparently derived from an earlier relationship with her mother. Before the session described below this latter fact had not been adequately recognized by the patient. Cameron states:

> The following statement was driven some 15 or 20 times: "That's what I can't understand—that one could strike at a little child." She has reference here to the fact that her mother used to take out all her own frustrations and disappointments and antagonisms on the patient during her early childhood—even going so far, when the patient was 7, as to tell her, "I tried to abort you, too, but you just wouldn't abort."
>
> After some 10 repetitions the patient said, "You know, that makes me feel dizzy and queer just to listen to it. I want to burst into tears—it makes the skin stand out on my arms; I am scared; my hands are wet." Then later, after it had been stopped, subsequent to some 15 repetitions, she said, "You know, I wanted to tell you to stop." In immediate discussion of the playback, she said, "I can see that it was really my mother who damaged me. I also see that not being able to trust my mother not to hurt me has made me mistrustful of everybody. It scares the hell out of me to think that my mother might be deliberately mean to me. It gives me one of those 'all gone' feelings just to think of it. I think my mother may have felt inadequate and taken it out on me." [Cameron, 1956, p. 504]

The type of reaction illustrated in the preceding excerpt is what Cameron describes as an immediately constructive response. That is, with relatively few repetitions the patient gains insight relative to the driven material. The next passage shows how the appropriate verbal cues continuously driven may result in rejection and blind escape responses:

> A girl in her early twenties came to us suffering from a severe character neurosis with marked immaturity, overt hostility toward her husband, underlying incestuous longings for her father, and much sex guilt. The incestuous longings were never accepted by the patient, although she was able, after therapy lasting several months, to identify the father's sexual desires for her. A passage in her psychotherapy was selected in which her own sexual longings for her father came close to the surface, and was set up as a playback circuit. After about 3 to 10 playbacks, she became progressively disturbed, grew very angry, called the therapist a fool, asked him what he meant by playing that stuff to her, and finally lept off the therapeutic couch and ran out of the Institute to her downtown apartment. She refused to return to therapy and had to be admitted at a later date as an in-patient, quite deeply disturbed. It should be underscored that this procedure is one of considerable potency, and care must be exercised in the selection of the playback circuit, otherwise the progress of the patient, as in this case, may be seriously retarded. [Cameron, 1956, p. 505]

SELF-STIMULATION

We have seen that verbal stimulation may serve to mediate both evaluative responses and action tendencies. However, what is even more important for human adjustment is the fact that the individual can both be enabled to solve problems by his own thinking, and be rewarded and punished by his own thinking. Furthermore, these internal events may be largely independent of current environmental stimulation. True, the individual's tacting facility and intraverbal repertoire are basically conditioned by people and events outside of his skin. However, once established, verbal operants play a major role in determining the way the individual reacts in life situations. On the positive side they may function to enable the individual to think, reason, plan, be creative, and reassure himself. On the negative side they may condition inappropriate thinking and reasoning; they may sustain anxiety, hostility, and other negative attitudes and feelings; and they may condition a shattered sense of mastery, recrimination, and self-blame.

Thinking and Planning

Through the mediation of other persons (culture) the learning of appropriate tacts and intraverbals may enable the individual to solve problems which have baffled the minds of the human race for centuries. Given the appropriate learning history (education), an individual may be able to cope with the problems of radiation surrounding the planet Mars; transplant a human kidney; write a movie script; or cope successfully with a problem of interpersonal relations.

Within our own culture there are wide individual differences in the "reasoning" which people engage in or accept. Contrast the intraverbal syllogistic chains of the professor of logic with those of the mass media of communication. The latter, even excluding advertisements and commercials, may acceptably employ incompatible contents within the same phrase, e.g., "armaments for peace."

Because of their learning histories, there are also wide differences among individuals in their ability to stop and think before making overt responses on an impulsive basis to the first intraverbal cue which becomes available. Responding by covert trial and error and generating intraverbal cues relative to the solution of a given problem is often extremely difficult. This is not only due to variations in the difficulty of problems which one must solve, but also because of the tension produced during the process of utilizing weakly conditioned verbal responses and arriving at novel associations.

This can readily be demonstrated in a laboratory. Students can be asked to think of novel uses for some article such as a coat hanger. After the first few responses even this relatively minor type of problem may be highly frustrating. When creative solutions are hit upon they are strongly reinforced because of the ensuing tension reduction. Children may also be taught to think relatively independently and in a problem-solving manner, or they may never be given the necessary encouragement for engaging in such initially difficult behavior.

Reassurance

Consider the role words play in giving reassurance. When the physician says, "You do not have cancer," the person may experience considerable relief. The student with professional aspirations and plans may similarly be relieved by the words, "You have been unconditionally accepted as a degree candidate in the school of...." The opposite of such remarks would, of course, cue off anxiety.

The individual may also alter his own emotional behavior by the appropriate verbalizations. The person who is plagued by financial difficulties may, after considering various aspects of his financial and life situation, come to solutions which offer considerable relief. One of the first, and perhaps most important, steps in successfully coping with emotions is the very recognition that the emotion exists. In the following example the tacting of emotional behavior came from an outside source, but the effect might have been the same if the patient had emitted the same self-tacts.

> During the second hour of therapy the patient became excited, began to weep and tremble, to lose his breath, and to show other signs of marked perturbation. The therapist understanding the behavior as a transferred fear reaction to himself said, "You realize that you are having an attack of anxiety?" This interpretation tended to stop the hysterical behavior and the patient slowly "cooled off." The therapist had explained to this intelligent patient that he was having an attack of "unreasonable fear." A statement from an authority that a fear is unreasonable tended to inhibit anxiety in this patient who had learned previously to be comforted by such statements. Furthermore, it showed the attack to be an event known to the therapist—experienced before and presumably manageable. The idea that helpful skill was at hand was also comforting. The patient had not at all realized that his behavior was motivated by fear. Labeling the emotion "fear" helped to discriminate the current safe situation from a really dangerous one. [Dollard & Miller, 1950, p. 297]

In order for the individual's emotions to acquire what Dollard and Miller term a "voice within him" and consequently be represented in reasoning, the individual must have been taught to tact his own feelings. Unfortunately, the early learning of the tacting of emotions is often not very elaborate or precise. Nowhere in our educational system is any attention given to formal training in tacting and understanding emotional behavior. Consequently the individual may be at the mercy of feelings of which he has no focal awareness and which have no verbal representations. Therefore he is unable to see himself as he really is.

The individual's verbal learning history may also result in his being able to tact his feelings appropriately when certain cues are present, but not when others are. Thus, he may be able to tact a hostile feeling at one time and a destructive act at another time but not be able to see a connection between the two. This phenomena is called *patterning*. For this particular example the sources of reinforcement might involve the individual's having been likely to be punished as a child if he performed a destructive act because he was angry, and less likely to be punished if he was angry without behaving destructively.

Anxiety

The individual may become reassured by words, but he may also become fearful in relation to purely verbal stimuli. Ellis (1963) has gone to great pains to underscore the fact that it is not what actually happens to the individual at point A which creates emotional disturbance, but rather what the individual tells himself at point B. Thus, the fact that the individual does not gain approval of others in certain situations does not directly cause anxiety; but the individual says to himself, as it were, "I do not have the approval of others (what Ellis would describe as a perfectly sane sentence) and that is terrible, catastrophic, etc." The latter phrase Ellis would describe as "insane."

Of course, this type of analysis is artificial. Undoubtedly the individual does not literally talk to himself in sentences in such cases. The thoughts are highly telescoped. They are attached by conditioning to certain discriminative stimuli. The reverse process could take place. The individual could be taught to think, reason, and plan in response to any given stimulus complex. That is, the same situation could, through conditioning, take on a discriminative function for very different intraverbal chains.

Although the individual originally learned to fear environmental occurrences at point A, the fear would not be expected to continue for periods of years because as the actual punishing contingencies changed from childhood to adulthood extinction would take place. However, the discriminative function of the situational cues causes the individual to continuously reindoctrinate himself.

This phenomenal description requires a further refinement which Ellis also recognizes: the intraverbal chains with which the individual continuously reindoctrinates himself are implicit, and only very rarely within the awareness of the individual. They function precisely like the "flower–stem–petal" implicit intraverbal mediating links of our earlier example.

This continual reindoctrination also involves the awakening of habit-family hierarchies of related implicit intraverbal responses. Thus, a variety of ideational and behavioral patterns may be elaborated in the absence of any significant further environmental conditioning. Positive patterns could also be mobilized. Salter (1964) refers to positive elaboration as the "reflexological will to health." This process is sometimes called "autokinesis," a term used by Gantt (1953) for the "ability of the organism to develop and acquire new responses on the basis of old stimulations and their traces, and to change its relationships to old stimulations without the aid of any new external stimulation."

This process acts in conjunction with the original learning of mands and tacts. First, early conditioning functions to bring behavior under the control of internal speech. In the trivial case when the child is manded, "Drink your milk," reinforcement is made contingent upon the child's behaving appropriately to the verbal stimulus. Again, when the child learns the echoic response, "I am going to drink my milk," he is reinforced if he does what he says he is going to do.

The process through which the child's internalized tacts and mands come to function as discriminative stimuli for behavior is a slow and arduous one. As Salter points out, using examples from the work of the Russian psychologist, Luria, it is difficult to get the young child to switch from one action to another. Try manding the child to put rings on a bar while it is taking them off. The result is that the ongoing action becomes intensified rather than altered. Here we see the connection between the conditioning whereby the individual comes to respond to his own internalized speech and the elaboration of patterns of self-stimulation. As Salter points out, again appealing to Luria's finding (which involve too many complexities to be discussed at length here), the child's speech system comes to involve considerable systematic organization. Salter concludes:

> With verbal patterns so fully permeating the cortex, and with the contents of the cortex in a large sense even built around these verbal patterns, the slightest self-stimulation of these cerebral traces can be conducive to auto-elaboration toward health, or toward sickness. [Wolpe, Salter, & Reyna, 1964, p. 31]

Ellis (1963) has detailed a number of categories of implicit verbal chains which are commonly learned in our culture, and which frequently result in emotional disturbance; he calls these "irrational ideas." Self-stimulation involving these irrational ideas tends to create false anxiety and unrealistic self-blame. They may be paraphrased as follows:

1. It is a *dire necessity* for the individual to be loved or approved of by virtually every other significant person; that almost everything the individual does must yield such approval; and that what others think of him is most important.
2. The individual *must* be especially adequate, competent, intelligent, and achieving; this must be true in all possible respects; and any evidence of incompetence is an indication that the individual is *not* a *worthwhile* person.
3. Others *should* be blamed for their mistaken or immoral behavior; it is *terrible* for others to behave in such a way; and others *must* be criticized and reformed.
4. If things are not the way one would like them to be, it is *awful* and *catastrophic*; other people should strive to make things easier for the individual, help him with life's difficulties, and gratifications *should* be immediate and not involve frustration.
5. A person *cannot* control his emotions and unhappiness, which are created by other people and events.
6. Because something is dangerous or threatening, continued deep concern and worrying are *necessary* and magically function to eliminate or alter the stimulus.
7. It is easier to avoid than face life's difficulties and self-responsibilities; one *should* rebel against doing unpleasant things.

8. One *must* totally accept the conditioning of parents or society and because certain people and events once strongly affected one's life, one's present behavior *must* continue to be affected indefinitely by these all-important determiners.

9. Other people's problems and disturbances are the *dire concern* of the individual, and, further, there is invariably a right, exact, and perfect solution to human problems; if the perfect solution is not found it is *catastrophic.*

We have noted that these categories of irrational ideas are artificial in the sense that the individual probably does not literally tell himself these sentences. However, both his behavior and subjective experiences indicate that some such implicit symbolic representations serve as mediating links which may become greatly elaborated and serve to continuously mobilize autonomic and motor functioning. Presumably other sets of ideas operate autokinetically to mobilize emotionally healthy patterns of behavior. At the present time these analyses are relatively crude and represent first approximations toward a more precise, valid, and complete scientific interpretation of the role of verbal behavior in complex human adjustment.

The reader will recall from our discussion of Arnold's theory of emotion in Chapter 2 that a stimulus or a stimulus complex must be perceived and its effects on the organism appraised before an emotional state can occur in reference to it; and further, that sensory impulses must be received and integrated in the cortex before the mechanisms involved in emotions become operative. This process of appraisal which Arnold speaks of may well involve learned implicit intraverbal chains of an evaluative nature such as those discussed in this section.

THE ROLE OF AWARENESS

It will be recalled that when verbal behavior is punished the emotional reactions generated by the punishment may generalize to the individual's covert verbal behavior. This negative reinforcement of verbal behavior results in avoidance responses so that the individual avoids or escapes from the anxiety-producing verbal cues. This was well illustrated in the second case of psychic driving presented earlier. Under these conditions the individual learns not to think certain thoughts. The thoughts may, however, function as implicit intraverbal chains outside of the individual's awareness. This has been called the process of repression. The mechanism of repression and the related experimental evidence for it will be discussed in detail in Chapter 10. It is important at this point to note that the mechanism of repression leads to a lack of awareness of those aspects of the individual's behavior in which the relevant tacts and intraverbals have been punished.

In either punishment or positive reinforcement both distortions and restrictions of awareness on the part of the individual may result in unrealistic behavior which is less adaptive and in the long run less satisfying than it would otherwise be. When the individual is in possession of all the relevant information he reacts maturely and less self-defeatingly. Psychotherapy with neurotics, whose behavior is stupid and self-defeating because of distortions and restrictions in awareness, involves both reducing anxiety in order to lift repression and teaching the individual new tacts in order to enhance his ability to discriminate and to generalize appropriately. Generally, the quality of the individual's adjustment is given by the range of awareness. Although awareness may emerge without specialized training in tacting, the process is greatly enhanced by making deliberate changes in the individual's verbal repertoire. As anxiety is reduced the individual gains the free use of his mind so that, in possession of all the variables, his decisions and behavior become less self-defeating and more mature. The process is a very complex one.

SUMMARY

A scientific interpretation of human personality and adjustment must include an analysis of human verbal behavior. What the individual says, how he says it, and the meaning of what is said involve habits based upon previous learning experiences. These habits appear to follow the principles of operant and respondent conditioning.

The many complex aspects of emitted speech have been classified into categories of verbal operants. These include the *echoic verbal operant*, the *intraverbal*, the *tact*, the *mand*, and the *autoclitic*. The first two of these are concerned with the individual's learning to repeat what he has heard and to associate certain words, sentences, and thoughts with one another. The tact involves his ability to label and describe objects and events in his experience. The category of verbal behavior called the mand is concerned with the individual's learning to make verbal demands upon other people or upon objects and events in his world. Finally, the autoclitic refers to a group of verbal habits that allow the individual to describe conditions surrounding his own verbal behavior and the reasons for it.

The process by which word meaning is established is analogous to classical conditioning. Word meaning appears to consist of conditioned sensory responses to spoken or written symbols. The sensory responses may involve seeing, feeling, hearing, autonomic, and emotional responses. Semantic generalization allows for autonomic reactions to become conditioned to objects or events through the mediation of conditioned sensory responses. This happens without those objects or events ever being directly paired with anything that would naturally elicit the autonomic responses in question. Hence emotional, attitudinal, and related behavioral responses can be learned on a purely verbal basis.

Verbal behavior is largely conditioned by an audience. It is sustained or increased by generalized reward and inhibited by punishment. The arousal of particular feeling states is directly related to the presentation of appropriate verbal stimuli.

Omissions and associations which occur in spontaneous speech are sometimes significant in that they reveal aspects of the individual's emotional experience and emotional conflicts. Slips of speech may do the same. An individual may also mishear, misinterpret, or mistakenly believe that remarks are directed towards him because his expectations are based more on his self-concept and upon his feelings than upon reality. Key statements that have to do with important aspects of clinical patient's problems have been purposely presented repetitively to patients. The procedure sometimes makes the patient tense and fearful, but sometimes it helps him develop insight and leads to eventual personality reorganization.

Words can be used both to reassure the individual and to increase his anxiety, making him tense and afraid. It may well be that certain irrational ideas and attitudes, along with their related behavioral manifestations, are sustained by continual self-indoctrination: by what the individual tells himself. One psychologist has suggested that disturbed behavior is learned and perpetuated in this way.

Implicit verbal chains may function outside the individual's awareness, mediating positive or negative feelings toward people and events in the individual's life. Awareness versus nonawareness may make the primary difference in whether or not the individual can control his behavior and cope effectively with his world.

REFERENCES

Bousfield, W. A. The problem of meaning in verbal learning. In C. N. Cofer, & B. Musgrave, (Eds.), *Verbal learning and verbal behavior.* New York; McGraw-Hill, 1961. pp. 81–90.

Cameron, D. E. Psychic driving. *Am. J. Psychiat.*, 1956, **112**, 502–509.

Dollard, J. *Victory over fear.* New York: Reynal & Hitchcock, Inc., 1942.

Dollard, J., & Miller, N. E. *Personality and psythotherapy.* New York: McGraw-Hill, 1950.

Ellis, A.E. *Reason and emotion in psychotherapy.* New York: Lyle Stuart, 1963.
Gantt, W. H. The physiological basis of psychiatry: The conditional reflex. In J. Wortis (Ed.), *Basic problems in psychiatry.* New York: Grune & Stratton, 1953.
Goldstein, M. J., & Palmer, J. O. *The experience of anxiety.* London and New York: Oxford Univer. Press, 1963.
Greenspoon, J. The effect of verbal and nonverbal stimuli on the frequency of members of two verbal response classes. Unpublished doctoral dissertation, Indiana Univer., 1950.
Jenkins, J. J. & Russell, W. A. Systematic changes in word association norms. *J. abnorm. soc. Psychol.*, 1960, **60,** 293–304.
Jung, C. G. Studies in word-association. Trans. by M. D. Eder. London: William Heineman, 1918.
Kent, G. H., & Rosanoff, A. J. A study of association in insanity. *Amer. J. Insanity,* 1910, **67,** 37–96, 317–390.
Miller, N. E., & Dollard, J. *Social learning and imitation.* New Haven: Yale Univer. Press, 1941.
Mowrer, O. H. The psychologist looks at language. *Amer. Psychologist,* 1954, **9,** 660–694.
Osgood, C. E. *Method and theory in experimental psychology.* London and New York: Oxford Univer. Press, 1953.
Phillips, L. W. Mediated verbal similarity as a determinant of the generalization of a conditioned GSR. *J. exp. Psychol.,* 1958, **55,** 56–62.
Rheingold, H. L., Gewirtz, J. L., & Ross, H. W. Social conditioning of vocalizations in the infant. *J. comp. physiol. Psychol.*, 1959, **52,** 68–73.
Russell, W. A., & Storms, L. H. Implicit verbal chaining in paired-associate learning. *J. exp. Psychol.,* 1955, **49,** 287–293.
Salter, A. Conditioned reflex therapy. In J. Wolpe, A. Salter, & L. J. Reyna, (Eds.), *The conditioning therapies.* New York: Holt, 1964. pp. 21–37.
Skinner, B. F. *Verbal behavior.* New York: Appleton, 1957.
Staats, A. W., & Staats, C. K. *Complex human behavior.* New York: Holt, 1963.
Verplanck, W. S. The control of the content of conversation: Reinforcement of statements of opinion. *J. abnorm. soc. Psychol.,* 1955, **51,** 668–676.
Wolpe, J., Salter, A., & Reyna, L. J. *The conditioning therapies.* New York: Holt, 1964.

Chapter 8

Social Learning and Imitation

INTRODUCTION

Most human learning has a social basis. Reward for its acquistion and maintenance is mediated by what significant people in our lives say and do. In the previous chapter we discussed many aspects of behavior rewarded and maintained by others. There it was seen that learning mediated by others involves no less than the entire realm of verbal behavior, including modes of thinking and emotional reaction patterns sustained by learned verbal responses. It was also noted that verbal cues learned in a social environment or produced in a social media may serve a discriminative function for a variety of instrumental and motivated behaviors.

A great deal of the individual's behavior is not taught verbally or by the direct manipulation of reinforcing and punishing contingencies. Both children and adults learn from others by observation. At times the characteristics, attitudes, and behavior of others are learned by the individual or influence him without any specific attempt on the part of others to teach or otherwise have an effect on him. When others serve as models for subsequent behavior psychologists refer to the processes as identification and imitation learning. In this chapter each of these processes will be discussed separately. We will then analyze a variety of reinforcement patterns which involve these processes, and which, in conjunction with the learning processes previously discussed, serve to shape and maintain significant aspects of adjustment. The learning of aggression, dependency and evaluative dependency, and sexual behavior will be discussed in some detail.

THE MODELING OF BEHAVIOR

Children frequently imitate adult models in their play. They typically assume the roles appropriate to adults of the same sex and may reflect parental patterns in their use of gestures, voice inflections, and attitudes. Older children and adolescents are exposed to both peer and adult models outside of the immediate family, models which may exert strong influence on their behavior. In fact, the parents may become less influential than peers, other adult models, or the mass media of communication in establishing the child's preferences, values, and behavior repertories.

A residue of parental mannerisms, postural and verbal habits, and attitudes may survive as the person matures. Occasionally one has the experience of meeting the parents of an acquaintance or a friend for the first time and being surprised by certain underlying similarities in attitudes and behavior between the individual and his parents. We are surprised to see that some of the behavior that we had previously thought to be uniquely his is actually imitative of his parents.

Laboratory Demonstrations of Modeling

Technically, modeling means that the observing individual may acquire new responses by imitating others. The modeling effect can be made to occur experimentally. For example, Bandura and Kurpers (1963) performed an experiment in which children participated with an adult in a bowling game. Prior to the game both the children and their models were given free access to a plentiful supply of candy. The situation clearly indicated that they could help themselves as much as they wished. In one experimental condition the model praised himself and helped himself to more candy whenever his bowling score exceeded a relatively high standard. However, whenever his score was lower than the standard he criticized himself and took no candy. In another experimental condition the model behaved in the same way, except that his standard was much lower. Following exposure to these models, the children played a series of games themselves, without the models present. Under these conditions the children showed a considerable amount of imitative behavior. For those children who had been exposed to the high-standard model the incidence of candy-taking was minimal, except when they achieved relatively high bowling success, whereas the children exposed to the low-standard model rewarded themselves generously with candy for minimal success. Moreover, the children tended to reproduce the self-approval and self-berating comments of the models.

Bandura and his associates (see Bandura and Walters, 1963) designed a series of experiments to test the modeling effect in the learning of deviant behavior—primarily aggressive responses. In one study (Bandura,

Ross, & Ross, 1961) one group of nursery school children was exposed to aggressive adult models and another group of nursery school children was exposed to models who showed inhibited and nonaggressive behavior. In the aggressive group the adults exhibited unusual, idiosyncratic physical and verbal aggressive responses toward a large inflated plastic doll. The inhibited, nonaggressive adult model for the other group sat quietly, completely ignoring the doll and other possible aggressive manipulanda in the room.

The children were then allowed to play freely and the behavior of the two groups was compared with a control group of children not exposed to either of these adult models. It was found that the children in the nonaggressive-model group displayed the inhibited behavior characteristic of their model to a greater extent than did the control children, and the children of the aggressive-model group displayed a great number of precisely imitative aggressive responses, which only rarely occurred in either the nonaggressive-model group or the control group. If the adult model kicked, swung, pushed, or hammered at the plastic doll, the children reproduced such gross behavior. In so doing they produced many of the fine-grain aspects of the model's postural adjustment, position, and mannerism as well.

An extension of these experiments in which real-life models, human film models, and cartoon film models were compared (Bandura, Ross, & Ross, 1963a) indicated that film models were as effective as real-life models. Figure 8-1 shows the results of one of these experiments. Notice that a greater amount of post-frustration aggression was obtained for both the film and cartoon than for the real-life model. Since in these experiments mild frustration was employed during the test period, they also served as demonstrations of how reactions to frustration may be shaped through modeling. Observation of parental and other models during the child's development undoubtedly provide him with many opportunities to observe their stress reactions and increase the probability that he will respond imitatively in frustrating situations.

Learning vs. Performance

Another experiment by Bandura (1962) has indicated that imitative behavioral repertoires may be learned, but not performed until the individual is rewarded directly for performing such responses. During acquisition in this experiment children were exposed to a film model who exhibited novel aggressive responses and distinctive verbalizations. The children were then divided into three experimental groups. One group was punished for imitating the model, one group was rewarded, and the third was neither punished nor rewarded. A later test of imitative behav-

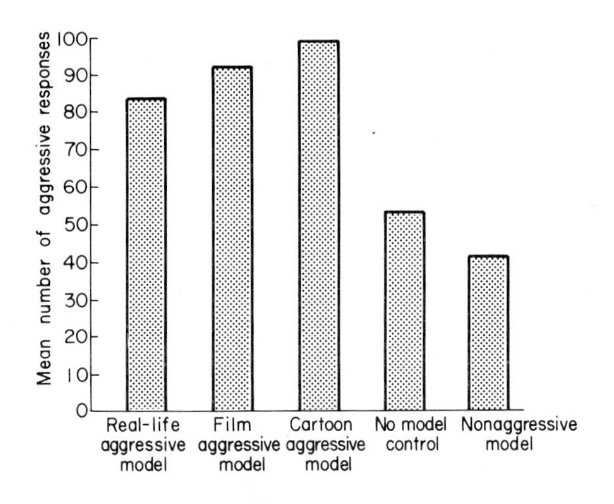

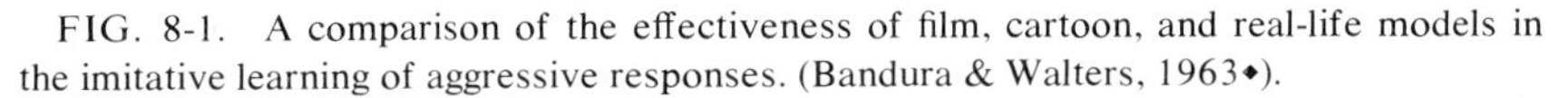
FIG. 8-1. A comparison of the effectiveness of film, cartoon, and real-life models in the imitative learning of aggressive responses. (Bandura & Walters, 1963•).

ior showed that children in the model-punished condition imitated less than both the model-rewarded and the no-consequence conditions. Then children in all three groups were offered attractive incentives if they reproduced model's responses. Surprisingly, the introduction of positive incentives for performing the model's responses resulted in the children of the model-rewarded, model-punished, and neither rewarded nor punished conditions all showing the same amount of imitative behavior. That is, previous performance differences among the groups were entirely eliminated. Thus, imitative learning may occur but not be reflected in performance until the appropriate reinforcing contingency is set up.

Parallel Nonlaboratory Observations

This is consistent both with clinical experience and with everyday experience. Children learn a great deal by observing, thinking, and drawing conclusions—none of which may be reflected in their behavior. This is exemplified by a commentary by Virginia Axline, excerpts of which are quoted below. They deal with the covert learning of Dibs, a severely withdrawn child who had originally been thought to be mentally retarded, but who actually possessed a very superior intelligence. Although this particular child is exceptional in that he is emotionally disturbed and unusually intelligent, the following comments from Axline's book, *Dibs: In Search of Self* (1964), point to observation learning and imitation processes by means of which a considerable amount of learning normally occurs throughout the child's development.

At one point during a play-therapy session Dibs tried to turn on a radiator. He was told that it didn't work and that there were men in the basement fixing it.

"What is wrong with it?" Dibs asked.
"I don't know," I replied.
"You could find out, you know," he said, after a short interval.
"I could? How?"
"You could go down in the basement and hang around out of the way on the edge of things close enough to watch them and hear what they have to say," he told me.
"I expect I could," I replied.
"Then why didn't you?" he asked.
"To tell you the truth, Dibs," I replied, "it had not occurred to me to do that."
"You can learn lots of interesting things that way," he said.
"I'm sure of that," I told him. And I was just as sure that Dibs had learned many, many things in that manner, hanging around, out of the way, on the edge of things, close enough to watch people and hear what they had to say. [Axline, 1964, p. 52]

During a later session Dibs discussed Marshmallow, a school pet rabbit, sang a marching song, and gave other evidence that his almost complete lack of communication with adults had not been a result of mental retardation. Axline, speaking as the therapist says:

Even though Dibs' outward behavior at school might not indicate it, school meant much to him. Even though his teachers might be baffled, frustrated, and feel a sense of defeat, they had gotten through to Dibs. He knew what was going on there. That marching song was probably one the children had been taught at school. Marshmallow was their pet—rather, their caged animal. But Marshmallow was a part of the school experience . . . .We never know how much of what we present to children is accepted by them, each in his own way, and becomes some part of the experiences with which they learn to cope with their worlds. [p. 58]

Dibs turned out to possess a wide variety of skills. He could read, write, spell, and draw. At one point Axline says:

According to all the existing learning theories, he should not have been able to have achieved any of these skills without having first mastered verbal language and without having had appropriate background experiences. Nevertheless, Dibs did possess these skills to an advanced degree.[P. 97]

Theories of learning aside, both experimental work and experience with children and adults indicate that behavioral repertoires can be acquired through observation learning and imitation without being reflected in actual performance.

Studies of Variables Determining Imitation Learning

The factors controlling these learning processes are at the present time only poorly understood in a scientific sense. There has, however, been some success in isolating the variables involved in imitation learning. Miller and Dollard (1941) suggested that cues which connote some degree of environmental competence (such as age, brightness, skill, etc.) function as discriminative stimuli for imitation. This was tested by Rosenbaum and Tucker (1962). Subjects were asked to choose the outcome of a series of horse races. They could see the choices of a supposed other subject who was predicting the outcome of a different series of races. Actually the responses of the supposed other subject were programmed into the apparatus by the experimenter. In three different experimental conditions the model was reinforced (given a "correct" green light) 80, 50, and 20% of the time, respectively. These varying amounts of reinforcement of the model were to be thought of as reflecting varying degrees of competence of the model in judging the outcome of his races.

The electrical circuitry of the apparatus was so arranged for half of the subjects that when they made the same choice as the model's immediately preceding one they were rewarded. For the other half of the subjects opposite-choice or nonmatching responses were rewarded. Thus, had the model's behavior alone been the significant cue for imitation the subjects should have learned to imitate (or not to imitate in the case of the nonmatching group) 100% of the time. This is because had they done so they would always have been rewarded (nonrewarded in the case of the nonmatching group). In contrast, if the amount of reinforcement received by the model was the significant cue, then the subjects should have learned to imitate in varying degrees in accordance with the 80, 50, and 20% model's competence conditions.

The results of this experiment are shown in Fig. 8-2, where the acquisition of imitative behavior can be clearly seen. Imitation was apparent as the experiment proceeded, with the final level of imitation matching somewhat the behavior of the model. Even though had the subjects in the matching group matched the responses of the model (and the nonmatching group not matched the responses of the model) they would have been correct 100% of the time, none of the subjects did this. Instead, the amount of reinforcement received by the model was the actual variable controlling their imitative behavior. When the model was imitated it was not because his *behavior* served as a cue. The *consequences* of his behavior determined the extent to which imitation learning occurred.

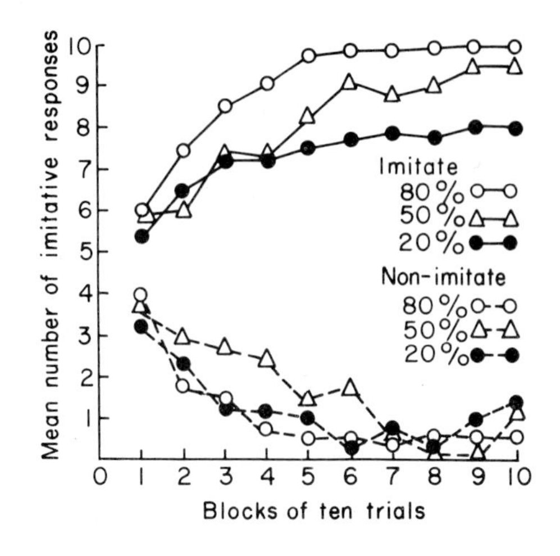

FIG. 8-2. Mean number of imitative responses per block of 10 trials for subjects observing models who were "correct" 80, 50, or 20 percent of the time. Those in the group labeled "imitate" were rewarded for agreeing with the model; the "nonimitate" group was rewarded for disagreeing with the model. (Rosenbaum & Tucker, 1962.)

Notice also that when nonimitation was learned, and a response which did not match that of the model would have been reinforced 100% of the time, imitation decreased as the experiment continued. It can be seen, however, that the matching of incompetent models is not as readily controlled by incompetence cues as the imitation of competent models is controlled by competent cues. The effect of early experiences in learning not to imitate may be quite complex: incompetent models may serve as useful guides to nonimitative behavior, early learning experiences provided by incompetent behavior may be primarily irrelevant to successful responding, or competence cues alone may be highlighted during early experiences so that all cues associated with competence are more salient than those related to incompetence.

In the Rosenbaum and Tucker experiment subjects imitated competent models or did not imitate incompetent models because of their previous life experiences, wherein they learned to imitate successful people and not to imitate unsuccessful people. Certain other characteristics of the individual also influence the extent to which his behavior will be imitative in any given situation. DeCharms and Rosenbaum (1960) have shown that individuals low in self-esteem, who have not received sufficient reinforcement, are predisposed to imitate a reinforced model. It has also been shown that when individuals believe themselves to have characteristics in common with models they are more likely to be influenced when they believe themselves to be different. For example, Stotland and Patchen (1961), some time after the administration of a measure of prejudice to students, led half of them to believe that they had characteristics in common with a highly prejudiced person whose case history was read to them. The remainder of the subjects were informed that the model was dissimilar to them in many ways. When the prejudice scale was readministered at a still later date (3 weeks later) the students who had been told they were similar to the highly prejudiced persons scored higher than they had on the first administration. The scale scores of those who were led to conceive of themselves as different from the prejudiced model did not change.

Many other experiments have indicated that very dependent individuals are predisposed to imitate reinforced models. Evaluative dependency, that is, a high degree of dependency on the positive evaluation of oneself by other people, has been shown to facilitate imitative learning, even when the individuals are not aware of the reinforcing contingencies. Crowne and Marlowe (1964) have devised a scale for measuring this factor, which they called the need for approval. In one of a series of experiments (see Crowne and Marlowe, 1964) they required subjects to observe the verbal conditioning of a student, who was really the experimenter's confederate. His responses were programmed so that his acquisition of the reinforced responses followed a typical verbal conditioning curve. In the second part of the experiment the students were subjected to the same procedure which they had observed, with the exception that the experimenter merely recorded their verbal responses without reinforcing any particular response category. The results indicated that no conditioning occurred during this period, but when the data were broken down and examined using only the responses of the high-need-for-approval groups, it was found that these subjects did, indeed, show a significant vicarious verbal-conditioning effect.

Finally, the degree of influence a model has on imitative behavior depends to some degree upon the emotional state of the individual. Schachter and Singer (1962) injected some students with adrenalin, and others with a neutral solution. Some of the adrenalin-injected students were informed about the harmless side effects of the drug, such as flushing, trembling, and palpitations, whereas others who had been injected with adrenalin were told nothing. Then the subjects were placed in a room with a confederate of the experimenter, who behaved in a highly aggressive manner. Compared to the placebo group, those who were injected with adrenalin and not informed of its effects, displayed a considerable amount of aggressive behavior. On the other hand, those injected with adrenalin and told of its side effects showed little imitative aggressive behavior. These results are very interesting. They indicate that models may be most influential when observers are emotionally aroused and are not able to tact the emotion or the emotion-arousing contingency. We saw in the last chapter that people may be unable to label and to represent their own emotional behavior in thinking. One consequence of this appears to be that they attribute their feelings to other stimuli in the environment and may imitate the emotional and operant behavior of models to justify, as it were, their own state of arousal.

Other experiments have shown that emotional arousal which is environmentally, rather than physiologically induced may increase the probability of imitative behavior, at least when the arousal is of moderate intensity. Typical of these studies is an experiment by Walters, Marshall, and Shooter (1960), who found that subjects placed in a stress situation tended to imitate fictitious arbitrary autokinetic judgements to a greater extent than subjects who had not experienced stress. Other experiments have shown attitude change is facilitated when the individual is emotionally aroused (e.g., McNulty & Walters, 1962).

IDENTIFICATION

Various writers have used the terms "imitation" and "identification" in different contexts. Imitation has been defined as matching behavior in the presence of a model and identification as matching behavior that occurs later in the model's absence (Mowrer, 1950). The term imitation has also been used to refer to the observational learning of specific behavioral acts, whereas identification has been reserved for generalized imitative learning (e.g., Parsons, 1951). In some instances the term identification has been reserved for imitative responses involving attitudinal and symbolic learning (e.g., Osgood, Suci, & Tannenbaum, 1957). Actually, little is gained by employing two different terms for the same process. Personality theorists tend to favor the term identification and the use of this term seems to be more appropriate because it is more common. This is especially true during discussions of the major emotional habit patterns and behavioral sequences that are learned on the basis of observation of others throughout the child's early development.

In the early developmental period the parents and other family members serve as powerful sources of unconditioned and conditioned reinforcement. The range of available responses for imitation is restricted primarily to these models. Later, of course, children are exposed to a variety of models outside the immediate family. The extent to which new modeling takes place on the basis of models provided by the peer group and other adults depends upon the diversity of the models (how different they are from the original family models), their availability, the degree of the model's relationship with the identifying person, and the extent to which each of them receives rewarding or punishing consequences for his behavior.

Older theories of identification (e.g., Freud, 1924) stressed the fear of punishment in the identification process. It was thought that when the child was unable to compete successfully with the adult for gratifications, primarily those of a psychosexual nature, fear of the more powerful competitor resulted in the child's avoiding punishment and gaining fantasy rewards by imitating the rival adult. This was primarily the Freudian thesis. It was thought to be especially true in the case of the boy child in competition with the father as the recipient of gratifications from the mother. A somewhat similar but more recent theory has been proposed by Whiting (1959; 1960). His theory involves defensive identification based on status-envy generally. Neither of these two theories has been strongly supported by empirical research.

Another thesis of earlier theories of identification (e.g., Freud, 1925) was that identification occurs when a nurturant parent begins to withhold reinforcement previously given in abundance. This aspect of the theory has found much more research support. It will be recalled that an organism can be taught to acquire new responses in order to avoid the withdrawal of reinforcement. If imitative responses are reinforced when the insecurity and frustration associated with the withdrawal of nurturance occurs, they would be expected to be learned.

In early childhood parental nurturance acquires strong reward value. Strong positive emotional components are attached to the nurturant activities and presence of the mother especially, so that the infant develops dependency to a high degree. Sears (1957) has pointed out that the consequent dependency—frustration as the mother begins to demand more independent behavior may lead the child to identify with the mother in order to reinstate the parental rewarding state of affairs. This may result from the fact that the child can administer positively conditioned reinforcers to himself by imitating his mother because the mother's presence and behavior have acquired secondary reinforcing properties (cf. Mowrer, 1950).

In this connection many studies have indicated an association between parental rewarding characteristics and the likelihood of identification. Sears (1953) found that boys whose fathers were affectionate tended to assume the father role in doll-play situations more often than boys whose fathers were emotionally distant. On the other hand, boys who tended to adopt the mother's role in doll-play situation were found to have mothers who behaved affectionately toward them, and who also

Children imitate adults whom they perceive as being affectionate toward them. (Photograph by Tor Eigland.)

tended to devaluate their husbands. Mussen and Distler (1959) reported that children with strong masculine-role preferences tended to perceive their fathers as powerful sources of both reward and punishment.

Studies of Identification

Bandura and Huston (1961) have shown by direct experimental manipulation that reward characteristics of the adult may be a relevant variable. In their experiment one group of nursery school children was exposed to a highly nurturant and rewarding female model. Another group of children was exposed to the same model who behaved in a nondemonstrative, nonrewarding manner. Later, during a game, the model exhibited a variety of novel responses—verbal, motor, and aggressive—responses which had nothing to do with the ongoing game. Children in the nurturant-model condition imitated these responses to a much greater extent than did children in the nonreward condition. Furthermore, the authors pointed out that the children exposed to a model possessing rewarding qualities not only tended to imitate specific behavior but also tended to emit responses falling within the same class as those made by the model but which the model did not in fact emit.

It is of special interest that more recent experiments have found that the controller of reinforcements rather than the consumer tends to become the imitated model. That is, research supports a "social power" theory of identification, rather than a "status-envy" theory. Five types of social power have been distinguished by French and Raven (1959). They are rewarding power, attractiveness, legitimacy, expertness, and coersiveness. Rewarding power and attractiveness may mediate wider imitative ranges of behavior because they strengthen approach and interactive sequences. Expertness may result in more specific modeling effects. Legitimacy and coersiveness may be associated with the elicitation of envy and inhibitory behavior, especially when the source of legitimacy or coersiveness is derived from the individual's ability to employ punishment. Therefore they come to control only limited aspects of behavior.

The experimental investigation of the power theory of identification has yielded promising results. In accordance with the social power theory of identification (Maccoby, 1959; Mussen & Distler, 1959; Parsons, 1955) Bandura, Ross, and Ross (1963c) predicted that children would reproduce more of the behavior of an adult who controlled positive reinforcers than that of a powerless adult model, and that power inversions on the part of male and female models would produce cross-sex imitation. Nursery school boys and girls were used in the experiment.

An adult (the consumer) who accompanied the child to the experimental room made positive comments about another adult (the controller) who owned the nursery school "surprise room" and a fabulous collection of play materials. A number of other techniques were used to establish one adult as the controller.

In one condition the adult consumer requested permission to play with the toys. The controlling adult agreed, and the child was left stranded at a table with only uninteresting toys. The controller and the consumer then continued to play enthusiastically for a 20-minute period. A dart game and a pinball machine, appropriate to adult play, were frequently used by the adult models. During this play session the consumer and the controller were differentiated. Verbal comments were made by the consumer about the controller's highly attractive resources. Half-way through the session the controller left and brought in a soda fountain dispenser and cookies. While the consumer was enjoying his snack the controller turned on a "TV" radio that played a melody while a revolving dial displayed storybook scenes. A male model served as the controller for half of the children in each treatment condition and a female model served as the controller for the other half. Thus the conditions were made to simulate the husband-dominant family and the wife-dominant family.

Following this play period all three participants engaged in a guessing game in which the adults had the first set of turns. The consumer and the controller of the prior sessions engaged in a variety of novel verbal and motor responses. For example, during the game the controller invited the others to join him in selecting a "thinking cap." The controller selected a green feathered hat, and remarked "feather in the front" and placed it on his head in that manner. The other model selected a yellow feathered hat, said "feather in the back," and wore the hat with the feather facing toward the back.

The imitation scores consisted of the frequency of occurrence of verbal, motor, and postural responses which were imitative of the consumer and controller models. The results showed that the children clearly identified with the controller of resources, the source of rewarding power, rather than with the rewarded model. Cross-sex imitation was also produced by power inversion on the part of male and female models, particularly in girls. That is, girls imitated male controllers and boys imitated female controllers, the boys less so than the girls. These were the general findings. It was further found that children also reproduced some of the elements of the behavior of the subordinate model. Imitation was not confined exclusively to the controller.

Thus, reward control, or rewarding power may influence identification in the nuclear family. More recent research has indicated that power as measured in terms of parental dominance may also be a significant factor in the identification process. For example, a study was performed by Hetherington (1965) in which 326 couples were visited in order to obtain 108 mother-dominant and 108 father-dominant families. Each parent was seen individually at home, read 12 hypothetical problem situations involving child behavior and asked how he would handle them when he was by himself. Then both parents were brought together and asked to arrive at a compatible solution on handling the hypothetical problems.

A standard dominance index devised by Farina (1960) was used. When six of the seven indexes indicated paternal dominance the family was classified as father-dominant, whereas if six of the indexes indicated maternal dominance the family was classified as mother-dominant. Equal numbers of boys and girls in the age ranges 4–5, 6–8, and 9–11 were employed in the experiment. They were first given a projective test of sex-role preference. Then a parent–child similarity measure was devised: the children were rated on an adjective check list, and the parents were asked to give the name of someone who knew them well enough to rate them on the same adjective check list.

For the imitation task the child was told that he and his parents were participating in a study attempting to evaluate what things people think are prettiest. As the child watched, the parent previously instructed by the experimenter, indicated the picture in each of 20 pairs of pictures which he thought was the prettiest. The child then went through the series making his selections. Some time later the same procedure was repeated with the other parent.

It was found that more appropriate sex-role preferences occurred when the father was dominant than when the mother was dominant. Boys from mother-dominant homes showed less masculine sex-role preferences. It was also found that whereas boys tended to develop a preference for the masculine role at ages 4–5 which continued and increased only slightly to ages 9–11, girls showed a significant increase in feminine sex-role preference in the 9–11 age group.

Children tended to be more similar to the dominant parent than to the passive parent. Boys from mother-dominant homes were found to have acquired more non-sex-type traits like the mother, as well as more feminine sex-role preferences. Since these differences were also present at ages 9–11, it appears that whatever later social pressures may exist for the boy to acquire masculine preferences they are not adequate to alter the early conditioning. Boys and girls from father-dominant homes differed. Whereas the boys identified more with the father the girls from father-dominant homes identified equally strongly with both parents.

The imitation data were analogous to the sexual preference and similarity data. The dominant parent was imitated more than the passive parent by children of both sexes. Girls were also found to imitate more than boys. It is interesting that girls showed no difference in sexual preference at any age whether they were from mother- or father-dominant homes. Hetherington suggested several possible reasons for this: that differences due to maternal dominance might have been attenuated by the fact that the feminine role in our culture is less well defined and less highly valued; that the mother could be more dominant than a passive husband and still not be dominant or "unfeminine" relative to other women; and that perhaps boys and girls initially both identify with the mother and the socially appropriate sex-role behavior of paternal dominance may be necessary to facilitate the shift in identification models for boys, whereas girls can imitate the opposite parent and still maintain their primary sex identification. It is likely that this latter factor is most significant. In homes where the mother is dominant and the father is weak the sexual identification of the boy is much more disrupted. In extreme cases of such disruption homosexuality may result, but the full development of a homosexual orientation and behavioral pattern depends upon other factors being operative as well, especially factors conditioning deep-seated fear of being rejected by women in conjunction with the learning of inappropriate sexual behavior by direct reinforcement.

ANGER AND AGGRESSION REINFORCEMENT

The learning of how and when to become angry is one of the most difficult developmental tasks imposed upon children. Various forms of restraint and of punishment for spontaneously motivated behavior appear to elicit a reflex rage reaction in the young child. Anger reactions also result from punishment in the form of the withdrawal of positive reinforcers. This is primarily because of the young child's state of being reinforced with love, affection, and attention in a variety of ways analogous to generalized crf programming. The angry and aggressive reactions of young children are natural in the sense that they represent either unconditioned respondents or predictable and lawful behavioral manifestations, given the reinforcement patterns which the very young child is likely to have experienced.

A variety of situations may provoke anger, resentment, and hostility throughout the individual's formative years. In general terms those situations likely to evoke anger and aggression involve the following: (1) Any occasion upon which a noxious stimulus is delivered to the child. This may involve deliberate aggression from other children or unintentional (unconscious) expressions of hostility from adults; but primarily it involves situations in which adults behave punitively toward the child. (2) The interference, restraint, or blocking of goal-directed activity. (3) The withholding or withdrawal of gratifications and needed objects. (4) Frustration other than those involving punishment or direct goal interference.

There are a great variety of specific anger-provoking situations represented by one or another of the above categories. Each new developmental task imposes some degree of frustration. The child must give up long-standing privileges and acquire new, sometimes difficult, behavioral repertoires. Frustrations ranging all the way from those involved in learning to tie one's own shoelaces to learning that other people have wills of their own may produce resentment and hostility.

The behavior of siblings provides fertile grounds for anger episodes. The mere existence of a younger child may be resented. The younger child may be envied and resented because he still enjoys the privileges of being a "baby," or he may be allowed to assume other privileges quickly which the older child had to wait and work for. There may be competition for evidences of parental love. At times when the older child is disappointed or frustrated the younger may be the most available and the weakest target for aggression.

The child may resent occasions when the parents are indifferent or nondemonstrative. Parental rejection or neglect are likely to evoke anger. Of course, in instances where the parents serve as stimuli for hostile feelings the child is likely to feel guilty and become anxious. The child may be afraid that if the parents know he feels hostile toward them they will reject, or otherwise severely punish him. Thus, in many instances involving hostility provoked by the parents or other significant figures a suppression or repression of the child's feelings occurs. Neither the adult nor the child himself may be aware of the underlying resentment.

Anger and aggression are not always punished. The child must be taught when and how to become angry. Anger and aggression are permitted in self-defense or in response to unfair treatment. After a history of punishment for being angry or aggressive it is sometimes very difficult for the individual to learn, let us say, to react to insults or aggression from others with appropriate anger and counteraggression.

Assertiveness of the "excuse me, but I was here first" variety; an untimid and vigorous approach to problem solutions; a degree of assertiveness appropriate to competitive situations, etc., must be generalized from early learning episodes in which the individual was rewarded for acting aggressively. But unless some degree of appropriate assertiveness occurs in the first place it cannot be rewarded. On the other hand, if aggressive behavior is rewarded and if aggressive patterns of thought and action and an attitude of aggressiveness toward the social environment come to be characteristic of the individual he is likely to be tacted as "pushy," and considered undesirable by others. It is easy to understand how such fine discriminations required of both the child and the adult represent an imposing learning task, particularly since the appropriate learning must involve underlying emotional habits and attitudes.

Anger, Anxiety, and Awareness

People upon whom the individual is dependent such as parents, and later his spouse, and his boss, are likely to evoke hostility from time to time, in that they are controllers of reinforcing stimuli such as money, affection, and sexual gratification. Many people are unable to express anger and assertiveness toward people upon whom they are dependent. In extreme cases any hostility which may arise toward people upon whom the individual is dependent evokes intolerable anxiety. Since society condemns hostility as being "bad" the individual comes to feel guilty and anxious when he is hostile. Since he is afraid that if other people know that he feels hostile they will reject him he may seethe with chronic, secret resentment. If the hostility is intense and cues off considerable anxiety, repression may occur as a personality defense. Then neither the other person nor the individual himself is aware of any interpersonal hostility. The individual may be unwittingly sarcastic, forget promises and intentions, and behave so as to undermine the self-confidence of others. At the same time he may not be able to criticize the person upon whom is dependent in any objective way. Figures upon whom he is dependent are likely to be idealized.

Sometimes the only cue available that the individual is hostile is the fact that he is on guard and shows caution in personal relationships. In the extreme case anger may be suppressed with protestations of love. In these latter situations the only available cue is the fact that the person behaves inappropriately because he is too nice. The reason that hostile individuals may appear to others as guarded, cautious, and timid is because of the learned mechanism of projection. The individual has learned to expect that when he shows hostility towards others they, in turn, will threaten or become angry toward him.

Unconscious feelings may be expressed in indirect ways. (Photograph by Tor Eigland.)

It is extremely difficult to get a person who has become afraid to express his feelings to learn that the expression of hostility in some measure is appropriate. It serves to let other people know and understand how their behavior is offensive. This allows for an impasse in interpersonal relations to be overcome and for an opportunity for growth. It is only when the individual is able to stand up for his emotional rights and express his feelings that others are able to react toward him appropriately and decide to maintain or change their attitudes and behavior. Moreover, it is only then that the individual's hostility can be subjected to reality testing. When the individual is able to express his feelings he is also able to experience and recognize them, and consequently to make use of his total faculties for judging whether they are appropriate or inappropriate under any given set of circumstances. When he is aware of his hostility he is able to cope with it rationally or take steps to remove or change the factors upon which it is based (e.g., excessive dependence, rigid standards, lack of real skill).

The following excerpt from Dollard and Miller (1950) illustrates the increased control which awareness, or being able to tact emotional experiences, provides for the individual. In this case a young man who had had a good deal of psychotherapy reports as follows:

> He had invited a girl whom he particularly liked but did not know very well to a formal dance. The girl had declined the invitation but said she would be glad to go to dinner before the dance. Trapped by her suggestion and his attraction to her he had agreed to take her to dinner. The boy found the dinner rough going. He was immediately attacked by a stomach ache so severe that he thought he would be forced to leave the table. He was thus urgently motivated to understand the situation. . . . He recalled that aggression against a woman frequently took gastric form in his case. Could it be that he was angry at the girl, and if so, why? He realized immediately that he *was* angry and had repressed his anger. He had felt exploited by her suggesting dinner when she could not go with him to the dance. He was just being used to fill in a chink of time before the dance. When these thoughts occurred, ones contradictory to them came up also. The girl did not seem like an exploitive type. Maybe she wanted to show that she really liked him. There would, after all, be another dance, so why not ask her then and there for another date. This he did and she accepted with evident pleasure. The combination of this lack of cause for aggression and hope for the future brought relief; the stomach ache disappeared. [Dollard and Miller, 1950, p. 443]

In this example it was a stomach ache which provided the cue to the individual that he was angry. Although physical complaints of this sort sometimes occur, more commonly the neurotic reacts with generalized anxiety. If the hostility itself is widely generalized, periodic depression may occur. The etiology of these reactions involves contingencies of reinforcement in which punishment for anger or resentment has resulted in the cues of the anger emotion itself becoming conditioned aversive stimuli. Hence, anxiety is produced. When the anxiety is severe enough, it tends to out-compete the anger response itself.

Just as, through learning, anger and resentment can become response-produced cues for anxiety, the response of anxiety can serve as a cue for assertive responses, or for anger and aggression. An individual may learn, perhaps by accident, that reacting in an anxiety-producing situation with an assertive response inhibits anxiety. The fact that assertive responses may, indeed, inhibit anxiety has been shown experimentally for both animals and humans (Wolpe, 1958; Wolpe, Salter, & Reyna, 1964). In fact, in some cases it is desirable for therapeutic reasons to teach people to condition themselves to make assertive responses in anxiety-producing situations.

Anger and anxiety involve many of the same respondents. Instead of a positive, assertive response becoming conditioned to cues of anxiety an anger response may occur. Although anger partially inhibits the experience of anxiety it intensifies emotionality—the physiological response pattern of anger. The anger response pattern under these circumstances may also contain varying degrees of the anxiety pattern. People who have learned to react with anger to the response-produced cues of anxiety are likely to criticize, belittle, and otherwise display hostility in interpersonal relations which contain for them an undercurrent of threat; or in response to verbal and ideational cues eliciting anxiety. Institutions, organizations, and generalized authority ("they") may have cue value for anxiety-hostility respondents.

Reinforcement may also occur for acting aggressively in anxiety-producing situations because the aggression serves to remove the source of annoyance. Anger and aggression may be instrumental in removing the feared stimulus.

The Reinforcement of Aggression

We have seen that aggression may be reinforced through modeling in the laboratory. Other data indicate that a modeling effect occurs under natural circumstances. Bandura (1960), for example, found that parents of inhibited children were generally more controlled and inhibited than parents of aggressive children. The latter were relatively more expressive and sometimes even impulsive. Other data tend to support this (e.g., Levin & Baldwin, 1959).

Parents can also reinforce aggression directly. Davis (1943) and Davis and Havighurst (1947) have reported that lower-class parents reinforce aggression more than middle-class parents by tolerating and showing approval of aggressive actions and imposing fewer frustrations on their children's impulsive behavior.

Experimental results indicate that the intermittent reinforcement of aggression produces effects similar to those produced by the intermittent reinforcement of other types of behavior. Walters and Brown (1963) have shown that intermittent schedules of reinforcement for aggression result in more persistent aggressive responding during extinction than continuous reinforcement. They have also provided evidence that intermittently reinforced aggressive responses are not only more persistent but tend to generalize more to new situations. Children reinforced on a 1:6 fixed-ratio schedule for hitting a doll, compared to children continuously reinforced, showed more aggression in competitive physical contact games during a later free-play session.

The reinforcement of aggression may be verbally mediated. Lovaas (1961) showed children two dolls, a clean doll and a dirty doll. Half of the children were reinforced for making "good doll" verbal responses and the other half were reinforced for making "bad doll" verbal responses. It was found that children who had been rewarded for verbal aggression displayed a significantly greater amount of nonverbal aggression during a test period than children who had been rewarded for making nonaggressive verbal responses. Patterson, Ludwig, and Sonoda (1961) reported that children who had been given generalized verbal reinforcement for hitting an inflated plastic doll hit the doll more frequently in a subsequent session compared to a control group. Generalized verbal reinforcement for aggression may have a strong effect on the child (or the adult). The same person who has learned to inhibit and suppress hostility for fear of being rejected for expressing his feelings might be enthusiastically aggressive and destructive when approval is available for such behavior.

Much of the intricate learning of when to become angry and what to do when one is angry is mediated verbally by significant people in the individual's life. Middle-class children are taught, primarily by verbal conditioning, to inhibit aggression generally, but to display aggression in competitive sports, military combat, and other special circumstances when the provocation warrants it. An excerpt from a case history of a normal individual by White (1961) illustrates how aggression can be verbally conditioned. A young Harvard undergraduate reported the following incident in which his older sister, Connie, taught him a lesson:

> I was thin and small when young. One day when I was about eight, I guess, I was playing with my cousin, a girl half a year older than I. One of the bullies of the neighborhood came over and began making unpleasant remarks about what he was going to do to me. Betty (my cousin) stuck up for me and held him off until I could make my escape. I ran home as fast as I could, Mother wasn't home, but Connie was. I told her the story, expecting sympathy. To my utter amazement she accused me of cowardice and shamed me for hiding "behind a woman's skirts." She told me if I had any courage I'd go back there and assert my rights. This I did, I must admit with fear and trembling inside of me, and when I called his bluff with no changes in my anatomy occurring I found that my terror of him vanished; and from that time on I never again feared him and never ran from fright in a fight. [White, 1961, pp. 31–32]

In instances such as this the pairing of certain words such as *cowardly* and *fleeing from a bully* may have a very strong conditioning effect. Consider what this individual would have learned relative to the expression of aggression had no significant person been present to administer such verbally mediated reinforcement. The example also underscores the enormous potency and flexibility which verbal discriminative functions may have.

DEPENDENCY REINFORCEMENT

Dependency involves a broad class of behavior directed toward the elicitation of ressurance, help, and attention from others. Dependency may also be directed toward physical gratification, as when the child seeks the physical warmth, protectiveness, and cuddling of the mother, or other signs of physical affection. Physical proximity, contact, and manifestations of affection are also reinforced, and therefore sought after, in the adult marital situation. In this case it is permitted, in fact expected, by the culture. Dependency behavior may be conceptualized as either person-oriented or task-oriented. In the former case the behavior involves a history of reinforcement for seeking emotional, symbolic, and nurturant support from others; whereas in the latter case it involves help in problem-solving and coping with objective situations.

Dependency behavior generally is taught, and sustained by parental solicitousness, nurturance, mothering, and other behavior directed toward its reinforcement. There is a classic experiment by Heathers (1953) in which 6- to 12-year-old children were blindfolded and requested to walk along a narrow plank, raised 8 inches from the floor. At the beginning of the walk the experimenter touched the back of the child's hand and waited for the child to accept or reject help. Analysis of parent profiles in the form of rating scales of various training procedures indicated that the children of those parents who had encouraged their children to depend upon others, rather than encouraging them to care for themselves were the ones who sought help in the experimental situation.

A great many other studies have indicated that the incidence of children's dependency behavior is related to parental patterns of reward and punishment for dependency responses in children when such responses were verbally approved, and a decrease in dependency behavior when they were mildly verbally punished. Adult nurturance, demonstrativeness, and warmth has frequently been shown to be related to the dependency of children. Sears, Maccoby, and Levin (1957) reported that the mothers of very dependent children indulged their dependent behavior and were generally affectionate. Bandura and Walters (1959) found that the parents of nonaggressive but dependent adolescent boys were more demonstrative and affectionate than those of aggressive boys who displayed a relative degree of independence.

Levy (1943) similarly reported that mothers who were oversolicitous and extremely indulgent in caring for their children had children who were very strongly dependent. These children were the victims of maternal overprotection. It would be expected that such children would develop into dependent adults who would be handicapped not only in facing the world independently, but also in their interpersonal relations. The following is a brief case history taken from Levy in which the overprotective behavior of the mother of an 8-year-old boy is described in some detail:

The mother had excessive contact with the child. She breast fed him much longer than was common (13 months). She tried not to leave him alone for an instant, and worried if he ever got out of her sight. She sleeps with him at night when he calls her even though he is now eight years of age. When walking in the street she holds his hand. The mother dresses him every day and takes him to and from the school, and arranges for special lunches. She refuses to let him help in housework for fear he might hurt himself.

The boy has only one friend to whom his mother takes him every two weeks. The mother always gives in to him and does everything for him. She in turn is dominated by him. When mad the boy strikes his mother and sometimes spits on her.

The individual is most likely to become what significant others have taught him. In a sense, it is not his habits which mediate his behavior, but theirs. In this case the operation of a crf reinforcement schedule is apparent. Notice the overemotional reaction of the child to presumed slight provocation from the mother.

The Lack of Parental Nurturance

In infancy and early childhood parental care, physical contact, and affection appear to be a necessity. One psychologist (Ribble, 1943) has argued that lack of adequate mothering is a causal factor in the development of infantile marasmus, a rare disease in which the infant literally wastes away and dies. Spitz (1945) has claimed that the development of infantile depression may be directly related to mother-separation. Harlow (1958) has demonstrated that "contact-comfort" is an important factor in the normal development of infant monkeys. He observed that monkeys separated from their mothers several hours after birth exhibited a strong attachment to folded gauze diapers in much the same way that human infants may exhibit an attachment for soft blankets, pillows, and stuffed animal toys. He constructed two mother surrogates, one of which was made of wood covered with a layer of sponge rubber and a layer of terry cloth. The other was made of wire mesh. The infant monkeys were fed by means of a bottle inserted in the chest region of the crude model of a mother. Both were warmed by radiant heat. If contact-comfort were required then infant monkeys were predicted to prefer the cloth mother, whereas if hugging and affectional responses depended upon conditioned reinforcement then either the cloth mother or the wire mother could serve as conditioned reinforcing stimuli. The results of Harlow's experiment clearly indicated that infant monkeys prefer the cloth mother.

Harlow discovered that infant monkeys found security and psychological relief from fearful stimuli by clinging to the cloth mother, and that this reinforcement allowed the monkey to gather enough courage to explore the novel stimulus. Harlow's data also indicated that infants who lacked support of a mother did not learn to explore their environment.

In extreme cases of nonsupport they exhibited signs of fear and assumed a fetus-like posture and rocked back and forth—a response indicative of severe emotional discomfort which is sometimes found in soldiers under the stress of combat. Thus, at least a modicum of early physical contact appears to be essential, and may be related to behavior other than dependency.

A lack of social responsiveness has been evidenced by children deprived of early mothering. Rheingold (1956) mothered and took care of institutionalized children for an 8-week period. Those children who had received such care-taking were found to be more socially responsive following the experiment than control children. Goldfarb (1943) has similarly reported that institution-raised children are severely handicapped in their ability to relate to others on an emotional and loving basis by the time they reach adolescence.

Firstborn children show a greater need of social companionship, especially under stress, than do second-born and later-born children (Schachter, 1959). A tendency for firstborns to join social organizations more often than later-born children has also been found. It may be that firstborns tend to receive more attention from their parents, especially when they are in trouble, so that other people tend to take on greater conditioned reinforcing value than they do for later-borns.

The Punishment of Dependency

The punishment of dependency has varying effects, depending upon the severity of punishment and the atmosphere or emotional climate in which punishment occurs. It will be recalled from our discussion of frustration and intermittent reinforcement that data by Sears, Whiting, Nowlis, and Sears (1953) suggest a curvilinear relationship between maternal frustration and punishment of dependency and the incidence of dependency in preschool children (see p. 209). Since even when dependency is punished generally it may be rewarded from time to time, intermittent schedules of reinforcement are likely to be operative. Also, since dependency is reinforced during infancy with very strong sources of gratification, the child's efforts to obtain these many gratifications is especially likely to be intensified as the parents begin independence training.

Parental rejection usually involves the punishment of dependency behavior. When dependency is intermittently reinforced by the rejecting parent an intensification of the behavior would be predicted to occur. However, under severe forms of rejection it would be expected that dependency behavior would be inhibited. This appears to be the case. Bandura and Walters (1959), for example, found that aggressive boys whose parents were severely rejecting behaved much less dependently than more accepted nonaggressive boys. Boys who felt rejected by their parents were also found to display a low incidence of dependency toward parents, teachers, and their peers.

Some parents behave neurotically toward their children. They strongly reward and actively punish dependency under the same circumstances. The inconsistent use of rewards and punishment is likely to develop strong dependency behavior. This has been shown experimentally by Fisher (1955), who compared the social responses of two groups of puppies. One group was affectionately handled by the experimenter whenever they approached him. The second group was given the same kind of treatment but in addition they were handled roughly and given an occasional electric shock for approach responses. After 13 weeks of training, tests of dependency indicated that puppies which had received both punishment and reward showed a greater degree of dependency (remained close to humans) than puppies who had been only rewarded.

Evaluative Dependency

Many individuals in our culture depend upon the positive evaluation of other people for their self-esteem. Actually, many people with a high need for approval appear to be well adjusted. They tend to place themselves in a good light on personality inventories, to hold the appropriate attitudes, to be conventional, not to display hostility; in short, they reflect many of the values of the American middle class. Closer analysis, however, indicates that their personal satisfactions may be meager and that when the approval motive is extreme, considerable defensiveness, conflict, and continued unsatisfactory solutions of emotional problems may result.

Crowne and Marlowe (1964) developed a scale for measuring evaluative dependency. In their book, *The Approval Motive*, they have detailed the procedure and results of a great variety of laboratory experiments employing students who score at the extremes of their scale. The data indicate the following characteristics for high-approval-dependent subjects: (a) When subjected to a dull and uninteresting task these subjects are likely to claim that it was enjoyable, presumably for fear of the disapproval of the experimenter. (b) In verbal conditioning experiments they conditioned faster than others. They also show vicarious conditioning. (c) Several experiments indicated that they are open to social influence, suggestibility, and that they use conventional norms as a frame of reference. (d) They tend to show perceptual defense. (e) They tend to suppress and not recognize their own feelings. For example, when angered experimentally they appeared unable to express the anger directly, but readily imitated a model who behaved euphorically. (f) In other experiments they were shown to have a low level of aspiration in a competitive situation. (g) Their projective test responses tend to be defensive, unrevealing, and unproductive. (h) Evaluative-dependent people in psychotherapy tend to leave early (presumably a defensive measure). (i) Finally, other people actually have a tendency not to like them, although they seek affiliation. Moreover, awareness of how others evaluate them appears to be tangential, emerging only in fantasy, as measured by projective tests.

Approval-dependent people, then, show a syndrome of traits: they show favorable attitudes in realistically unfavorable situations; they lack a frank acknowledgment of their feelings and appropriate assertiveness; in competitive situations they tend to anticipate failure; and they tend to justify their behavior and validate their worth defensively.

The fact has been well established that anxiety leads to inhibitory behavior and an increase in behavior with an avoidance history. It has also been established that anxiety results in a decrease in positively reinforced, spontaneous behavior and courage, or behavior which is reality-confronting. Individual growth and positive development involves the ability to be somewhat autonomous and self-directing. But people who have been anxiety-conditioned, including those with an extreme degree of evaluative dependency, tend to be "alienated from their real selves"—unable to obtain a high degree of personal security and productive capabilities. One of the patterns which emerges from anxiety conditioning and consequent self-alienation is the evaluative dependency pattern. It may be expressed in authoritarianism, a pattern in which the individual feels lost unless he is subordinate to strong authority, or it may be expressed by a loss of the sense of being self-directing and by striving to do what others do in a given situation.

When the individual has a dire need for approval and dependency he is forced into situations which promote suppression and defensiveness. He must do nothing in order to incur any displeasure on the part of those upon whom he is dependent. Interpersonal dependency and mutual cooperation are desirable as long as they do not necessitate undue suppression. However, when the individual feels that unless he is approved and admired he is not worthwhile then unhealthy results may follow. Success means continual striving for approval, a goal which is unrealistic and usually impossible to attain for any length of time. Therefore, feelings of inadequacy and depression, along with suppressed hostility, are likely to result. Sometimes the major factor responsible for the maintenance of growth-interfering relations with others—relations in which the individual may submit to continual humiliations and frustration—is the belief that he is unable to obtain satisfactions elsewhere.

SEXUAL BEHAVIOR REINFORCEMENT

The evidence is that in higher animals hormonal and unconditioned environmental stimuli decrease in importance as controllers of sexual behavior and sexual responsiveness. In man sexual excitement can readily be conditioned as a response to previously neutral cues (e.g., Beach, 1947).

Through the socialization process the individual is taught both how and when to behave sexually. The learning of appropriate sexual control and of the appropriate expression of sexuality is another extremely difficult task imposed upon our children. The difficulty is partly due to the very strong sources of reward inherent in the natural expression of sexuality; it is partly due to the very long developmental period during which our society demands that the expression of sexuality be restrained; and partly due to the fact that the mores relative to sexual behavior are not always clearly delineated, are subject to some degree of change, and even to some degree of conflict.

Relative to this latter point Dollard and Miller assert the following:

> Despite recent trends of tolerance and enlightenment, our society is far from having a truly positive attitude towards sexuality. We have not learned how to tame sex during the formative years while preserving its full force for expression during years of marriage. The surviving climate is one of disapproval and nervous neglect. The expression of sex in conventional ways wins a certain reluctant assent but the person who is fearful because of early training is likely to perceive the reluctance and secrecy surrounding sex as signs of hostility toward it. Neurotic persons, alert for punishment, cannot discriminate between this social coolness and real danger. They find it hard to believe that the culture will tolerate behavior which it will not openly encourage. [Dollard and Miller, 1950, p. 347]

Because sexual excitement is readily conditionable the sources of reinforcement associated with sexuality are manifold. The climax, or the orgasm, is normally intensely pleasurable. Although there may be minor variations in the pleasure experienced from sexual excitement and climactic release during the individual's life, at least the possibility for strong physical reward exists from the time of early rudimentary sexual awakening in the young child until late in life. Newman and Nichols (1960) have reported that people in their eighties and nineties may maintain sexual activity and experience sexual pleasure.

Aside from physical pleasure as a source of reinforcement, through conditioning the sexual act may become associated with love and affection, esteem, and a great variety of specific and generalized positive conditioned reinforcers. In short, the sexual act may become a symbol of love, a ritual cementing the smallest social unit (husband and wife) together. In fact, the development and expression of sexual behavior is so closely related to the learning of loving and affectionate behavior that the separation of the two topics must be, to some extent, artificial. Sexuality is also obviously related to other personal and social factors such as the conception of children, economic, ideological, and moral considerations.

Under some conditions sexual behavior may be associated with fear-reduction, rather than affection and love. Sexual behavior may serve as an escape, or as a first-line defense against anxiety, much in the same way as does excessive reading, television viewing, or movie going. Sexual conquest may serve to assure the individual that he or she is adequate, attractive, or desirable. Again, the fearful individual may learn to gain some measure of attention and affection, however meager, by behaving sexually.

Generally, sexual behavior associated with various aspects of fear-reduction is less acceptable, less mature, less integrative, and less desirable than sexuality associated with affection, tenderness, respect, and other positive sources of pleasure. Unfortunately healthy sexual behavior appears to be relatively rare in our society. The reasons for this will become clearer when we have considered the factors which may operate to adversely influence the learning of sexual behavior and expression.

When sexual behavior functions as a trade for affection, as a means to bolster the individual's esteem-income, or as an escape from emptiness, boredom, and other thwarting circumstances of life it is primarily immature and undesirable. Security and anxiety-reduction associated with sexual behavior under such circumstances are likely to be superficial and short-lived because the underlying feelings of worthlessness and inadequacy from which the sexual behavior serves as an escape are not essentially diminished. In our culture sexual expression can be associated with considerable anxiety, guilt, and disgust and, consequently, with mild forms of frigidity in the female and impotence in the male. Although various forms of sexual behavior are counter to the cultural mores (masturbation, premarital intercourse, homosexuality, etc.), they may serve as compensatory or defensive measures against dissatisfactions and frustrating experiences in the individual's life, but they may also serve to increase underlying feelings of guilt, inferiority, and self-hate. Increased sexual activity and other mechanisms of escape and avoidance may result. The real satisfactions are likely to become more and more meager. Hence, a vicious circle may be created.

The Freudian Stages of Psychosexual Development

One can hardly discuss human sexual behavior without reference to Freudian psychology and psychoanalysis. Freud felt that adherence to the sexual morality of his time contributed to "nervousness" and devoted much of his professional life to creating an elaborate theoretical system of complexly related concepts, many of which, unfortunately, were formulated in a manner which rendered them incapable of being tested. Aside from his theory, which is decidedly unscientific (primarily because of the way in which it was formulated), many of his observations of infantile and pathological manifestations of sexuality appear to have been astute, valid, and repeatable by anyone who is willing to make the same observations with some measure of candor and objectivity. However, psychologists overwhelmingly hold that learning theory and the established principles of learning and conditioning better account for the complex facts of sexual behavior.

In brief, Freud postulated stages of psychosexual development through which the individual normally progresses, unless severe fixation occurs at any given stage. The progressive stages are the oral, anal, phallic, latency, and genital periods.

During the *oral* stage pleasure is derived from sensual experiences associated with the oral cavity: sucking, swallowing, biting, etc. This "sexual" pleasure may, in the adult, be fused with pleasures from other stages and is postulated to take highly generalized forms, such as preoccupation with traveling ("drinking in the scenery"), extreme sarcasm ("biting"), etc.

During the *anal* stage of development, pleasure associated with elimination and bladder tensions become salient for the child. In combination with the learning resulting from toilet training, generalized carryovers from this stage of development are postulated to result in certain response patterns in the adult of an anal character, e.g., sadism, stubborness, excessive neatness, and punctuality.

During the *phallic* stage, which occurs from 3 to 5 years of age, the child becomes preoccupied with exploring his genitalia and with a sexual attachment for the opposite-sexed parent and a rejection of the like-sexed parent. This results in the Oedipus complex, or conflict (see below).

Following the phallic stage a *latency* period occurs up to the age of 12 or 13 during which, because of repression, the child has little interest in the forms of sexual pleasure learned in the earlier stages. However, with the onset of puberty the individual normally progresses to the final, genital stage of adult sexuality. Although the genital stage represents the end of the normal progression, not everyone is able to attain complete adult genital sexuality. In this stage the primary source of pleasure is associated with complete heterosexual behavior and with an integration of sexual pleasure with other values, especially tenderness, affection, and love. The genital character is able to love and work effectively. It is particularly characteristic of this stage that there is little need for repression because of guilt or anxiety. The individual has little unresolved sexual conflict, and is able to reconcile sex with tenderness and esteem.

The Freudian theory of sexual development is extremely complex. The term sexuality is so inclusive as to be applicable to an inordinate variety of life situations. Thus the classical analyst may see a relationship between a man's toilet training and his attainment of wealth, or between his love for big game hunting and his large family.

Learning theorists have attempted to reinterpret certain aspects of the Freudian theory and to account for the learning of sexual behavior and its manifestations in a manner consistent with responsible observation and scientific validity. It appears that infantile sexuality does, in fact, exist, although because of their own anxieties regarding sexuality adults may minimize or ignore the many instances in which it occurs. Candid observation reveals that young children do show strong sexual curiosity: they may inspect each other's bodies when the opportunity arises; learn to masturbate; carry on sex play with siblings; and display a decided preference for the opposite-sexed parent.

The normal middle-class adult may be horrified by episodes which are viewed by the psychoanalyst as being significantly related to the child's psychosexual development. For example, the very young child may bring his feces to the adult as a present. In fact, it is *his* production, something he can be creatively proud of. The reader can imagine the uneasiness and disgust with which the average middle-class adult might react in such a circumstance. The attitudes of the parents, cultural practices, and specific contingencies of reinforcement and punishment result in development of the child's, and later the adult's, behavior and emotional reactions toward sexual and other physiologically based tensions.

Masturbation

The evidence is that masturbation is nearly universal in our culture, but it is counter to the social mores. When very young children are discovered exploring and manipulating their genitalia parents frequently become anxious and may even punish the child severely. At times the infant is merely in the process of discovering its own body and there is little grounds for alarm. When the child is severely and inexplicably punished considerable anxiety relative to sexual excitement and his sexual behavior (masturbation) may be generated. Many adolescents in our culture, and even adults, return to masturbation as a source of satisfaction. Unfortunately, few are able to do so without feelings of conflict, misery, and guilt. Others can maintain self-esteem only if they abstain.

Unless infantile masturbation is excessive, parents would be well advised to avoid making an emotional issue out of it. During the child's development excessive masturbation may occur as a compensatory mechanism, or a defense against anxiety and may be a symptom of the child's inability to get satisfactions from relationships with the family and others. Masturbation may be indicative of the immaturity of the child or it may evidence an unsatisfactory relationship between the parents and the child. If it is taken as a sign of degeneracy or worthlessness the parents may avoid facing the real cause and the child may be irrationally punished.

Generally the later expression of appropriate sexual control and sexual behavior is contingent upon a background which is relatively free of fears and guilt regarding sex. This can be promoted if the parents are able to show kindness, understanding, and to give explanations which the child can understand when the child deviates from their own sexual ethics. The child need not to be threatened with untruthful stories about the results of masturbation. If the relationship between the parents and the child is healthy they need only indicate that they would like him to stop. The child will eventually acquire the standards and ideals the parents would like him to have.

The Oedipal Situation

It is the ideal case when the parents are able to display loving behavior, be a rich source of satisfaction, and represent models which the child derives considerable satisfaction from imitating. More commonly, parents punish their children's infantile masturbation and sexual play with siblings and other children. This results in a sex-fear conflict for the child. Dollard and Miller point out that this early learning, in conjunction with sex-typing, sets the stage for what the psychoanalyst terms the Oedipus complex. Sex-typing begins with male and female names, toys, clothes, patterns of play, and continues into later life where specialized sexual roles for men and women are defined. Much of this learning is, of course, verbally mediated. It is also probable that parents tend to favor the child of the opposite sex. According to Dollard and Miller:

> The ultimate love object of the child is defined as a member of the opposite sex. The nascent sexual reactions of the child are directed towards stimuli of the opposite sex. The child is led to expect eventual sex rewards from persons of the opposite sex. [Dollard and Miller, 1950, p. 143]

In the Oedipus situation the little boy begins to fantasy possession of the mother and to exclude the father from her affection and attention. Similarly, the little girl wants to be the exclusive recipient of daddy's attentions and affections. The matter is further complicated by the fact that each child has some positive feelings toward the rejected parent, and also by the fact that each child fears some kind of retaliation from the rejected parent. The classic Oedipus complex for the male child has been described by the learning psychologists Dollard and Miller as follows:

The anxiety which adolescents, and often adults, show at the prospect of heterosexual contact must be explained. It does not arise by chance. It arises rather in the family situation which is the child's most important early learning situation . . . the boy child turns to his mother in fact or thought in the hope of getting sex rewards partly by generalization (Miller, 1948; Selsinger, 1948) of expectation of reward—that is, by analogy to the many rewards the mother has already given him—and partly from the fact that by sex-typing he has learned to expect sex rewards from a woman and his, mother is a woman at hand. Doubtless some of the anxiety already learned in connection with masturbation generalizes to the sex impulse when it begins to show itself towards family women.

A new source of anxiety appears, however, that is, fear of the father. The five year old boy knows his father is the head of the house, the symbolic source of punishments and discipline. He also knows that his father is the husband of his mother and has some unique relationship with her. This rivalry of the father does not exist merely in the boy's mind. It is often made very concrete in the father's behavior. The father may complain that the little boy sleeps in the mother's bed when it is "too old" for such behavior. The father may object to the fact that the child or children sop up so much of the mother's time and leave so little to him. The father may impose certain restrictions about entering the parent's room which leaves the child with a mystery on his hands. Whenever the male child makes emotional demands upon the mother, the father may become more critical of him in other more general aspects, saying that the boy talks too much, that he does not work enough, and so forth. If the boy reacts with fear toward his father as a rival, it is because the father, consciously or unconsciously, is acting in a way that seems fearsome and rivalous. The child is usually unable to discriminate between opposition on ground of sexual learnings and that evoked by its other claims on the mother. The whole thing may be played out as a kind of dumb show. The sexual strivings of the boy toward the mother may be behaviorally real and active but not labelled in the boy's mind. On the other hand, the opposition of the father, though active and effective, may be oblique and unconscious. [pp. 144–145]

The normal resolution of the Oedipus conflict appears to be repression, resulting in the latency period. With the onset of puberty a reawakening of sexual interest occurs and, unless the fear, guilt, and anxiety learned in the Oedipal situation are too great, the individual shows an increasing capacity for mature, adult sexuality. If, however, excessive fear and anxiety were conditioned in the early Oedipal situation the individual may show undue inhibition and find the prospect of heterosexual behavior either too frightening to engage in or unpleasurable. In the case of the boy, if the father was excessively threatening, he may not be able to comfortably accept adult authority in later life.

Notice that Dollard and Miller described the Oedipal situation as a "dumb show." Aside from the possibilities of learning that sexual ex-

The oedipal situation. The family drama is largely outside of the awareness of the participants. (Photograph by Tor Eigland.)

citement, thought, and expression are dangerous and frightening on a semiconscious or unconscious basis such as might occur in the Oedipal situation, the individual may be deliberately taught that sex is "undesirable," "dirty," "disgusting," or "immoral" by direct verbal conditioning from the parents throughout his development. External factors combine complexly to determine the learning for any given individual. Thus, the individual raised to have a prudish and inhibited attitude toward sex might be capable of normal sexual expression as an adult if his peers, circle of friends, or perhaps other adults do not share the same repressive attitudes. On the other hand, the child might be exposed only to other children and adults who share the same attitudes of the parents. The complexity of factors which may determine what gets learned becomes apparent when the causes of deviant sexual patterns, particularly homosexuality, are delineated (see below).

Some individuals who suffer from neurotic sexual conflict have been taught that sex is a degrading activity. If the individual has been given somewhat rigid moral training in regard to sex and particularly if he has learned to fear the father, what the Freudians call a "split imago" may result. That is, the individual can feel tenderness and affection for "nice" women, but is unable to feel sexual excitement toward them. On the other hand, the same individual may be easily sexually excited by "low" or "bad" women, but be unable to feel any tenderness or affection for them. In short, since sex is believed to be a degrading activity, and since the individual has been taught to honor and respect "good" women, he becomes incapable of fusing sex with love in marriage. In the extreme case the individual may become impotent.

Affection, Love, and Sex

When sex is integrated with love, tenderness, and affection, it may be an important condition for happiness. Sexual behavior is complicated by the fact that sexual relationships are interpersonal relationships involving the transfer of affectional habits and loving behavior. Appropriate and adequate sex appears to allow the individual to function better in his work and in his relationships with others. This is because appropriate sexual behavior occurs in an atmosphere of relaxed and effective love.

Learning to love is difficult because so many conditions must be met before the learning can occur and because it involves the acquisition of a complex of skills. Thus, loving behavior appears to be an outgrowth of having been loved by the parents and by significant others. Under these conditions the individual learns to be concerned with his own growth and happiness and to implement these values. He acquires actual practice in respecting, caring for, and knowing himself. Such an individual is not likely to have developed unreasonable expectations and demands of himself. He is likely to have a measure of inner security and freedom from anxiety. Since he is not all wrapped up in himself, he can devote some time, energy, and skill to the loved object. Moreover, freed from the necessity to interpret others in terms of his own needs he is able to acquire an accurate knowledge of the needs, feelings, and causes of the behavior of the loved object.

Aside from the skills involved in knowing, caring for, and promoting the growth and happiness of the object the individual must have acquired a sense of the value of loving others. If he has acquired a personal sense of worth he is predisposed, as it were, to realizing the importance of achieving a sense of respect, worth, and security for others. Lacking a dire need for approval himself and being unafraid and unashamed to convey his deepest feelings, needs, and desires to others, the individual is able to bring emotional freedom and reinforcing power to the relationship. It is in this context that the richest sex satisfaction can occur.

Loving behavior at times requires the ability to tolerate deprivation of one's own needs in the process of devoting oneself to the needs of the loved object. It also requires an ability to accept in turn the concern, understanding, and affection of the loved object. It appears that practically everyone has a need to be loved, but if early experiences in interpersonal relationships have been fraught with punishment, hurt, and frustration the individual may not have conscious longings for love and may be incapable of active love. Such early learning may produce the so-called psychopathic personality. This generally refers to a syndrome involving antisocial behavior, egocentricity, callousness, insincerity, shallowness of affect, lack of conscience, extreme sensitivity, pity-seeking, and related traits. Usually a human relationship of sincere involvement can be found in the background of even the most hardened psychopath, but other crippling experiences have precluded normal development.

Less severe than psychopathic development, many individuals are simply inhibited and aloof in interpersonal relations. They find involvement aversive and unpleasant. On the other hand, in most neurotics there tends to be what Levy (1937) termed a "primary affect hunger," a longing for affectionate relationships which must be repressed. Love–sex relationships for these individuals are likely to involve passivity and dependence, rather than mutual active love.

Persons who have very strict consciences or who have come to fear sex in general are likely to repress sexual feelings, deny any sexual intent, and lack conscious sexual wishes and desires. Just as in the case of the repression of other feelings, sexual desire is likely to become manifested in unpredictable and uncontrollable ways. The individual may show a prudish and condemning attitude toward any open expression of sexuality. His own behavior may show unwitting overtones of an interest or preoccupation with sex. He may become anxious or guilty without knowing why or experience impulses and inappropriate sexual episodes which serve to reinforce fright and defense.

In contrast, the individual who has been appropriately sex-trained does not feel threatened by his own feelings and does not need to deny their existence. He has learned, however, that sexual behavior is appropriate and satisfying only when certain interpersonal, ethical, and perhaps esthetic conditions are met. He may be aware of his own sexual feelings even when the conditions are inappropriate, but he is capable of controlling himself because he has learned that sexual intercourse, when the conditions are inappropriate, is not fundamentally satisfying. Hence, he is not driven to immediate release at the jeopardy of many important values.

Sexual Role Inversion

For some individuals life experiences combine to produce a homosexual, rather than a heterosexual adjustment. Since in our society homosexuality is taboo and engaging in sexual relationships with members of the same sex is subject to fine, imprisonment, social derision, perhaps loss of job, and other threats, individuals who practice homosexuality can be thought of as unrealistic and self-defeating. Because of the strong direct sexual satisfaction which homosexuals receive and because other neurotic gains may be associated with the practice of homosexuality many clinicians regard the treatment of homosexuals, and sex deviates generally, as especially difficult (e.g., Buckle, 1949; Gurvitz, 1957; Rickles, 1950; Wolfenden, 1957; etc.). Some psychologists, such as Ellis (1965) seem to feel that the majority of homosexual individuals are severely disturbed—an additional reason for a poor prognosis. According to Ellis:

> Although it is theoretically possible for homosexuals to be accidentally conditioned to a strong attachment to members of their own sex and then merely to maintain this kind of conditioning, while leading undisturbed lives in other respects, I have not actually found such an individual among the scores of sex deviates I have seen for psychotherapy . . . In every case I have seen, irrational fear played the leading role either in inducing the individual to become homosexual in the first place or inducing him to maintain his early acquired homophilic conditioning in the second place. The main fears homosexuals have include fear of rejection by members of the other sex, fear of heterosexual impotence, fear of intense emotional involvement, fear of marital responsibilities, fear of competition, among others. Basically, they are terribly afraid of failure and of what other people think of them if they do fail in any way; and they consequently withdraw from heterosexual relations which (peculiarly enough!) are relatively difficult to initiate and maintain in our society, into the "gay life," where sexual conquest, at least, is much safer. [Ellis, 1965, p. 80]

These comments are not meant as an indictment against the homosexual, but rather as a considered interpretation of homosexuality. They underscore the fact that sexual behavior cannot be divorced from other behavior with which the individual has learned to cope in interpersonal relationships and with life circumstances.

An elaborate discussion of the etiology and manifestations of homosexuality is beyond the scope of this text. However, it is instructive to consider briefly some of the many factors that may operate in the individual's life to condition a homosexual orientation. Thus, the learning of homosexuality may center around an unhealthy solution to the Oedipal situation. A strong fear of competing with the same-sexed parent may function as a basic predisposition. The individual may magically identify with the same-sexed parent in attempts to gain favor and admire members of his own sex whom he feels are much worthier and heroic. He may feel so inept that he believes he cannot live adequately without a strong protector who will care for him and run his life for him.

Another predisposing factor appears to be identification with a dominant opposite-sexed parent. We have seen that the dominant parent tends to be imitated and identified with. Wife-dominant families are not uncommon in our middle-class society. The mass media of communication often foster the image of daddy as inept, bungling, and passive. Daddy is a "jerk"—certainly not someone who can command respect and serve as a model to be imitated. The boy may therefore find his behavior and attitudes modeled after his mother and other females, both within and outside the family. In other instances, extreme anger and rebellion against either or both parents may predispose the individual to degrade himself and them through the practice of homosexuality. This may be especially true when the resentment and anger have been deeply repressed.

Many other learning experiences may serve as predisposing factors. The individual may be raised almost exclusively with members of the opposite sex, or with his own sex so that he learns their tastes, interests, and emotional patterns. He may retreat from heterosexual relations because he finds them unsatisfying or frightening. Fear of failure, shyness, and withdrawal may prevent him from making overtures to members of the opposite sex and he may consequently fail to gain any skill and experience; or seduction by a member of his own sex may provide the first satisfying experience for him.

Ellis points out that the anxious and hostile individual may find himself rejected by members of the opposite sex, and may be accepted by severely disturbed homosexuals who will make allowance for the hostility of others. Hostile individuals may find they can victimize others more easily in the disturbed homosexual world. Other neurotic trends may predispose the individual to becoming homosexual: a dire need to be loved; inability to reject prudish sex attitudes as antihuman and to enjoy normal sexuality; a refusal to face adult responsibilities and a desire to maintain childish grandiosity, rebelliousness, immediate gratification, etc.

On top of the predisposing factors given above, the individual may chance to have childhood or adolescent homosexual experiences. Although, except in large cities, he rarely has adult models whom he can imitate directly, his curiosity may be stimulated by articles and discussions about homosexuality in the mass media of communication, by the mimicry of homosexual behavior in the entertainment world, and by jokes and stories. Individuals who have learned to withdraw, and those who are isolated from normal human relationships because of economic, social, and family circumstances, as well as those who are lonely and unable to cope with the frustrations of their lives, are likely to seek out homosexual relationships. They may pursue homosexuality as excitement, as a pathetic gesture of establishing sound human relationships, as protection, as a rebellious gesture, and for related reasons. The individual need not appear disturbed in his normal relationships with others. Many disturbed individuals develop a facade of normalcy and conventionality, yet they are actually lonely and frightened amidst ostensibly normal relationships with people.

Once homosexuality has been learned some measure of affection and some experience in the learning of loving behavior, however meager and confounded with neurotic needs, may be open to the individual. Probably a psychotic break with reality is prevented in some cases and in others a modicum of security may be gained. There may also be secondary gains in that the individual can avoid the expenses, difficulties, and amenities of courtship; the threat of pregnancy; and deep emotional involvement. The female homosexual may similarly avoid conforming to sexual restrictions: the necessity to continuously act ladylike, dressing up, etc. For some individuals fear has been attached to the idea of normal heterosexual relationships, but they have not been equally warned relative to their own sex. Antiheterosexual ideas, such as the ugliness of sexual organs, disgust about sex odors, and menstrual functions, may serve as minor sources of reinforcement.

Even when given all of the possible sources of reinforcement it appears that, aside from the fact that such behavior is self-defeating and unrealistic, truly satisfying homosexual relationships rarely, if ever, occur. When fear and repression are alleviated through psychotherapy, normal heterosexual adjustment appears to be far more satisfying.

SUMMARY

In this chapter imitation learning and the learning of identification have been discussed, along with behaviors that are based on social learning, and that mediate certain significant aspects of the individual's adjustment. These latter categories of behavior include aggression, dependency, evaluative dependency, and sexual behavior.

Both laboratory and nonlaboratory observations indicate that children will imitate the behavior of adults or of other children. Certain aspects of the individual's experience with others provide cues for the imitation of them. In general, it is not the model's behavior that determines whether or not imitation will occur, but rather the kind and extent of imitation is related to the amount and kind of reinforcement received by the model. Rewarded models tend to be imitated to the extent that their behavior is rewarded. Models who are not rewarded tend not to be imitated.

Older theories of identification stressed the fear of punishment in the identification process. It was also theorized that identification occurs when a nurturant parent begins to withhold reinforcement previously given in abundance. This latter theory has been somewhat substantiated by research. There is a marked tendency for children to identify with parents whom they perceive as nurturant and who behave affectionately toward them. Other experiments have shown that the social power of an adult in the sense of his or her being in control of rewards increases the likelihood that children will use him (or her) as a model for identification. Research has also indicated that children tend to be more similar to the dominant parent than to the passive parent. There is some evidence that girls in our culture are better able to identify with the opposite parent without a disruption of their primary sexual identification than is the case with boys.

The learning of how and when to become angry is one of the most difficult of developmental tasks for children. Countless situations evoke anger and resentment throughout the child's development. If the child's anger is punished too severely and/or indiscriminately he may learn to be afraid of his own feelings of anger. Situations that would normally arouse anger and resentment then come to arouse anxiety, which, in turn, leads to repression. Repression is unfortunate because it is only when the individual is aware of his real feelings that he is able to cope with them rationally, to judge their appropriateness, or to take steps to remove or change the factors upon which they are based.

Some people learn to react with anger when they are made anxious. The response of anxiety is not recognized for what it is, but instead it cues off anger and aggression. Under some circumstances reacting with assertiveness may be integrative. However, insofar as the reaction is unrealistic and insofar as it prevents the person from recognizing his true feelings and the reason for them it may be maladaptive.

Direct verbal instruction accounts for a great deal of the learning of how and when to become angry. Anger and aggression can also be learned imitatively. In general people in our culture are taught to be aggressive under special circumstances of provocation, in military combat, and in competitive sports.

Dependency, behavior directed at the elicitation of reassurance, help, and attention from others, is generally maintained by parental solicitousness and indulgence. Parental rejection and punishment, when it is not very severe and when it occurs intermittently, also appears to produce strong dependency behavior. Experimental animals that have been treated affectionately *and* shocked showed marked dependency compared to those treated affectionately only.

Many people in our culture have a high need for the approval of others, for the maintenance of their self-esteem. This is called approval dependency. They tend to be optimistic, yet lack appropriate assertiveness and the ability to frankly appraise their feelings. Interpersonal dependency and a desire for mutual cooperation are desirable, so long as undue suppression or repression does not result. Otherwise, the individual may become alienated from his true feelings and capabilities.

The learning of appropriate sexual control and appropriate sexual expression is another difficult learning task. The sexual act and sexual behavior can become associated with love, affection, and esteem. On the other hand, under certain conditions it may become associated with anxiety, disgust, and guilt. Freud believed that there were stages of psychosexual development. Unfortunately, the evidence for his theory of psychosexual development is largely equivocal. However, the Oedipus complex proposed by him does seem to occur in some family constellations.

There are many possibilities for the distortion of healthy sexual attitudes and behavior. For many shy, inhibited people love–sex relations involve passivity and dependence, rather than mutual active love. Because of fear and prudishness many are unable to enjoy sexual rewards or integrate them with personal happiness. For some individuals life experiences combine to produce a homosexual, rather than a heterosexual adjustment. There are strong social and legal taboos against this. Some authorities suggest that this form of sexual adjustment is frequently secondary to the fact that the individual is emotionally disturbed.

REFERENCES

Axline, Virginia M. *Dibs: In search of self—personality development in play therapy.* Boston: Houghton, 1964.

Bandura, A. Relationship of family pattern to child behavior disorders. Progr. Rep., U.S.P.H. Res. Grant M-1734. Stanford Univer., 1960.

Bandura, A. Social learning through imitation. In M.R. Jones (Ed.), *Nebraska symposium on motivation.* Lincoln, Nebr.: Univer. of Nebraska Press, 1962. Pp. 211–269.

Bandura, A., & Huston, Altha C. Identification as a process of incidental learning. *J. abnorm. soc. Psychol.*, 1961, **63**, 311–318.

Bandura, A., & Kurpers, Carol J. The transmission of patterns of self-reinforcement modeling. In A. Bandura & R. H. Walters. *Social learning and personality development.* New York: Holt, 1963.

Bandura, A., Ross, Dorothea, & Ross, Sheila A. Transmission of aggression through imitation of aggressive models. *J. abnorm. soc. Psychol.*, 1961, **63**, 575–582.

Bandura, A., Ross, Dorothea, & Ross, Sheila A. Imitation of film-mediated aggressive models. *J. abnorm. soc. Psychol.*, 1963, **66**, 3–11. (a)

Bandura, A., Ross, Dorothea, & Ross, Sheila A. A comparative test of status envy, social power, and the secondary-reinforcement theories of identificatory learning. *J. abnorm. soc. Psychol.*, 1963, **67**, 527–534. (b)

Bandura, A., Ross, Dorothea, & Ross, Sheila A. Vicarious reinforcement and imitative learning. *J. abnorm. soc. Psychol.*, 1963, **67**, 601–607. (c)

Bandura, A., & Walters, R. H. *Adolescent aggression.* New York: Ronald Press, 1959.

Bandura, A., & Walters, R. H. *Social learning and personality development.* New York: Holt, 1963.

Beach, F. A. Evolutionary changes in the physiological control of mating behavior in mammals. *Psychol. Rev.*, 1947, **54**, 297–315.

Buckle, D. The treatment of sex offenders. *Int. J. Sexol.*, 1949, **3**, 1–8.

Crowne, D. P., & Marlow, D. *The approval motive: Studies in evaluative dependencies.* New York: Wiley, 1964.

Davis, A. Child training and social class. In R. G. Barker, J. S. Kounin, & H. F. Wright (Eds.), *Child behavior and development.* New York: McGraw-Hill, 1943. Pp. 605–619.

Davis, A., & Havighurst, R. J. *The father of the man: How your child gets his personality.* Boston: Houghton, 1947.

DeCharms, R., & Rosenbaum, M. E. Status variables and matching behavior. *J. Pers.*, 1960, **28**, 492–502.

Dollard, J., & Miller, N. E. *Personality and psychotherapy: An analysis in terms of learning, thinking, and culture.* New York: McGraw-Hill, 1950.

Ellis, A. *Homosexuality: Its causes and cure.* New York: Lyle Stuart, 1965.

Farina, A. Patterns of role dominance and conflict in parents of schizophrenic patients. *J. abnorm. soc. Psychol.*, 1960, **61**, 31–38.

Fisher, A. E. The effects of differential early treatment on the social and exploratory behavior of puppies. Unpublished doctoral dissertation, Penn. State Univer., 1955.

French, J. R. P., & Raven, B. The bases of power. In D. Cartwright (Ed.), *Studies in social power.* Ann Arbor, Mich.: Inst. Soc. Res., 1959. Pp. 150–167.

Freud, S. The dynamics of transference. In E. Jones (Ed.), *Collected papers.* Vol II. London: Hogarth, 1924. Pp. 312–322. (First published in 1912, *Zentralblatt*, **II**)

Freud, S. Mourning and melancholia. In E. Jones (Ed.), *Collected papers.* Vol. IV. London: Hogarth, 1925. Pp. 152–170. (First published in 1917, *Zeitschrift, Bd.*, **IV**)

Goldfarb, W. Infant rearing and problem behavior. *Amer. J. Orthopsychiat.*, 1943, **13**, 249–265.

Gurvitz, M. Sex offenders in private practice: Treatment and outcome. Paper read at the Amer. Psychol. Assoc. Conv., Sept., 1957.

Harlow, H. F. The nature of love. *Amer. Psychologist*, 1958, **13**, 673–685.

Heathers, G. Emotional dependence and independence in a physical threat situation. *Child Develpm.*, 1953, **24**, 169–179.

Hetherington, M. E. A developmental study of the effects of sex of the dominant parent on sex-role preference, identification, and imitation in children. *J. Pers. Soc. Psychol.*, 1965, **2**, 188–194.

Levin, H., & Baldwin, A. L. Pride and shame in children. In M. R. Jones (Ed.), *Nebraska symposium on motivation.* Lincoln, Nebr.: Univer. of Nebraska Press, 1959. Pp. 138–173.

Levy, D. M. Primary-affection hunger. *Amer. J. Orthopsychiat.*, 1937, **94**, 643–652.

Levy, D. M. *Maternal overprotection.* New York: Columbia Univer. Press, 1943.

Lovaas, O. I. Interaction between verbal and nonverbal behavior. *Child Develpm.* 1961, **32**, 329–336.

Maccoby, Eleanor E. Role-taking in childhood and its consequences for social learning. *Child Develpm.*, 1959, **30**, 239–252.

McNulty, J. A., & Walters, R. H. Emotional arousal, conflict, and susceptibility to social influence. *Canad. J. Psychol.*, 1962, **16**, 211–220.

Miller, N. E., & Dollard, J. *Social learning and imitation.* New Haven: Yale Univer. Press, 1941.

Mowrer, O. H. *Learning theory and personality dynamics.* New York: Ronald Press, 1950.

Mussen, P. H., & Distler, L. Masculinity, identification, and father-son relationship. *J. abnorm. soc. Psychol.*, 1959, **59**, 350–356.

Newman, G., & Nichols, C. R. Sexual activities and attitudes in older persons. *J. Amer. Med. Assoc.*, 1960, **173**, 33–35.

Osgood, C. D., Suci, G. J., & Tannenbaum, P. H. *The measurement of meaning.* Urbana, Ill.: Univer. of Illinois Press, 1957.

Parsons, T. *The social system.* New York: Free Press, 1951.

Parsons, T. Family structure and the socialization of the child. In T. Parsons and R. F. Bales (Eds.), *Family, socialization, and interaction process.* New York: Free Press, 1955. Pp. 53–131.

Patterson, G. R., Ludwig, M., & Sonoda, Beverly. Reinforcement of aggression in children. Unpublished manuscript, Univer. of Oregon, 1961.

Rheingold, Harriet L. The modification of social responsiveness in institutional babies. *Monogr. Soc. Res. Child Develpm.*, 1956, **21**, No. 2 (Serial No. 63).

Ribble, M. A. *The rights of infants.* New York: Columbia Univer. Press, 1943.

Rickles, N. K. *Exhibitionism.* Philadelphia: Lippincott, 1950.

Rosenbaum, M. E., & Tucker, I. F. The competence of the model and the learning of imitation and non-imitation. *J. exp. Psychol.*, 1962, **63**, 183–190.

Schachter, S. *The psychology of affiliation.* Stanford, Calif.: Stanford Univer. Press, 1959.

Schachter, S., & Singer, J. E. Cognitive, social, and physiological determinants of emotional state. *Psychol. Rev.*, 1962, **69**, 370–399.

Sears, Pauline S. Child-rearing factors related to played sex-typed roles. *Amer. Psychologist*, 1953, **8**, 431. (Abstract)

Sears, R. R. Identification as a form of behavioral development. In D. B. Harris (Ed.), *The concept of development.* Minneapolis: Univer. of Minn. Press, 1957. Pp. 149–161.

Sears, R. R., Maccoby, Eleanor E., & Levin, H. *Patterns of child rearing.* New York: Harper, 1957.

Sears, R. R., Whiting, J. W. M., Nowlis, V., & Sears, Pauline S. Some Child-rearing antecedents of aggression and dependency in young children. *Genet. Psychol. Monogr.*, 1953, **47**, 135–234.

Spitz, R. A. Hospitalism: An inquiry into the genesis of psychiatric conditions in early childhood. *Psychoanal. Study Child*, 1945, **1**, 53–74.

Stotland, E., & Patchen, M. Identification and changes in prejudice and in authoritarianism. *J. abnorm. soc. Psychol.*, 1961, **62**, 265–274.

Walters, R. H., & Brown, M. Studies of reinforcement of aggression. II. Transfer of responses to interpersonal situations. *Child Develpm.*, 1963, Pp. 563–571; also see Bandura and Walters (1963).

Walters, R. H., Marshall, W. E., & Shooter, J. R. Anxiety, isolation, and susceptibility to social influence. *J. Pers.*, 1960, **28**, 518–529.

White, R. W. *Lives in progress: A study of the natural growth of personality.* (2nd ed.) New York: Holt, 1961.

Whiting, J. W. M. Sorcery, sin, and the superego. In M. R. Jones (Ed.), *Nebraska symposium on motivation.* Lincoln, Nebr.: Univer. of Nebraska Press, 1959. Pp. 174–195.

Whiting, J. W. M. Resource mediation and learning by identification. In I. Iscoe and H. W. Stevenson (Eds.), *Personality development in children.* Austin: Univer. of Texas Press, 1960. Pp. 112–126.

Wolfenden, Sir John. *Report of the committee on homosexual offenses and prostitution.* London: H. M. Stationery Office, 1957.

Wolpe, J. *Psychotherapy by reciprocal inhibition.* Stanford, Calif.: Stanford Univer. Press, 1958.

Wolpe, J., Salter, A., & Reyna, L. (Eds.) *The conditioning therapies.* New York: Holt, 1964.

Chapter 9

The Self-Concept and the Learning of Anxiety

INTRODUCTION

In the previous chapter attention was focused on the learning of sexual behavior, aggressive behavior, the expression of anger, and dependency. There are many significant aspects of human adjustment which remain to be considered.

One of the most important factors underlying the quality of the individual's adjustment is the manner with which he learns to cope with anxiety. The great variety of adjustive patterns revealed in the clinic, ranging from relatively mild emotional disturbances to severely disorganized states, have a common denominator: the experience of anxiety.

It is difficult to do justice to the significance of the concept of anxiety without first forming an appreciation of how people learn to perceive themselves. The individual's perception of himself is shaped by environmental influences. Various self-evaluations are learned as a by-product of the socialization process and of the human relationships that are a part of it. Once learned, the image the individual has of himself is important in determining his behavior. How he comes to perceive and evaluate himself as an object (his self-concept) and its relation to the learning of anxiety will be discussed in the first part of this chapter. The latter part of this chapter will be concerned with an elaboration of the topic of anxiety and with personality defense.

THE RELEVANCE OF CONDITIONING TO HUMAN BEHAVIOR

The learning psychologist Mowrer (1960) has conceptualized the conditioning process in a way which will help the student appreciate the role of conditioning in our discussions of the development of the self-concept, anxiety, and personality defense. It will be recalled that when we discussed respondent conditioning in Chapter 4 it was said that highly controlled and, in a sense, artificial, laboratory situations were employed in the study of the respondent-conditioning process. In the early research of Pavlov the amount of salivary glandular arousal was almost exclusively employed as the dependent variable, primarily because it was so easily quantifiable. The number of drops of saliva secreted could be taken as a sensitive measure of the effect of the manipulation of various independent variables. Actually, however, in these simplified laboratory situations the salivary glandular response represents only a minor aspect of the animal's reaction pattern in response to the conditioned stimulus.

Simple motor reflexes such as the leg-flexion reflex, finger-withdrawal, or eyelid reflexes are extremely difficult to condition. On the other hand, the more the involvement of the autonomic nervous system and emotional circuits, the faster and more effective is conditioning, and the greater resistance it has to extinction. Presumably this is because the autonomic nervous system and the emotional circuits reinforce the CS-UCS associations. In our discussions of the r_g–s_g and the r_f–s_f mechanisms it was pointed out that rewarding or aversive events generalize by means of conditioned antedating responses that serve as a source of reinforcement underlying lengthy operant or instrumental behavioral chains. That is, respondent conditioning was viewed as "embedded within" operant, or instrumental conditioning (see pp. 113–118).

Mowrer (1960) refers to these underlying reinforcing events as "hope" and "fear," depending on whether they are rewarding or aversive in nature. In the Pavlovian situation the salivary response is a simple glandular response which is part of autonomic and emotional reaction patterns mediating the conditioned disposition of the animal. The more important fact is that the sound of the bell as a conditioned stimulus makes the animal "feel good," or "feel hopeful." Its nervous system reacts with the learned expectancy that something good is about to happen. Mowrer has labeled the conditioned expectancy involved in reward conditioning as one of hope. On the other hand, when the unconditioned stimulus is aversive or painful the conditioned expectancy becomes an emotional reaction pattern of shock, dread, or fear.

This view of learning stresses the fact that the acquisition and maintenance of behavior involves feeling components. According to Mowrer we *only* learn to hope and fear. Thus, when the child learns the tact "Daddy," the ability to produce it is, as we have seen, already within the child's repertoire. Learning the particular response "Daddy" occurs because the child is made to feel good whenever he emits it appropriately. Analogously, verbal or other behavioral responses may be inhibited when the child is punished or made to feel fear.

A discussion of abstract principles of conditioning is not always enough to allow the concepts of conditioning to have sufficient meaning for the student so that he can make use of them in interpreting complex behavior. More examples might be helpful in this respect. Bugelski (1964) has employed Mowrer's terms to describe effective teaching. Following the example given below, the reader might consider the ways in which the parents may teach hope and fear in the process of employing normal child-rearing techniques. Bugelski says of the teacher:

> He is the US for emotion. If he is flat and dull, the words that he uses (arithmetic words, grammar words, geography words) will become conditioned stimuli for boredom and apathy. If he is mean, sarcastic, irritable his words will create emotional reactions of resentment, and the content words of the subject will become conditioned stimuli for such emotional reactions. The job of the teacher is to arouse emotional reactions of "feeling good" (hope) in connection with the content he is attempting to teach. Creating such positive feelings as responses to the content of a course is the job of the teacher. In effect, his purpose is to get the student to fall in love with the subject—just as a talking bird is said by Mowrer to fall in love with the sound of the human voice. If the teacher cannot do this, he is not a teacher, no matter what his job may be called. It is probable that only a few people occupying teaching posts actually are genuine teachers. The great majority probably succeed in killing interest in their subjects. In some cases they may not *completely* destroy such interest, and other people (parents or friends) repair the damage done by "teachers."
>
> The successful teacher is one who finds his students reading books in his subject, taking more courses in the area, "majoring" in the field. The teacher does not "instruct." The student learns because he "feels good," saying (to himself or to others) the kind of words about which he "felt good" when he heard them from the teacher. In summary, the teacher, through his appearance, actions, manner, "personality," or other characteristics, directly arouses (as an unconditioned stimulus) favorable (hope) or unfavorable (fear) reactions. The words he uses and the things he talks about become conditioned stimuli for these same reactions. These stimuli later guide and control the behavior of the student in other situations. He will, in a sense, seek to get more or less of such words, saying them to others or refraining from saying them, depending upon the conditioning undergone. [Bugelski, 1964, pp. 121–122]

This is certainly an oversimplification. Perhaps it is an unpleasant one for some readers. Yet, however simplified, Bugelski's example attempts to communicate what many psychologists feel to be the case: that some such reinforcing processes underlie all complex behavior. In many human situations the precise contingencies of reinforcement are obscured, relatively complex, and usually not obvious, especially to the participant. Even the reinforcing contingencies involved in unusually dramatic instances, such as instances of truly effective teaching, are not always apparent except through careful analysis and research.

It must be remembered that the individual is ordinarily not conscious of his behavior as being influenced by environmental "contingencies of reinforcement." He perceives himself as "wanting to major in geography," with perhaps only a vague awareness of the stimuli that actually shaped his interest. The culture does not particularly teach people to tact the factors influencing their behavior. Even if it did, and even if our behavioral sciences were advanced enough to enable a complete and unequivocal account of the factors influencing the individual's behavior under any given set of circumstances, the experience, perception, or feeling the individual would have of himself as possessing various volitions (wanting, deciding, initiation of action, etc.) would probably remain much the same as when he was ignorant of the factors influencing his behavior. We might think of the individual's experience of "wanting to major in geography" as a condition that is present when the person has been exposed to certain environmental contingencies of reinforcement.

THE SELF-CONCEPT

A person's life experiences result in relatively complex sets of expectancies. Experiential reality includes certain aspects of the inner world of his own feelings, wishes, and thoughts. These covert sources of stimulation are very important in mediating complex human adjustment. The concept formations which result from the individual's capacity to be aware of and therefore responsive to his own internal and symbolic processes, as well as events in the external world, allow for varying degrees of organization of his experiences. As a by-product of his experiences, especially his experiences with people, he develops a perception of himself in the sense of "who he is" and what sort of person he is.

Those patterns of stimulation which produce feelings of well-being, hope, mastery, and control mediate recognizable patterns of perception, variously called the *self-structure* or the *self-concept*. The self-structure is the individual's perception of his general habits of thinking and acting, whereas the self-concept is perhaps better reserved to mean the individual's way of looking at himself in an evaluative sense. Both of these evolve from the individual's ability to respond to himself as an object.

Ordinarily he perceives his thoughts and acts as if he were responsible for his own behavior. He perceives himself as being master of his thoughts, feelings, wishes, and acts. This sense of mastery, control, and identity appears to be correlated with the development of hope and related, in the final analysis, to positive experiences in his life.

Objects, events, and people which have become rewarding for the individual may also become part of the matrix of experience perceived as the self-structure. Events which affect the person's loved ones may be experienced to some degree as if they had happened to him. Depending upon how rewarding they are, aspects of his values, country, business organization, club, family, etc., may be similarly experienced. Under certain conditions material possessions may also be incorporated into the self-structure and perceived as extensions of the self.

The Early Development of the Self-Concept

There is little reason to believe that the neonate or the very young infant has any basic awareness of himself as a separate entity. It is possible that he has a vague awareness of proprioceptive feedback from his own bodily movements, but these are probably not distinguishable from other sensory experiences of colors, lights, sounds, painful discomforts, and bliss. During the first year of life the infant can be seen developing a differentiation among objects in his environment. With the development of eye–hand coordination infants begin to explore their own bodies and to differentiate themselves from their surroundings.

Probably the developing awareness that he is a separate entity occurs as part of the process of the development of conditioned stimuli. The appearance of his mother is paired with a variety of comforts and release from unpleasant hunger, pain, cold, and other sources of discomfort. Events over which he has little or no control happen to him as an object. Eventually an increasing capacity to locomote, push, grab, grasp, and manipulate appears. There is a growing capacity to explore his environment and an increasing variety and organization of expectancies. Some things are hard, soft, wet, alive, pleasant, or fearful. Gradually the world becomes differentiated into "me" and "not me."

Language learning introduces new dimensions. Among the first words the child is taught are the pronouns "me" and "mine." Cultural practices with regard to possession are taught early. The child is given toys, clothes, etc., which are said to belong to "him." The child's learning of these tacts enhances his ability to discriminate. When baby sister does not drink her milk she is "bad." In contrast people with good table manners and a hearty appetite (like "me") are "good."

As the child's capacity for speech and motor coordination develop he shows an increasing capacity to cope with his environment. When the child has been encouraged, presented with, and praised for accomplishing tasks that are within his capacity to learn at successive stages of his development, then a generalized feeling of confidence and well-being—hope—develops as part of his self-experience. Many aspects of the world become discriminative stimuli for exploration and attempts at mastery, thus creating an extended sense of self.

Socialization and the Self-Concept

During the 1st and 2nd years of life the child's freedom and spontaneity must be greatly restricted by the socialization process. In very early infancy, although he may suffer some discomforts, he is typically the recipient of an unlimited supply of care, attention, and affection. Difficulties may arise during the weaning period, but adults are likely to cope with them by techniques other than direct punishment. As we have noted previously, however, with the advent of toilet training the adult's motivation to control the child's behavior is so strong that direct punishment is frequently initiated as a disciplinary technique. With the child's increasing capacity to use and respond to language, verbal and physical forms of punishment are increasingly combined. For the child the 2's and 3's are a period of considerable confusion and emotional reactivity.

The behavior of parents toward the child provides him with information about his adequacy, goodness, and worth. If he is given a basic feeling of being loved and accepted as a person the transition from unconditional affection to acceptance and affection that is given only when he behaves as adults want him to behave may be relatively painless and normal. The lack of unconditional acceptance may at first be disturbing, but as the child gets to know the parents better he comes to want to model his behavior after theirs and behave as they would like him to. Living up to parental expectations becomes part of his self-concept. Parental standards of behavior become his own standards and he begins to admire and have a satisfactory concept of himself when he behaves accordingly. New possibilities of reward and a sense of security and well-being become available to him.

On the other hand, the child may learn anxiety. The manner in which he has learned to appraise punishing situations appears to be the most significant factor in determining whether or not anxiety will be learned, and the quality of the general relationship the parents have developed with the child appears to be most important in determining his appraisal of punishing situations. If the parents frequently punish the child, but do not demonstrate warmth and attention at other times, he may come to feel that he is being rejected or abandoned. He may feel a loss of support and be overwhelmed by feelings of helplessness. The realization of the possibility of being rejected and unloved as a person, inaccurate thought it may be, may create feelings of fright and anxiety.

The behavior of the parents toward the child provides him with information about his goodness, adequacy, and worth. (Photograph by Tor Eigland.)

When the child is hurt—for example, by being punished for scribbling on the walls with a crayon—he will withdraw from people and attempt to cope with his feelings. How should the experience be interpreted? Does it mean that *he* is bad, or that his *behavior* is bad? If the anxiety experienced by the child is greater than he can easily cope with he will seek evidences of acceptance and security. He may cling to adults. He may bring the parents presents. He may seek acceptance and recognition for something he has done. The parents can indiscriminately reinforce this security-seeking behavior so that it becomes greatly exaggerated. That is, of course, undesirable. However, if the adults fail to recognize that the child is actually attempting to gain emotional support, they may see all such behavior as a further annoyance and inadvertently punish it. When that happens his ability to cope with his feelings may be seriously impaired.

By observing the behavior of adults and of other people towards him the child's concept of himself begins to take on evaluative characteristics. Real or imagined evaluative attitudes of the parents appear to be especially important in determining the self-concept. For example,

Helper (1958) compared evaluative ratings which parents gave their children with evaluative ratings the children gave themselves and found a significant positive correlation between the two sets of ratings. Jourard and Remy (1955) felt that the actual parental attitudes were not as important in determining self-evaluation as the person's perception of the parents attitudes, e.g., what the person thinks they are. They found a significant and extremely high relationship between the way college students see themselves and their beliefs about the way their parents see them. These correlations were slightly higher for the mother than for the father.

The Self-Concept and the Interpretation of Behavior

Consider the following experiment:

> A small group of college men agreed to cooperate in establishing a shy and inept girl as a social favorite. They saw to it . . . that she was invited to college affairs that were considered important and that she always had dancing partners. They treated her by agreement as though she were the reigning college favorite. Before the year was over she developed an easy manner and a confident assumption that she was popular. These habits continued her social success after the experiment was completed and the men involved had ceased to make efforts on her behalf. They themselves had accepted her as a success. What her college career would have been if the experiment had not been made is impossible to say, of course, but it is fairly certain that she would have resigned all social ambitions and would have found interest compatible with her social ineptitude. [Guthrie, 1938, p. 128]

This experiment illustrates that when a person is rewarded by others he tends to develop a positive self-concept. The self-concept is then reflected in behavior. Since this is the case, the reader might ask, Why do we need to consider the self-concept at all? Why isn't a knowledge of the kind of social experience the person has had sufficient to interpret and predict the person's behavior?

There are at least three reasons why a consideration of the self-concept is of some importance. In the first place, although feelings of hope, mastery, or anxiety are originally learned under specific rewarding or punishing conditions, they are not *maintained* by a continuation of them, but by the individual's continual, often unconscious, reindoctrination of himself. In this connection the reader will recall Ellis' theory (see p. 245) that fear and anxiety may be chronically reactivated by the intraverbal chain—(a) X person rejects me; (b) therefore I am not a lovable, worthwhile person and that is terrible!—where portion (b) of the chain may be unconscious. There appear to be wide individual differences in learning to respond in this manner. Whether or not these responses are made depends upon other factors in the person's experience, factors which will be clearly delineated only by future research.

In the second place, relevant information concerning the person's past experiences may not be readily obtainable, or, if obtainable, it may be of questionable validity for any number of reasons. However, if a reasonably valid picture of the individual's self-concept can be obtained, interpretation, prediction, and alteration of his behavior may be possible without knowing the details of his past learning history.

Finally, the continual development of self-report scales, projective personality tests, and other techniques designed to measure various aspects of the self-concept has given rise to a significant body of research. Relationships are being established between individual differences on such tests and a wide variety of behavioral differences in life situations. These findings are not only important in themselves. They also have considerable bearing on the development of theory, and, in turn, an expansion of our ability to interpret complex human behavior. Many of these findings and their implications will be discussed throughout this and the following chapters.

The Unconscious Self-Concept

Individuals rarely have a neat, homogenous concept of themselves. The matrix of self-perception may contain contradictory perceptions many of which may be only partially sensed by the person or completely outside of his awareness. A person may not be fully conscious of his own self-evaluations any more than he is fully aware of other factors influencing his behavior. Thus, an ambitious, driving man may give an exaggerated impression of a sense of confidence and well-being. However, closer inspection may reveal that he feels anxious and uneasy when things are going well. It is as if he had to have challenging situations or problems so that he could continuously prove his adequacy.

Consider the woman who thinks of herself as the ideal mother and wife, or even as self-sacrificing. Yet objective observation reveals that she spends much more time in community-service work than she does with her family, and that she has actually hurt her children's chances for emotional maturity by overprotecting them to a degree that they are overly dependent and helpless. Her real inability to be affectionate and loving may be betrayed by her sudden revulsion and readiness to turn against members of her family when their behavior or aspirations run counter to her wishes.

Consider the individual who gives the impression of being indifferent to others, who thinks of himself as a person who is completely detached and self-sufficient. Yet close observation reveals that he is actually lonely and unhappy much of the time.

In instances such as these, certain aspects of the person's behavior indicate the existence of a self-concept of which he is unaware. The hard-driving business or professional man may conceive of himself at the bottom as someone who is basically helpless and weak. The self-sacrificing mother may be actually quite domineering and vindictive. The aloof individual may be showing a compulsive disregard of his own feelings. It is as if he were really contemptuous of himself, or had no interest in his true well-being.

People are rarely totally unconscious of the contradictory trends in their self-structure. Suppression or repression may occur, but when the individual is allowed to express himself without fear of being judged or punished in some way, he may reveal an awareness of these aspects of himself. However, psychotherapists commonly note that their patients are rarely aware of the extent to which these suppressed aspects of their self-concept may influence their behavior, decisions, and total pattern of living.

Research on the Unconscious Self-Concept

Measuring Instruments. Until the recent attempts at rapprochement between experimental psychology and clinical psychology, clinical psychologists had found the simpler behavioristic models too limited to account for the phenomena they were observing. Because of the pressing practical problems of dealing with emotional disturbances, and because of the paucity of relevant research, they were more willing to speculate on the basis of their experience without attempting to rigorously determine how certain antecedents of behavior should be tied to observable events. Hence, the unconscious self-concept has been considered by some to be more important than the conscious self-concept in determing behavior (e.g., see Fisher & Cleveland, 1958). In this connection a great deal of clinical research has employed projective tests and techniques. The most widely used projective tests are the Thematic Apperception Test (TAT) and the Rorschach test.

The TAT consists of a series of ambiguously drawn pictures, mostly of people alone or in social situations. The subject is asked to make up imaginative stories about the pictures. The theory is that the person will describe many of his own unconscious feelings and tendencies. It is as if he projected his private wishes and strivings into the imaginary characters and events of his stories. The theory underlying the Rorschach test is the same. In this test the subject is asked to respond to a series of highly ambiguous smudged ink-blot figures.

Experimental Examples. A few examples of research designed to measure the unconscious self-concept might be worth our consideration.

Friedman (1955) assumed that a lack of correlation between the conscious self-concept and TAT indexes of the self-concept would indicate "a large portion of personal experience out of awareness." He predicated that correlations between conscious self-ratings and ratings made on the basis of the projected self would be higher for normal subjects than for neurotics and paranoid schizophrenics. Although the expected rank order among the magnitudes of correlation was found, only the paranoid schizophrenic group differed significantly from the other two. The inference underlying this study was that the degree of pathological behavior is related to the degree to which the self-concept is conscious.

Davids, McArthur, and McNamara (1955) scored the amount of aggression and the direction of aggression ("anger-out" vs. "anger-in") from the TAT protocols of 20 male college students. The amount and direction of aggression was also rated by a clinical psychologist who had known these subjects for approximately 18 months. No correlation was obtained between self-evaluations of aggressiveness and TAT or clinician's ratings of aggressiveness. Relative to the direction of aggression it was found that subjects showing anger-out rated themselves as high on aggression, whereas subjects showing anger-in rated themselves as low on aggression.

It has been a common finding that there is little or no correlation between the consciously avowed possession of a given trait by subjects and the amount of that trait attributed to ambiguous stimuli in certain situations. For example, Goldings (1954) found no correlation between subjects' self-rated happiness and the amount of happiness attributed to photos of persons with ambiguous facial expressions. On the other hand, there is some evidence that unconscious self-evaluations are related to specific personality variables. For example, A. H. Rogers and Walsh (1959) found that defensive subjects judged their own picture, superimposed on other photographs quickly at a low illumination level to be less attractive than nondefensive subjects did under these same conditions of unwittingly evaluating themselves.

Research on the Conscious Self-Concept

Measuring Instruments. Most research on the self-concept has dealt with the subjects' consciously avowed self-concept. Typically, subjects are asked to check adjectives from lists containing several hundred words as applying to them or not applying to them. Words implying favorable self-evaluation and those implying a nonfavorable self-evaluation are distributed randomly throughout the list. Indexes of the person's self-acceptance may be derived by subtracting the number of unfavorable adjectives checked by the subject from the favorable adjectives which the subject checks as applying to him or by comparing the subject's profile with a reference group, composed of a large number of people taking the test.

Another technique is the Q sort devised by Stephenson (1953). In this technique the subject is given a large number of cards (e.g., 100) with a different statement typed on each card. He is then told to sort the cards into categories ranging from "least like him" to "most like him." The number of cards that the subject may place in each category is prearranged by the experimenter according to an approximation of a normal distribution, so that only a few statements are placed in the extreme categories and the majority are placed in the middle categories. Presumably, distortions which may occur by forcing the subject to place his self-ratings into an approximation of a normal distribution are more than offset by the wide number of statistical procedures which can be readily employed with such data. There is considerable flexibility in the technique. The subject may be asked to describe himself, another person, or his ideal self. Correlation coefficients can readily be calculated between the various Q sorts in order to determine whether, and to what extent these various self-descriptions are related.

With such paper-and-pencil rating techniques a problem immediately arises, namely, whether or not the subject is willing to reveal his evaluation of himself. If he knows he is being assessed he may behave defensively and try to create a favorable impression, or distort his communicated self-image in other ways. Investigators have had little control over this fact except by way of instructing the subjects that the results would remain anonymous, and by encouraging individuals to give as valid a subjective picture as possible. Tests have also been devised which measure the individual's concern with "social desirability" (i.e., with the favorableness of his responses). Considerable research has been concerned with how this factor influences the individual's test-taking responses (see Edwards, 1957). We will have more to say about social desirability in the following chapter. Individual differences on social desirability scales have been found to be related to many other aspects of personality and behavior.

The Significance of Self-Ratings. Several studies (e.g., Bass, 1953; Mussen & Porter, 1959) have found that people who are rated as leaders by others and those who are rated as more effective in group discussions or work situations tend to have significantly more positive self-concepts than nonleaders, or than those who are rated least effective.

There is also a relationship between the degree to which the person tends to accept himself and his acceptance of other people (e.g., Phillips, 1951; Sheerer, 1949; Stock, 1949). Generally, significant correlations have been found between scales proporting to measure acceptance-of-self (AS) and acceptance-of-others (AO). Since these correlations are not extremely high, considerable individual variability may occur. Thus, whereas many individuals high in AS are also high in AO, some high AS subjects are found to be low in AO. On the other hand, although low AS individuals are likely to have low AO scores, some may have a high index of AO. Some exceptions to this general relationship have been found. For example, Gordon and Cartwright (1954) reported that an increase in the AS did not involve a corresponding increase in the AO.

Fey (1955) included the subject's estimate of his acceptance by others (EA) and a measure of his actual acceptance by others (AA) in a similar study. He found in general that the person's estimate of his acceptability was not related to his actual acceptability. Those individuals who were most accepted by others had significantly less discrepancy between their AS and AO scores than did individuals least accepted by others. Those with high AS but low AO scores tended to be rejected by others, although they apparently felt that others accepted them (i.e., high EA scores were obtained from this group). Fey suggested that these individuals probably had attitudes of superiority which resulted in their actually being disliked.

An interesting finding of this study was that those subjects scoring high in both AS and AO were neither better liked nor more disliked. This is surprising in that these individuals would be generally thought of as comprising the most well-adjusted group. Fey explained this by saying:

> Perhaps the average person cannot identify easily with the overt paragon of emotional good health. Such a person may not appear to "need" friendship, or to repay it. It may be that his very psychological robustness is resented, or perhaps it is perceived and rejected as a Pollyanna-like facade. [Fey, 1955, p. 275]

Self-reports cannot be relied upon to give us the same description of personality that is obtained from outside judges. Chodorkoff (1954) obtained self-concept data from 30 college students by the Q-sort technique. Four clinical psychologists then made a Q sort for each subject using the same items that the subjects had used. This latter Q sort was based upon biographical information, projective test results (Rorschach and TAT), and word association test data, all of which were available to the judges for each subject. The judges also made a series of ratings on each student of the degree of adequacy of his adjustment. Chodorkoff found, as he had predicted, that the greater the agreement between the person's self-description and the appraisal of him made by the judges, the higher was his rated personal adjustment.

Another interesting finding of this study was that the greater the agreement between the individual's self-description and the judges' description of him, the less was the perceptual defense manifested by the subject in a laboratory perceptual-defense test. Threatening and neutral words were exposed subliminally, and the exposure speed was gradually increased until the subject was able to recognize all of the words. The difference between the recognition thresholds for neutral and threatening words constituted the index of perceptual defense. Thus, subjects who showed less perceptual defense and whose self-concepts were more in agreement with judges ratings were rated clinically as being better adjusted.

The Self-Ideal

If the person is asked to describe himself "as he would like to be," he may give a description which contrasts markedly from the one he would give if he were simply asked to describe himself. Just as in the case of the self-concept, the image the individual has of his ideal self may not be well delineated in the person's mind and, analogously, the raw material from which the self-ideal is formed comes from many sources. The ideas which the individual has about the kind of person he would like to be result from those sources of reinforcement which contribute to identification, modeling, and imitation learning (for a discussion of these topics see pp. 253–266) and from the verbal instruction of the parents and other people significant in the individual's life. Parents are usually not fully aware of the extent to which they teach the child their own ideals of behavior. Their influence is often indirect (e.g., by way of their own practices of admiration). The mass media of communication and the major institutions of society are also influential in shaping an image of personality and behavioral ideals.

The self-ideal may be associated with occupations the person would like to enter or roles he would like to play. It also involves characteristics and values which the person would ideally like to possess. Ordinarily, when the individual's self-concept approximates his self-ideal self-esteem is high.

The self-concept and the self-ideal are not independent. When ideals are too unrealistic, or when they take the form of "shoulds," instead of merely wishes, they may contribute to a negative self-concept and a lack of self-esteem. Many personality theorists, particularly those in the neopsychoanalytic tradition (e.g., Horney, 1950; Sullivan, 1954) have been impressed by the fact that when the individual is exposed to conditions producing anxiety, especially those producing the subjective feeling of not being lovable, he may form an idealized image of himself as a defensive measure. The distinction between his idealized self and the concept he has of himself as he really is becomes blurred. Horney (1950) sees the entire style of life of neurotic persons as an attempt to actualize an idealized image, irrespective of the person's true feelings and capabilities. These dynamics result in an alienation of the person from himself. Such theories must be considered in the light of the clinical data they are designed to interpret. They present a considerable challenge for future research.

Research on the Self-Ideal. Havighurst and his associates (Havighurst, Robinson, & Dorr, 1946) asked 1147 subjects, ranging in age from 8 to 18, to describe the person they would most like to be like when they grew up. They were told that the person could be real, imaginary, or a combination of several people. They found that the younger children based their descriptions on parental figures, whereas the older children tended to favor glamorous or romantic figures. The late-adolescent group, however, tended to describe a composite of desirable characteristics, or referred to specific young adults to whom they had been exposed in one way or another. Other findings from this study included the fact that children from families of low socioeconomic status continue their preoccupation with glamour figures as ideal models longer than children of higher socioeconomic status families. The children did not choose teachers as models for the description of ideal very often, but it was found that a teacher whom the child or adolescent finds especially attractive may have an unusually strong influence on the child's description of his ideal. This is also occasionally reflected in case history data: an encouraging high school biology teacher may be the reason given by an individual for his initial interest in becoming a doctor; a kindly college professor may do more than is absolutely required by his position to encourage a young writer, etc.

A great deal of research has employed paper-and-pencil tests, and/or the Q-sort technique to measure the individual's self-ideal. Carl Rogers and his associates (see C. R. Rogers & Diamond, 1954) have frequently employed the self-ideal measure in research programs designed to study the psychotherapeutic process. Butler and Haigh (1954) made a test of the hypothesis that people who come for counseling are dissatisfied with themselves, and that successful therapy significantly changes the individual's degree of self-dissatisfaction in a positive direction. In order to test this hypothesis they asked each client prior to the beginning of counseling to describe himself using the Q-sort method. They also asked each person to describe his self-ideal ("the person you would most like within yourself to be"). When they correlated the distributions of the two sortings for each person, they found that the average correlation between the self-concept (SC) and the self-ideal (SI) for the group of subjects in the experiment was zero. That is, for the group as a whole there was no relationship between their self-concept and their self-ideals. This was a marked contrast to a control group of subjects who were not interested in being counseled. The average correlation between the self-sort and the ideal-sort for this groups was .58. The subjects in this group thus appeared to be much better satisfied with themselves than the group seeking therapy.

After the completion of an average of 31 sessions of counseling the average correlation between SC and SI for the therapy group had significantly increased to .34, whereas the SC–SI correlation for the control group had not changed after an equivalent interval of time. Follow-up studies of clients within 6 months to 1 year following therapy indicated that the average SC–SI correlation remained about the same as it had been at the end of therapy. Butler and Haigh concluded that self-esteem (congruence between the self- and ideal-sorts) increased as a result of this particular type of therapy (client-centered counseling).

The question arises whether the significant increase in the average SC–SI correlation following therapy was a result of the self-ideal being lowered in the direction of the self-concept during the therapy, or a result of the self-concept being altered so that it became more in line with the self-ideal. Research bearing on this point has indicated that both occur. Rudikoff (1954) found that the SI changed in the direction of the SC during therapy, whereas Raimy (1948) found evidence for the self-concept changing toward increased self-approval during therapy. Raimy used a content analysis of transcribed records of therapy sessions in his study.

Experiments such as these led to the hypothesis that the SC/SI ratio could be taken as an index of adjustment. The lower the SC/SI ratio, the poorer the individual's adjustment. This was largely substantiated with data from patients (e.g., Chase, 1957) and data from college students (e.g., Turner & Vanderlippe, 1958). However, other findings indicate that a high SC/SI ratio must be interpreted with caution. Thus, Friedman (1955) found an SC–SI correlation of .63 for normal subjects, .03 for neurotics, and .43 for psychotic patients. Would that warrant the conclusion that psychotics show more self-esteem, and are therefore better adjusted than neurotics?

The evidence is beginning to accumulate that very high SC–SI correlations may be indicative of maladjustment. A defensive person is likely to try to give a flattering picture of himself, so that the correlation between the self-concept and the self-ideal will be high (see Block & Thomas, 1955). Individuals may present only socially desirable responses in order to create a flattering picture of themselves, or because they are not able to differentiate between their idealized self-image and themselves as they really are. Devising instruments and research designs to discriminate between these two groups is a challenging problem for future research.

One further finding relative to the SC/SI ratio is of particular interest. Along with SC and SI Q sorts, Levy (1956) had college students describe their home town, both as it is and as they would like it to be. He predicted a positive relationship between the way people evaluate themselves and the way they evaluate their home towns. The actual correlation he obtained between the two sets of Q sorts was .70, suggesting a very strong relationship. Thus for many individuals there appears to be a consistency between their evaluation of themselves and their evaluation of their environment. The attitudes the individual has toward himself may generalize widely to other aspects of his experience.

ANXIETY AND THE SELF-CONCEPT

The self-structure depends upon those sources of stimulation which give rise to feelings of hope, mastery, and well-being. Conditions producing the experience of anxiety may seriously disrupt the person's organized perception of himself and result in a feeling of helplessness and disorganization. People suffering from relatively severe anxiety attacks express the fact that they feel as if they are "going to pieces"—an extremely unpleasant experience.

The same conditions that produce a negative self-concept also produce anxiety. If the tension cannot be controlled in some way, it results in a subjective experience of disorganization and helplessness, and a lack of perception of the self as the center of things. Severe anxiety is apparently unbearable. Very primitive mechanisms such as fainting, emotional constriction, rigid behavioral inhibition, or blind rage reactions occur as a means of coping with the intolerably painful experience.

Since clinical workers characterize anxiety as a basic phenomenon and a central problem in neurosis and psychopathology, it is important that we discuss certain aspects of it. We will begin with some comments about the subjective experience of anxiety, continue with a discussion of objective techniques used to define and measure it, and then elaborate upon the conditions under which anxiety is learned.

The Subjective Experience of Anxiety

A distinction is frequently drawn between fear and anxiety. Many people prefer to use the term fear to describe the organism's reaction to an objectively painful stimulus or event; whereas anxiety is used to refer to similar unpleasant emotional tensions which occur in the absence of any objective, adequate stimulus for the fear reaction.

In terms of the individual's subjective experience it is probably not necessary to make a distinction between the experience of anxiety and that of fear. An intense state of anxiety is certainly subjectively inseparable from the experience of fear. Consider the following classic description of the experience of severe anxiety and panic given by William Ellery Leonard. In the excerpt which follows he is standing on a bluff overlooking a placid lake. He had left his walking companion behind in the woods. The recent suicide of his wife, for which the community had blamed him, had been causing a mounting distress. As he stood on the bluff a train passed along the opposite shore of the lake. An incident had occurred when he was a child in which he had barely escaped being run over by a locomotive.

> I stand looking out over the silent and vacant water, in the blue midday. I feel a sinking loneliness, an uneasy, a weird isolation. I take off my hat; I mop my head; I fan my face. Sinking . . . isolation . . . diffuse premonitions of horror. "Charlie" . . . "Charlie" . . . no answer. The minutes passed. "Charlie, Charlie" . . . louder . . . and no answer. I am alone, alone in the universe. Oh, to be home . . . home. "Charlie." Then on the tracks from behind Eagle Heights and the woods across the lake comes a freight-train blowing its whistle. Instantaneously diffuse premonition becomes acute panic. The cabin of that locomotive feels right over my head, as if about to engulf me. I am obsessed with a feeling as of a big circle, hogshead, cistern-hole, or what not in the air just in front of me. The train feels as if it were about to rush over me. In reality it chugs on. I race back and forth on the embankment. I say to myself (and aloud): "it is half mile across the lake—it can't touch you, it can't run you down—half a mile across the lake." And I keep looking to make sure, so intensely in contradiction to what the eye sees is the testimony of the feeling of that cabin over head, of that strange huge circle hovering over me . . . .
>
> Meanwhile the train chugs on toward Middletown. I rush back and forth on the bluff. "My God won't that train go; my God won't that train go away!" I smash a wooden box to pieces, board by board, against my knee to occupy myself against panic. I an intermittently still shrieking, "Charlie, Charlie." I am all the while mad with terror and despair of being so far from home and parents. I am running around and around in a circle shrieking, when Charlie emerges from the woods. [Leonard, 1927, pp. 304–307]

The emotional distress of this example was originally learned on the basis of cues which naturally elicted fear. When the original stimulus is intense enough, the learning of rather strong anxiety in response to cues associated with the traumatic experience can occur as the result of a single episode. Intense anxiety can also be experienced as the result of less severe, or even relatively mild, noxious stimulation occurring repeatedly over a long period of time.

The example above also illustrates the fact that the anxiety response can be cumulative. Perhaps the train appearing on the opposite shore would have only caused some apprehension and uneasiness (as opposed to panic) had it not been for the fact that the narrator had recently been exposed to other anxiety-producing experiences. The cumulative results of mild anxiety producing episodes will be discussed below.

Chronic Anxiety and Its Manifestations

The anxiety response may be chronic in that it is experienced repeatedly in many situations. Suppose the individual has learned to be anxious over the loss of approval by others. Perhaps the original conditions for learning to be anxious in the absence of signs of approval involved direct verbal instruction, or they may have involved imitative learning based upon parental models. In any event a person who has been taught to have a marked degree of anxiety at the prospect of his not being approved of by others may become upset in a great variety of interpersonal situations which present cues indicating lack of approval. The result is a more or less chronic experience of anxiety and tension. Another example of chronic anxiety would be that resulting from unrealistic perfectionistic strivings, with accompanying symptoms of irritability, hostility, etc.

Mild manifestations of chronic anxiety occur quite frequently. Others may note that an individual is hypersensitive or oversensitive to criticism. He may seem preoccupied and show an inability to concentrate or attend fully to the world around him. He may fail to comprehend simple statements unless they are repeated. Anxiety may be apparent in the tonal quality of the individual's speech. It may be tremulous or it may show a lack of modulation. His speech may be characterized by halting and blocking. Other signs of tension may be present. He may be fidgety or restless much of the time. He may nervously manipulate objects or parts of his body with his hands and fingers. He may show the startle reaction to incidental stimuli that are sudden or abrupt. He may urinate frequently, suffer from chronic insomnia, perspire in objectively mild stress situations, etc.

The Measurement of Anxiety

The direct physiological measurement of anxiety using indexes such as blood pressure, heart rate, muscle action potential, and electrical resistance of the skin may be possible in the future (see Martin, 1961). Some progress has been made in differentiating anger and anxiety (or fear) by means of these indexes. For example, the diastolic blood pressure appears to increase more during anger, whereas the heart rate increases more during fear or anxiety. It appears that a given individual has a pattern of autonomic responses peculiar to him which are elicited by threatening stimuli. At the present time, however, physiological instrumentation and research have not provided us with a reliable and accurate technique for measuring anxiety. Even when this becomes available, bringing together the equipment and creating the appropriate situation will be a difficulty.

Measures based on rating scales are more convenient and have the advantage of not disrupting whatever subject–environmental behavioral sequence is occurring at the time the person's anxiety level is being measured. In the typical rating scale procedure, judges are asked to rate the absence, presence, or degree of severity of the occurrence of various overt signs of tension or anxiety (restlessness, speech disturbances, irritability, etc.).

Aside from the rating of specific behavior, an overall estimate of the degree of overt anxiety is typically obtained by having the judge or judges rate the individual on seven- or nine-point scales. The degree of overt anxiety of a normal individual in the particular situation in which the ratings are being made is used as the midpoint for "less than" or "greater than" judgments.

Another common technique for measuring anxiety is the self-report scale. The Taylor Manifest Anxiety Scale (MAS) is typical of these. Taylor (1953) had five judges rate a large number of items (simple true–false statements) as to whether or not they reflected overt anxiety. A final scale was made up of those items which four of five judges agreed reflected anxiety. Another commonly employed self-report scale is the Welsh Anxiety Scale. Its derivation is too complex to consider in detail here. However, it is composed of items reflecting aspects of personality included in the categories: hypersensitivity, negative emotional tone, difficulties in thinking and thought processes, pessimism, and lack of energy.

Anxiety Scales and Behavior

The problem of the subject's willingness to give a valid self-report is as pertinent to anxiety scales as it is to other self-report scales. Nevertheless, from a scientific point of view the significance of the scale must be judged according to the degree to which individual differences in responses to such scales are related to individual differences along other dimensions and in other behavioral situations. Self-report measures of anxiety such as the MAS and Welsh scale have, in fact, been shown to be related significantly to other aspects of behavior. Highly anxious people, as indexed by the Welsh scale, have been found to be pessimistic, to lack confidence, and to be hesitant and easily upset in social situations (Dahlstron & Welsh, 1960). They have been found to be less accurate and more uncertain than low anxiety scorers in a psychophysical experiment of line-judging (Riedel, 1961). They are measurably less effective in communication than low-anxiety scorers (Chance, 1956). High-anxiety-scoring college students have been found to omit more items on a true–false course examination as compared to low-scoring students. They also failed to supply answers they knew were correct (Sherriffs and Boomer, 1954). Psychotherapy patients have been shown to have higher average anxiety scores than control groups of subjects (e.g., Dahlstrom & Welsh, 1960; Welsh & Dahlstrom, 1956).

MAS scores have also been shown to be significantly related to behavior. Thus, subjects who score high on the MAS learn more rapidly than low-anxious subjects to avoid noxious stimulation (e.g., Spence & Beecroft, 1954; Spence & Taylor, 1953; etc.). High-MAS subjects perform verbal-learning tasks more poorly than low-MAS subjects when the material is difficult, whereas the reverse is true with easy material (e.g., Lucas, 1952; Montague, 1953; etc.) Anxiety ratings of psychiatric patients and people undergoing clinical counseling have consistently been significantly correlated with MAS scores (e.g., Gleser and Ulett, 1952; Hoyt & Magoon, 1954; etc.). College students' grades have also been found to vary inversely with MAS for students of average intellectual aptitude. On the other hand, the poor grades of low-aptitude students were found to be unrelated to their MAS scores. Students with high aptitude scores get good grades irrespective of their MAS scores (Spielberger and Katzenmeyer, 1959).

Unconscious Anxiety

In general there is a correspondence between the individual's self-report on anxiety scales and the ratings of his anxiety level by others. However, it is very rare for the degree of correspondence to even approach a perfect correlation. The self-report of individuals and ratings of them by others are never perfectly related when the correlation coefficient is computed for a group of randomly selected subjects.

Part of this lack of correspondence is because some people try to create a socially desirable impression by claiming to be less anxious than they either perceive themselves to be, or than others perceive them to be. Another factor that may contribute to this lack of correspondence between self-ratings and other ratings of anxiety is that people may be anxious without consciously experiencing any anxiety. Therefore they are unable to report it. How many people during their wedding ceremonies report that they "feel just fine," or "really feel very calm," when in fact the groom may chain-smoke, put a sock on wrong side out, and give the general appearance of a trapped, frightened animal—while the bride may have a tight grip on her handkerchief, perspire, laugh too loudly, etc.

The question still remains, Are these self-reports merely socially desirable responses? Clearly, this is not always the case. The individual may feel quite well controlled and be completely oblivious to many general signs of tension. Even after the event, after hearing various anecdotes about his nervousness from his friends, he may admit by inference that he must have been tense, without actually recalling any subjectively felt anxiety.

A third reason for the lack of correspondence between self-reported anxiety and level of anxiety as judged by objective observers is that anxiety may be inferred when *neither* overt manifestations nor the subjective experience of it are present. Consider the case of hysterical paralysis. Let us say the patient is a soldier recently returned from the combat zone because of a paralyzed arm. No organic conditions, lesions, etc., are indicated. The clinical diagnosis is that the paralysis is functional, not organic. That is, the paralysis is viewed as a symptom which functions to reduce intense anxiety. Aside from the paralysis the patient does not appear anxious, nor does he have any subjective experience of anxiety.

Consider an individual with strong obsessive-compulsive characteristics. Instead of the type of symptom which might gravely interfere with his normal living, such as compulsive hand-washing or the compulsion to count telephone poles, suppose the individual's compulsivity is in line with cultural expectations, except that it is exaggerated. Suppose he must be scrupulously neat, always punctual, and must give careful attention to minute details of his work. Suppose he maintains an extremely rigid schedule and functions somewhat like a well-oiled machine. Again, the clinical inference would be that the compulsive rigidity and fixedness of his behavior functions as a defense against anxiety. Subjectively, however, he may report no anxiety whatsoever. Objective observers may report that he is unusually well controlled, poised, and unanxious.

This raises several important questions. In the first place, What allows the clinical inference of the presence of anxiety? The answer to this varies with the particular case in question. In the case of hysterical paralysis clinical experience has been that inevitably the symptom represents a drastic attempt to cope with unbearable anxiety. Under hypnosis or sodium pentothal severe anxiety may be evidenced as the patient begins to recall the traumatic event responsible for the symptom. The case of the obsessive-compulsive neurotic is somewhat different. There the inference of anxiety is based upon the fact that the person may experience severe anxiety if he is prevented from acting out his compulsive ritual.

Assuming that the behavior was learned on the basis of anxiety-reduction, if it is effective in reducing anxiety, then anxiety should not be present. It can also be argued that anxiety is present, and that it serves to maintain the anxiety-reducing behavior. The individual is not aware of the anxiety. He is only aware of his behavior. Irrespective of these conceptual problems, the inference of unconscious anxiety may be meaningful and useful as long as the observations upon which is it based are specified.

Conditions under Which Anxiety Is Learned

When the original fearful event is traumatic, anxiety may become strongly conditioned to stimuli surrounding the fearful event on the basis of a single instance. When we talked about Diven's experiment in Chapter 4 in which a single word from a list of words was shocked (the word "barn"), and the anxiety response generalized to associated words in the list, we noted that even in such simple situations anxiety tends to incubate. It increases and generalizes with the mere passage of time. Thus, if the Group Commander can arrange to get his pilot into a new aircraft and into the air again immediately after a serious accident, severe conditioned anxiety associated with flying may be prevented. On the other hand, with the passage of time severe anxiety may become attached to any number of cues in the situation: the aircraft, the airfield, or any associated symbolic cue.

Phobias

Irrational fears attached to specific cues such as these have been called phobias. Although phobias are relatively rare, when they occur they are likely to be especially impressive, because when the individual comes in contact with a stimulus cue that serves to elicit his fear he is thrown into a severe anxiety attack. One of the remarkable features of this type of reaction is that the person may come to fear stimuli remotely associated with the original frightening event rather than the event itself. The original frightening event is almost invariably related to experiences with people. For example, in very young children a school phobia occurs occasionally. Analysis of the situation usually reveals that the child has an overwhelming fear of separation from the mother. However, the fear becomes attached not to the prospect of separation, but to objective cues available in the situation—the teacher, other children at school, school buildings, etc.

One of the distinguishing features of the phobia is that it is a relatively focal fear. Although it is true that the person suffering from a phobia is thrown into panic by specific stimulus cues, as opposed to suffering from diffuse anxiety, phobias do show generalization. Thus, the man who shows a fear of riding in elevators may eventually come to fear other small, enclosed spaces. The fear may eventually generalize to the point where he finds being inside any building for any length of time insufferable.

Anxiety Learned by Verbal Instruction

We have also noted from time to time that anxiety may be learned solely on the basis of verbal cues. In the chapter on verbal learning (see pp. 244–245) we discussed the effect that direct verbal instruction had in producing anxiety about thoughts associated with sex and pregnancy in one patient. This is a very typical example of the way in which anxiety can be created on the verbal level. Telling the child that if his belly button is pressed his legs will fall off may be amusing. The child soon learns that it isn't so. Telling him that Santa Claus comes down the chimney and leaves presents for good little boys and girls is relatively harmless. On the other hand, telling him that masturbation will result in insanity; that if he is not good no one will love him; that he must be good at everything he does; that others are more important, richer, better; that other people, races, nationalities, are inferior; that the world is a jungle and others will try to take advantage of him—setting up rigid, unrealistic moral and ethical standards—these create undue emotional disturbance.

Verbal instruction which leads to emotional disturbance is likely to have the following features: First, it is of such a nature that it is likely to cue off anxiety, perhaps severe anxiety. Second, the individual has been strongly reinforced in the past for basing his attitudes and behavior on what the source says (e.g., what parents say is likely to have a strong emotional coloring for the child). Third, the content of what is said is not easily able to be validated or subjected to by testing reality. Finally, since the person cannot consider what is being told to him calmly and objectively without experiencing anxiety, techniques for coping with anxiety become mobilized.

Verbalizations of the parents which produce anxiety need not be in the direct form of criticism, discipline, or instruction. Comments of a negative evaluative nature which are made casually and unwittingly by the parents in the presence of the child may have a significant effect. The frequent overhearing of parental quarrels appears to be especially bad for children. Along with exposure to angry words, parental dissension, and possible loss of support, the child himself may be mentioned or implicated in some way. When the child is very young such episodes may produce more stress than he is capable of coping with effectively.

Anxiety Learned by Repetition

The one-trial learning of anxiety is relatively rare, occurring as it does under conditions of intense stimulation. More commonly anxiety is learned by way of relatively mild aversive stimulation over a period of time. Typically—a variety of cues—objective, verbal, behavioral, and intrapsychic—combine to produce a repetition of certain distressful experiences over a long period of time.

The National Film Board of Canada has a series of excellent documentary films which illustrate the development of very typical neurotic patterns resulting from this kind of episodic learning. One of these films is entitled, *The feeling of rejection:*

Margaret. The central figure, Margaret, is shy, noncompetitive, and afraid to take independent action. She feels rejected by others. She appears to be afraid to express herself in any way. She feels that her own feelings and thoughts are unimportant. Afraid to assert herself, she tends to acquiesce to the feelings and wishes of others. It is as if her own wishes did not really matter.

In one scene she is shown buying a blouse for herself. She is obviously delighted by a particular garment. However, with pressure from the salesman, she ends up buying one she doesn't really like—but one she knows her mother would approve. When she takes the blouse home and inspects it in front of the mirror she develops a headache.

She tends to take too much abuse from others. In one scene her sister asks her to do her share of the household chores, because she has a date that evening. Margaret doesn't really want to but she must comply in order to avoid the terrible anxiety she would experience if she were to assert her own rights. While doing the chores she develops a headache.

She is shy with others and feels rejected much of the time. In still another scene she is shown walking by a group of acquaintances who are talking excitedly. They obviously do not see her. She feels, however, intentionally rejected and snubbed. She relates to the psychotherapist that they ignored her and did not speak to her. He asks, "Did you speak to them?" She replies hesitantly, "Welll . . . no."

The latter part of the film switches to scenes of her childhood. Again and again the audience is shown mild episodes in which Margaret learned to fear her overprotective mother's disapproval. Apparently her normal child's play is disturbing to the parents. They tell her not to be such a nuisance, to sit down and be a "good little girl." Attempts to gain support from her father only result in rebuffs, e.g. "Can't you see I'm busy?" "Now sit down there and be quiet." Her early life is filled with "don'ts." In various scenes the audience is exposed to the mother's excited admonitions: "Don't go near that dog. Dogs bite." "Don't play with scissors. You will cut yourself." "Good little girls don't climb on fences," etc. In short, through repeated experiences over a long period of time the child learns to be afraid to assert herself. She learns to gain parental approval by being "good."

The cumulative result of these experiences is shown not only in her adult behavior, but during the formative phases as well. Thus, in one vivid scene she is shown rehearsing in her room for a possible part in the school play. It is the part of the "Princess." The audience sees a delighted, happy child. She knows her lines perfectly and recites them in front of the mirror with a childlike charm. The following day she is shown in the school room. The teacher is making the selection of the child to play the part. Another child recites the lines reasonably well, with some prompting. Then it is Margaret's turn. Margaret stands up, very tense. She is clutched with anxiety, totally unable to recite the lines which she knew so perfectly. The child had learned to anticipate rejection and, frightened of the prospect of not gaining approval, she had learned to fear being evaluated.

In order that the reader may more fully appreciate how anxiety is learned over time let us consider one more example, from another documentary film in the same series entitled *Over-dependency:*

Jim. Jim, the central character in this film is a talented young married man who, despite actually being completely physically healthy, suffers periods of illness which interfere with his work. Without his being aware of it, he has become overdependent upon his wife and his overindulgent mother. The film shows how as a child he was reinforced for obtaining comfort from his mother whenever he had to face any difficulties.

But that is only one side of the story. Jim is anxious in the presence of his boss or other authority figures. In one scene he is shown going for a job interview. While waiting in the outer office he experiences an anxiety attack. The anxiety mounts to the point where he has to give the man's secretary a feeble excuse and rush out into the street.

In other scenes Jim is shown as a child, "hanging around" his father's workshop. He begs his father to let him go fishing with him. The father tells him that he is not big enough. It can be seen that the attitude of the father is one of (unintentional) rebuff. When he tries to handle his father's tools he is again told that he is "not grown up enough yet."

Recalling incidents such as these later in psychotherapy Jim comes to the realization that he had gradually learned to become afraid of his father. To lessen his anxiety, then he would seek comfort and protection from his mother. Later in life, taking on any responsibilities resulted in his feeling inadequate, weak, and afraid. This, in turn, helped to reinforce his pattern of overdependency.

Defensiveness

Most of our anxieties are learned responses to social situations implying competition, criticism, rejection, hostility, guilt, and many other possible threats to our self-esteem. Minor defensive patterns such as attention-getting, unruly conduct, displaced aggression, or compulsive stealing, sometimes serve to protect the individual from anxiety associated with feeling of worthlessness, rejection, and inferiority. These behaviors may be associated with relatively minor and temporary frustrations. The student who does poorly on an examination may belittle the subject matter, criticize the professor, or proclaim loudly that he did not study. The person who has been insulted may brag too strongly, fantasy revenge, or make disparaging remarks about the person who insulted him.

Some people defend themselves against anxiety or punishment by asserting the exact opposite of what they feel. (Photograph by Tor Eigland.)

On the other hand, defensiveness may be chronic and generalized. People who have been continuously exposed to punishment, threat, or criticism during childhood respond with anxiety in many social situations implying competition, criticism, or almost any kind of evaluation of them as a person. Merely the presence of other people or situational cues may elicit an intense conditioned emotional response for these unfortunate individuals.

An anecdote from the writer's experience is pertinent here:

> As a young professor I was often mistaken for a student. One cold morning I was driving to the University in an expensive, luxurious automobile which I had temporarily acquired, quite by accident. While I was stopped at a traffic light a neatly dressed man came up, knocked on the window and asked for a lift. We chatted amiably for a few blocks, but soon he began to become tense and agitated. He glanced furtively around the car and at me. He noted the University parking sticker, my casual mode of dress (I was not teaching that day), the pushbutton windows. He obviously took me for a student. He soon began a boisterous conversation about his daughter who was a student too, his fine house, his automobiles. When I let him out of the car a number of blocks later he was talking very loudly and flailing his arms in the air, saying: "You'll meet my daughter. Her name is Susan K——. You'll see her. She drives a big Lincoln. You'll see her . . . you'll see her . . . ." I wondered briefly about him and his daughter. Had she, too, learned that her worth as a person depended upon her owning an expensive piece of machinery?

The reaction of this man is somewhat atypical. He was probably under stress in certain other areas of his life at the time. Overdefensiveness is common, but usually when defensive habits have been established the individual is not obviously anxious, although he may be egotistical and overbearing.

Through early experiences with being unloved, rejected, criticized, made to feel quilty, etc., the individual may develop a generalized, chronic concept of himself as inadequate or unworthy. This is generally called an *attitude of inferiority.* Karen Horney has referred to the early development of this type of generalized self-concept as *basic anxiety* (e.g., 1950). Horney believes that basic anxiety is developed very early in childhood, and serves as a basis for greatly elaborated defensive patterns as the individual develops. According to Horney:

> Through a variety of adverse influences, the child may not be permitted to grow according to his individual needs and possibilities. Such unfavorable conditions . . . boil down to the fact that people in the environment are too wrapped up in their own neuroses to be able to love the child, or even conceive of him as the particular individual he is; their attitudes toward him are determined by their own neurotic needs and responses. In simple words, they may be domineering or overprotective, intimidating, irritable, overexacting, overindulgent, neurotic, partial to other siblings, hypocritical, indifferent, etc. It is never a matter of just a single factor, but always the whole constellation that exerts the untoward influence on the child's growth.

> As a result, the child does not develop a feeling of belonging, of "we," but instead a profound insecurity and vague apprehensiveness, for which I use the term *basic anxiety*. It is his feeling of being isolated and helpless in a world conceived as potentially hostile. The cramping pressure of his basic anxiety prevents the child from relating himself to others with the spontaneity of his real feelings, and forces him to find ways to cope with them. He must (unconsciously) deal with them in ways which do not arouse, or increase, but rather allay his basic anxiety (Horney, 1950, pp. 3–4).

The following chapter deals exclusively with specific, or focal, defensive patterns. Defense patterns of the more pervasive sort described by Horney and by others are discussed in Chapter 11.

SUMMARY

The individual's perception of himself is shaped by social influences. Once it is learned, it is significantly related to his further adjustment. The perception of the self is inseparable from those environmental influences that produce feelings of goodness, worth, and mastery, at one extreme; and inadequacy, self-contempt, emotional disturbance, and anxiety, at the other. Some psychologists believe that the emotional elements in learning are of fundamental importance. It may be that all learning involves emotional components. Those events that cause the person to feel good or feel hopeful mediate positive behavior change. Those that cause him to feel disturbed and fearful mediate behavior characterized by escape and avoidance.

Parental attitudes toward the child strongly influence the concept he develops of himself. The way he perceives attitudes of parents and significant others appears to be an even more crucial factor than their actual attitudes. Once a child or a person has learned to evaluate himself in a positive or in a negative way the conditions of his life can change yet he may continue to reindoctrinate himself so that the original self-evaluation persists. By knowing how the individual sees himself it is often possible to interpret and predict much of his behavior without a detailed knowledge of his past history.

Sometimes the concept of the self which can be inferred on the basis of observation of the person's behavior is quite different from his consciously avowed self-concept. As a result the notion of an unconscious self-concept has been postulated. Various projective tests and devices have been used to measure it. Unfortunately, research using these instruments has not tended to produce highly reliable results and has had limited success.

A large number of self-report inventories, rating scales, and other paper-and-pencil techniques have been devised to measure people's consciously avowed self-concept. Surprisingly, these simple devices have led to a wide range of more or less integrated findings.

People can be asked to describe themselves "as they would ideally like to be" as well. That is, their "self-ideal" (SI) can be measured in the same way that their self-concept (SC) can. An index of the discrepancy between the two measures, the SC/SI ratio, has been shown to be significantly related to adjustment.

Stimuli associated with severely traumatic, fear-producing events can become strongly conditioned in a single trial to elicit many of the respondents of fear and anxiety. During severe anxiety states people often report feelings of unreality and a loss of a sense of self. Experiences of mild distress over a period of time can have a cumulative effect, creating quite strong generalized anxiety.

Very often these distressful experiences involve episodic mild fear that occurs within the framework of the person's relationships with others. Cumulative anxiety-producing experiences of this sort may have thematic content such as fear of loss of approval, feelings of rejection, or fear of authority; or they may produce more generalized, chronic, anxiety.

Physiological instrumentation has not as yet provided us with techniques that are accurate enough or flexible enough for measuring anxiety. However, as is the case with the self-concept, self-report inventories of anxiety have been fruitful in experimental personality research.

It appears that anxiety may be unconscious under some conditions in that the person's anxiety may be obvious to others, yet he himself may not be aware of it. It also appears that certain adjustments represent an attempt to cope with severe anxiety. Although they are reinforced by anxiety reduction, as long as the symptoms, or anxiety-reducing behavior, are effective there may be no obvious signs of anxiety present.

Anxious individuals often use defensive maneuvers to cope with anxiety. They may be relatively minor and transient, such as attention-getting or unruly conduct, or they may involve complex patterns of intrapsychic and behavioral responses to a generalized feeling of inferiority. These are discussed in more detail in Chapter 11.

REFERENCES

Bass, B. M., McGehee, C. R., Hawkins, W. C., Young, P. C., & Gebel, A. Personality variables related to leaderless group discussion behavior. *J. abnorm. soc. Psychol.*, 1953, **48**, 120–128.

Block, J., & Thomas, H. Is satisfaction with self a measure of adjustment? *J. abnorm. soc. Psychol.*, 1955, **51**, 254–259.

Bugelski, B. R. *The psychology of learning applied to teaching.* New York: Bobbs-Merrill, 1964.

Butler, J. M., & Haigh, G. V. Changes in the relation between self-concepts and ideal concepts consequent upon client-centered counseling. In C. R. Rogers & Rosalind F. Dymond (Eds.), *Psychotherapy and personality change,* Chicago: Univer. of Chicago Press, 1954. Pp. 55–75.

Chance, June E. Some correlates of affective tone of early memories. *J. consult. Psychol.,* 1956, **21**, 203–205.

Chase, P. H. Self concepts in adjusted and maladjusted hospital patients. *J. consult. Psychol.,* 1957, **21**, 495–497.

Chodorkoff, B. Self-perception, perceptual defense, and adjustment. *J. abnorm. soc. Psychol.,* 1954, **49**, 508–512.

Dahlstrom, W. G., & Welsh, G. S. *An MMPI handbook: A guide to use in clinical practice and research.* Minneapolis: Univer. of Minnesota Press, 1960.

Davids, A. H., McArthur, C. C., & McNamara, L. F. Projection, self-evaluation and clinical evaluation of aggression. *J. consult. Psychol.,* 1955, **19**, 437–440.

Edwards, A. L. *The social desirability variable in personality assessment and research.* New York: Holt, 1957.

Fey, W. F. Acceptance by others and its relation to acceptance of self and others: A revaluation. *J. abnorm. soc. Psychol.,* 1955, **50**, 274–276.

Fisher, S., & Cleveland, S. E. *Body image and personality.* Princeton, N. J.: Van Nostrand, 1958.

Friedman, I. Phenomenal, ideal and projected conceptions of self. *J. abnorm. soc. Psychol.,* 1955, **51**, 611–615.

Gleser, Goldine, & Ulett, G. The Saslow screening test as a measure of anxiety-proneness. *J. clin. Psychol.,* 1952, **7**, 279–283.

Goldings, H. J. On the avowal and projection of happiness. *J. Pers.,* 1954, **23**, 30–47.

Gordon, T., & Cartwright, D. The effect of psychotherapy upon certain attitudes toward others. In C. R. Rogers & Rosalind F. Dymond (Eds.) *Psychotherapy and personality change.,* Chicago: Unimzr. of Chicago Press, 1954. Pp. 167–195.

Guthrie, E. R. *The psychology of human conflict.* New York, Harper, 1938.

Havighurst, R. J., Robinson, Myra Z., & Dorr, Mildred. The development of the ideal self in childhood and adolescence. *J. educ. Res.,* 1946, **40**, 241–257.

Helper, M. Parental evaluation of children and children's self-evaluations. *J. abnorm. soc. Psychol.,* 1958, **56**, 190–194.

Horney, Karen. *Neurosis and human growth.* New York: Norton, 1950.

Hoyt, D. P., & Magoon, T. M. A validation study of the Taylor Manifest Anxiety Scale. *J. clin. Psychol.,* 1954, **10**, 357–361.

Jourard, S. M., & Remy, R. M. Perceived parental attitudes, the self, and security. *J. consult. Psychol.,* 1955, **19**, 364–366.

Leonard, W. E. *The locomotive-god.* New York: Appleton, 1927.

Levy, L. H. The meaning and generality of perceived actual-ideal discrepancies. *J. consult. Psychol.,* 1956, **20**, 396–398.

Lucas, J. D. The interactive effects of anxiety, failure and inter-serial duplication. *Amer. J. Psychol.,* 1952, **65**, 59–66.

Martin, B. The assessment of anxiety by physiological behavioral measures. *Psychol. Bull.,* 1961, **58**, 234–255.

Montague, E. K. The role of anxiety in serial rote learning. *J. exp. Psychol.,* 1953, **45**, 91–96.

Mowrer, O. H. *Learning theory and symbolic processes.* New York: Wiley, 1960.

Mussen, P., & Porter, L. W. Personal motivations and self-conceptions associated with effectiveness and ineffectiveness in emergent groups. *J. abnorm. soc. Psychol.*, 1959, **59**, 23–27.

Phillips, E. L. Attitudes toward self and others: A brief questionnaire report. *J. consult. Psychol.*, 1951, **15**, 79–81.

Raimy, V. C. Self-reference in counseling interviews. *J. consult. Psychol.*, 1948, **12**, 153–163.

Riedel, W. W. Scores on the Welsh A Scale as predictors of the number of "doubtful" judgments in a psychophysical experiment. Unpublished M.A. thesis, Univer. of Delaware, 1961.

Rogers, A. H., & Walsh, T. M. Defensiveness and unwitting self-evaluation. *J. clin. Psychol.*, 1959, **15**, 302–304.

Rogers, C. R., & Dymond, Rosalind F. (Eds.) *Psychotherapy and personality change: Co-ordinated studies in the client-centered approach.* Chicago: Univer. of Chicago Press, 1954.

Rudikoff, E. G. A comparative study of the changes in the concepts of the self, the ordinary person, and the ideal in eight cases. In C. R. Rogers and R. F. Dymond (Eds.), *Psychotherapy and personality change: Co-ordination studies in the client-centered approach.* Chicago: Univer. of Chicago Press, 1954. Pp. 85–89.

Sheerer, Elizabeth T. An analysis of the relationship between acceptance of and respect for self and acceptance of and respect for others in ten counselling cases. *J. consult. Psychol.*, 1949, **13**, 169–175.

Sheriffs, A. C., & Boomer, D. S. Who is penalized by the penalty for guessing? *J. educ. Psychol.*, 1954, **45**, 81–90.

Spence, K. W., & Beecroft, R. S. Differential conditioning and level of anxiety. *J. exp. Psychol.*, 1954, **48**, 399–403.

Spence, K. W., & Taylor, J. A. The relation of conditioned response strength to anxiety in normal, neurotic and psychotic subjects. *J. exp. Psychol.*, 1953, **45**, 265–272.

Spielberger, C. D., & Katzenmeyer, W. G. Manifest anxiety, intelligence, and college grades. *J. consult. Psychol.*, 1959, **23**, 278.

Stephenson, W. *The study of behavior: Q-technique and its methodology.* Chicago: Univer. of Chicago Press, 1953.

Stock, Dorothy. An investigation into the interrelations between the self-concept and feelings directed toward other persons and groups. *J. consult. Psychol.*, 1949, **13**, 176–180.

Sullivan, H. S. *The psychiatric interview.* New York: Norton, 1954.

Taylor, Janet A. A personality scale of manifest anxiety. *J. abnorm. soc. Psychol.*, 1953, **48**, 285–290.

Turner, R. H., & Vanderlippe, R. H. Self-ideal congruence as an index of adjustment. *J. abnorm. soc. Psychol.*, 1958, **57**, 202–206.

Welsh, G. S., & Dahlstrom, W. G. (Eds.) *Basic readings on the MMPI in psychology and medicine.* Minneapolis: Univer. of Minnesota Press, 1956.

Chapter 10

Repression-Sensitization and Mechanisms of Personality Defense

INTRODUCTION

In the last chapter we discussed some of the general aspects of personality defense. Primarily from psychoanalysis, psychology has inherited formulations of specific unconscious mechanisms which enable the individual to cope with anxiety. Many different kinds of anxiety-reducing activities which are unconsciously motivated have been labeled and described as *mechanisms of defense.* Standard textbooks in the psychology of adjustment or in introductory psychology list anywhere from 6 to 30 or more.

These classifications are not always satisfactory. For one thing, given behavior may be classifiable in a number of different ways. For another, the particular mechanism postulated to underlie a given sequence of behavior may represent concepts that are impossible to validate scientifically. Finally, the very word "mechanism" implies that the antecedent conditions, or the determinants, of the behavior in question are known. However, in spite of the central importance defense mechanisms are assumed to have, there has actually been little research either attempting to demonstrate the operation of particular mechanisms or specifying the conditions under which they occur. On the other hand, the evidence that does exist is promising. It represents one of the most exciting areas of investigation within psychology. Recent experiments dealing with the mechanisms of defense will be discussed throughout this chapter.

Advances in experimental personality research have suggested that the mechanisms of defense can be meaningfully viewed within a broader context. The classification employed in this chapter will relate the mechanisms of defense to established principles of anxiety-oriented behavior. Such mechanisms as repression, denial, reaction-formation, and sweet-lemon rationalization, will be classified as *repression-type mechanisms.* On the other hand, projection, intellectualization, compartmentalization, compulsivity, sour-grapes rationalization, and related defenses will be classified as *sensitization-type mechanisms.* Unfortunately, because of the present state of our knowledge, it will have to be pointed out from time to time that certain aspects of this classification must be viewed as tentative. However, the broader aspects of the classification appear to be sound.

THE REPRESSION-SENSITIZATION DIMENSION

Cumulative observation and research over a number of years have established that there are characteristic differences among people in their responses to threatening stimuli or to anxiety-producing situations. These anxiety-coping behaviors can be ordered along a continuum that might be briefly described as an approach–avoidance dimension. At one extreme of this continuum, people tend to react to threatening situations with avoidance, denial of threat, and withdrawal. These people are called *repressors.* At the other extreme, people tend to admit anxiety, approach threatening situations, and attempt to control the anxiety response and the anxiety-producing stimulus. People who behave this way have been termed *sensitizers.* Both of these extremes of anxiety-oriented behavior are thought of as unconsciously motivated attempts to reduce anxiety. They appear to be basic patterns of reactivity which underlie the mechanisms of defense.

Individual differences along the anxiety approach–avoidance dimension had been established by experimenters in the area of perception; notably, those at the Laboratory of Social Relations at Harvard (e.g., Bruner & Postman, 1947). Other investigators had related these individual differences to the concept of personality defense (e.g., Ericksen, 1950; Gordon, 1957). The development of instruments to measure repression-sensitization was soon to follow. Several were devised (e.g., Altrocchi, Parsons, & Dickoff, 1960; Ullmann, 1962), but the one which has enjoyed the most popularity was developed by Don Byrne at the University of Texas (Byrne, 1961; Byrne, Barry, & Nelson, 1963). He called it the *Repression-Sensitization Scale,* or simply, the RS scale.

It will be worth our while to review, at least briefly, some of the repression-sensitization research, most of which has employed the RS scale. Some of this research has not been published, but has been presented in a review by Byrne (see Maher, 1964). A knowledge of the relationships between individual differences on the RS scale and behavior will enable the reader to appreciate the variety of behavioral situations from which inferences concerning repression and sensitization are made. Most of this research has dealt with extreme scorers in both directions on the RS scale and has not included those scoring in the middle (normal) ranges. We will first discuss data relevant to low scorers on the RS scale (repressors), followed by a discussion of the high-scoring individuals (sensitizers).

Repressors

Those scoring low on the RS scale tend to show three broad behavioral trends: avoidance and denial of threat, a tendency to give socially desirable responses, and a tendency toward emotional stability.

A brief survey of research relative to the first of these tendencies involves findings such as the following: (a) Repressors tend to forget learned material when it is associated with anxiety-producing experiences (Gossett, 1964). (b) They also show higher recognition thresholds for failure-connected material (Tempone, 1962). (c) In objectively threatening situations, such as exposure to sexually arousing literary material, they deny experiencing any anxiety (Byrne & Sheffield, 1965). (d) In contrast to the fact that they do not admit any disturbance in objectively threatening situations, when physiological measures are taken increased autonomic responsiveness is indicated (Lazarus & Alfert, 1964).

The body of research indicating that repressors respond to test situations by giving socially desirable responses includes such findings as the following: (a) In responding to various kinds of ambiguous stimuli repressors are likely to give conventional, stereotyped answers and they tend to avoid the expression of sexuality and hostility (Carpenter, Wiener, & Carpenter, 1956; Lazarus, Ericksen, & Fonda, 1951). (b) They also present themselves positively on self-report scales designed to measure the self-concept (Altrocchi *et al.*, 1960). Since this is the case, they tend to have relatively small self-concept–self-ideal discrepancy scores. (c) This tendency to put themselves in a positive light on personality tests is readily apparent on scales designed to measure socially desirable responding (Edwards, 1959). (d) On rating scales designed to measure hostility, they tend to attribute relatively little hostility to themselves (Altrocchi, Schrauger, and McLeod, 1964). On the other hand, hostility may be indicated in their responses to personality tests designed to measure unconscious hostility (Lomont, 1964).

Finally, let us indicate some of the research findings relative to the relationship between repressor tendencies, as measured by the RS scale, and emotional stability: (a) In line with their tendency to give socially desirable responses, they score high on paper-and-pencil tests of emotional stability (e.g., Weinberg, 1963). (b) Other data indicate that they tend to show concern for maintaining friendly relations with others (Joy, 1963).

Sensitizers

Sensitizers (individuals who score high on the RS scale) show a pattern of behavior which is in many ways in direct contrast to the behavior of repressors. Let us review briefly the research findings in three areas: anxiety-oriented behavior; responses on self-report inventories, and indexes of adjustment.

First, relative to anxiety-oriented behavior the following findings are typical: (a) In contrast to repressors, sensitizers show a significant tendency to *recall* material associated with the experience of failure or anxiety (Gossett, 1964). (b) When they are presented with threatening material such as sexually arousing literary passages, they tend to *report* anxiety (Byrne & Sheffield, 1965). (c) As opposed to their verbal reports, physiological indexes are likely to indicate low autonomic responsiveness during such stress situations (Lazarus & Alfert, 1964). (d) There is some evidence that they are more appreciative of humor than extreme repressors. However, they are not as responsive as normal individuals, where "normal" refers to those scoring in the middle ranges on the RS scale (O'Connell & Peterson, 1964).

On self-report inventories sensitizers also show response patterns which are the opposite of those shown by repressors: (a) They tend to admit anxiety, failure, and inadequacies. (b) They have a somewhat negative self-concept, which results in their having a greater self-concept–self-ideal discrepancy than repressors. (c) On self-report scales of hostility they attribute more hostility to themselves than do repressors. (d) Finally, in contrast to repressors, on personality tests they tend to score in the "emotionally unstable" or "maladjusted" ranges.

Repression-Sensitization and Emotional Stability

Research findings to date suggest that sensitizers are more maladjusted than repressors or those scoring in the middle ranges between the two extremes. As the sensitization score increases, various indexes of maladjustment also tend to increase. Such a conclusion, however, must be made only on a highly tentative basis. Much of the research demonstrating this relationship may be open to distortions or reflect nothing more than the fact that sensitizers are willing to admit anxiety and unacceptable impulses on self-report inventories.

Nevertheless, there is evidence other than just that based on self-report measures that sensitization is related to adjustive difficulties. In studying the relationship between personality differences in dealing with stress and various obstetrical complications, McDonald, Gynther, and Christakos (1963) found a significant relationship between sensitization and the occurrence of various irregularities associated with childbirth. One study (Byrne, Steinberg, and Schwartz, 1968) concluded that sensitizers reported significantly more physical illness, and visited the university health clinic more often than either middle-scoring subjects or repressors. Neuropsychiatric patients have also been found to have significantly higher sensitizing scores than college students, although the psychiatric sample included some low-scoring repressors as well (Ullmann, 1962). Another kind of indicator of maladjustment is suggested by the fact that in experimental group discussions among three men – one repressor, one sensitizer, and one subject scoring in the middle of the scale – it has been found that sensitizers are chosen less often as desirable partners than those scoring in either of the other two categories (Joy, 1963).

The suggested relationship between sensitization and maladjustment does not appear to be in line with clinical experience. Clinical theory and practice strongly indicate that the denial and avoidance of anxiety is as maladjustive as the anxiety-approach behavior. Consider the extreme case of the mild-mannered, overcontrolled, nonaggressive person who suddenly and unexpectedly commits a violent criminal act. The dynamics of such cases are most apparent when the individual is a young person. Parents and acquaintances are frequently shocked with disbelief, because the child has been, insofar as they know, a model child. The clinical hypothesis is that such individuals appear less aggressive in their everyday behavior because assertiveness, anger, or aggression has been associated with anxiety. The denial or repression of unacceptable impulses, therefore, prevents the individual from being overwhelmed by anxiety. The cumulative effect of repeated denial and reality-distortion then results in eventual increased tension and a violent outbreak. Byrne has reported a study by Megargee (1964) in which a significant correlation was found between degree of criminality and repression as measured by the RS scale. However, there are few other data available to support the clinical concept of the relationship between repressor-type mechanisms and maladjustment.

Perhaps it can be seen more clearly how repressor patterns are related to adjustment if we consider a study, based on interview data, of personalities whose functioning could be described as above normal. Personal histories were collected on a sample of 100 intermediate-grade military officers by Renaug and Estess (1961). Renaug and Estess were not directly interested with repression–sensitization. Their main interest was in showing that "One-hundred men who, as a group, functioned at above average levels, and who were substantially free of psychoneurotic and psychosomatic symptomology, reported childhood histories containing seemingly as many traumatic events or pathogenic factors, as we ordinarily elicit in history-taking interviews with psychiatric patients who are in varying degrees disabled by their symptoms." Their conclusion, one with which the author would certainly agree, was that emotional disturbance cannot be predicted on the basis of any simple account of the number of disruptive events in the individual's life. Much more important considerations involve the situations and contexts in which frustration and conflict occur, and their relation to adjustive sequences in the person's life.

But let us take a closer look at the adjustive behavior of these men, as seen through Renaug and Estess' interviews with them. Considerable evidence can be found for the operation of repression, as opposed to sensitization patterns.

Thus, Renaug and Estess reported:

> With few exceptions these men's family backgrounds were characterized by appreciable degrees of tension and conflict. Roughly three-quarters of them, however, *seemed to be unaware of the fact of such conflict* even though they might at the moment be speaking of its results. [Renaug and Estess, 1961, p. 788]

Further comments which related to the handling of possible anxiety-producing situations, for example, the topic of sex, indicate the existence of repressor-type patterns. Relative to sexuality Renaug and Estess state the following:

> The writers were impressed with the amount of repression or suppression concerning childhood sexual memories. Involuntary contradictions, confusion of ages, or impossible replies ("Well, I just never thought about anything like that until I was thirteen"—on a question concerning masturbation) characterized roughly one-third of the interviews. [p. 790]

At another point in the study the authors stated:

> This (Sexual Relationships) was a difficult topic for most of the men to discuss. In many interviews, cliches, nervous sighs or over-talkativeness belied the subject's disclaimers of uneasiness. [p. 794]

Later on they say:

> In short, there seemed to be less exuberance or pleasure in sexuality than might have been expected in this age group. [p. 795].

Now let us turn our attention to the handling of hostility by these men. On this subject Renaug and Estess reported as follows:

> Hostility or disagreement seemed to be handled mainly by avoidance. To be a "complainer," "an arguer," or "guardhouse lawyer" was a distasteful role. The men placed a premium on being "average" and "normal," and described themselves by such phrases as "just about like everybody else," or "just average." [p. 792–793.]
>
> Much of what we learned about these men can be summarized in the statement that they had a very low tolerance for negative feelings or disagreement. [p. 793]
>
> It appeared to the writers that repression was the most commonly used mechanism. It may be stated in general that the subjects felt comfortable about making "complaints" only after prolonged difficulties, and it is of speculative interest to note that, typically, the mothers of these men were most able to express opinions or wishes in the form of grievances. [p. 793]

A final point to note, bearing on repressive patterns is that:

> Few men had strongly developed intellectual, theoretical, esthetic, or cultural interests. [p. 795]

The fact that these people were functioning at high levels in responsible jobs with little evidence of anxiety, tension, and emotional disturbance needs to be especially stressed. By ordinary standards they must be considered as people who have made a superior adjustment. Little, if any, sensitization trends were apparent in the interviews.

Studies such as these would seem to indicate that repressive patterns are maladaptive only when they are very extreme, and that repressive defenses against anxiety have considerable value in reducing emotional disturbance and allowing the individual to function on a responsible level.

On the other hand, although these people tended to be gregarious, sociable, and skilled in their jobs, their outlooks appeared generally to have a philistine quality. They were conventional and submissive to authority; somewhat passive in their marital relationships; and they tended to fear, deny, and avoid expression of sexual or hostile impulses.

Why would not such behavior eventually result in an increase in tension? For one thing, although the interviews were relatively thorough, a more intensive appraisal might have turned up more underlying tensions than were reported. However, perhaps the more important factor was that these individuals were receiving considerable reinforcement in their lives. Probably their status, success in their work, and their facility in dealing with others created a situation in which they could accept the facts of their social environment and repress certain of their needs without feeling inadequate, inferior, or guilty. A little positive reinforcement from peers and from superiors—that is, the very fact of their way of life being accepted by others—could go a long way toward reducing tension and enabling the repressive mechanism to function effectively. "Good adjustment" ordinarily implies that the individual is able to function well in his daily life, and that he is free from anxiety and hostility. Of course, many other kinds of value judgments are possible when other criteria are employed.

REPRESSION-TYPE DEFENSE MECHANISMS

The Mechanism of Repression

Repression, the avoidance learning of anxiety-producing thoughts and feelings, has been discussed from time to time in various contexts in the preceding chapters. Now, let us turn to a consideration of the mechanism itself. The repressive process may occur as the result of either of two types of anxiety-producing situations. One type occurs when the precipitating events are unusually traumatic, such as severe physical injuries or the witnessing of particularly horrifying events, such as combat scenes. These single-incident traumatic events may be so destructive that they take away the individual's sense of security and produce inordinate anxiety. Thoughts and feelings associated with such events undergo abrupt and severe repression because of their potential anxiety-generating power.

Another type of situation which results in repression is the learning of anxiety over a period of time. This is a much more typical occurrence. Impulses such as striving for mastery and independence, sexual curiosity, or hostility towards parents, may only very gradually be checked because of parental discipline. When the child changes so that the admittance of prohibited impulses or anxiety-producing thoughts no longer occurs, sometimes it is in response to relatively minor experiences in his life. This is because of the symbolic, or verbal nature of the repressive process. A single anxiety-producing experience may represent a condensation of a series of damaging events. For instance, Wolberg (1954), an experienced psychotherapist, points out that repression in childhood frequently occurs following an experience that convinces the child of the possibility of real danger. Consider an insecure child who has retained certain rebellious tendencies, and who happens to witness the whipping of a dog that has done something to anger its master. The child may be frightened by such brutal treatment and, by way of associating himself with the animal, may fear that he will be injured in some way if he continues to openly defy his parents. Sharp repression and denial of his own aggressive impulses may then ensue, possibly along with a phobia of dogs.

Dollard and Miller (1950) have stressed the fact that the mechanism involved in repression is very similar to that of the simple suppression of thoughts and feelings. The primary difference is that repression occurs automatically and unconsciously, whereas suppression occurs "intentionally," or at least, on a conscious level. Suppression refers to the conscious avoidance of unpleasant or anxiety-producing thoughts and feelings. It is a very common experience for all of us. We do not like to seriously consider topics such as death, atomic war, possible loss of loved ones; nor do we like to discuss events which have been unpleasant for us. Simple suppression occurs when we sit down to do a difficult job. Thoughts of the possibility of failure, of approaching deadlines, or even relatively minor intruding associations—events of the day, an invitation to visit friends, anything that competes for our attention—all of these thoughts have to be forced out of consciousness.

It appears that repression also does not have to occur abruptly and automatically. With the repeated suppression of certain thoughts the suppressive process itself may gradually become unconscious, just as any other overlearned habit does.

In this connection it is interesting to note that if repression is, indeed, learnable in this way then it ought to be teachable. Sensitizer personality types tend to worry and ruminate a lot. It is possible that deliberate training in suppression would be of some value in enabling the individual to cope with minor, but for him, anxiety-producing, aspects of his life. Sometimes in treating severely disturbed psychiatric patients it is advisable to strengthen certain defenses, particularly repression and not to discuss topics which mobilize anxiety. This is done when the patient is judged unable to handle any escalation of anxiety over his present state.

A further difference between suppression and repression is that material which has merely been suppressed can be freely recalled, although it might be of such a nature that the individual becomes uneasy and tense when he does so. On the other hand, the automatic, unconscious process of repression completely inhibits recall except under unusual circumstances in which the anxiety associated with the repressed thoughts or feelings becomes extinguished so that the thoughts are free to enter the person's mind.

An Experimental Analogue of Suppression and Repression

The mechanisms underlying both repression and suppression can be clarified by an experimental analogue of these processes. Let us consider an experiment by Eriksen and Kuethe (1956) in which the avoidance conditioning of verbal behavior occurred both with and without awareness.

The concept of repression rests upon the fact that thoughts which have become conditioned stimuli for anxiety tend to be eliminated or replaced by other thoughts. Therefore, these investigators set out to test, first, whether verbal associations which had been formed during the individual's life experiences could be changed by punishment (in the form of electric shocks), and second, whether these changes could occur without the subject's becoming aware that his behavior had been modified.

A word association test was used in this experiment. Fifteen stimulus words were employed, words like, "polite," "small," "cheerful," and "rural." Pilot studies had shown that when the same word list is repeated a number of times, subjects generally repeat their original associations to the words. In this experiment the subjects were asked to associate to the words on the list during ten repetitions of the list, or ten trials. Before the experiment began the experimenter randomly chose five words which were to be the critical stimuli for that particular subject. During the first presentation of the list the subject's associations to those five critical words were followed by a strong electric shock. Thereafter, on other trials, with the words presented in varying sequences, the subject received a strong shock every time he responded with one of the five associations that had been shocked during the first presentation of the list. The subjects were misinformed as to the nature of the experiment. They were told that the purpose of the experiment was to determine the "maximum speed of mental associations." They were told that shocks would occur if their associations to a stimulus word were too slow, and also that shocks would occur for another reason that the subjects could not be told about in advance, but they might be able to discover on their own, and hence avoid receiving the shocks. After these instructions the learning phase was begun. The subjects were instructed to respond to each word with the first word that came to mind, and to respond as quickly as possible.

At the end of the experiment it was found that most of the subjects could be divided evenly into a group that was able to give the real basis for the shocks and to describe what they had done to avoid them, and a second group of subjects who were not able to state any reasons for the shocks other than the misinformation they had been given at the beginning of the experiment. The respective labels of "insight" and "no-insight" were given to these two groups.

Figure 10-1 shows the percentage of associations that were repeated by the subjects in each respective group on succeeding trials. Notice first the upper two curves. They represent the percent of nonshocked associations that were repeated by the subjects on each trial: the percent

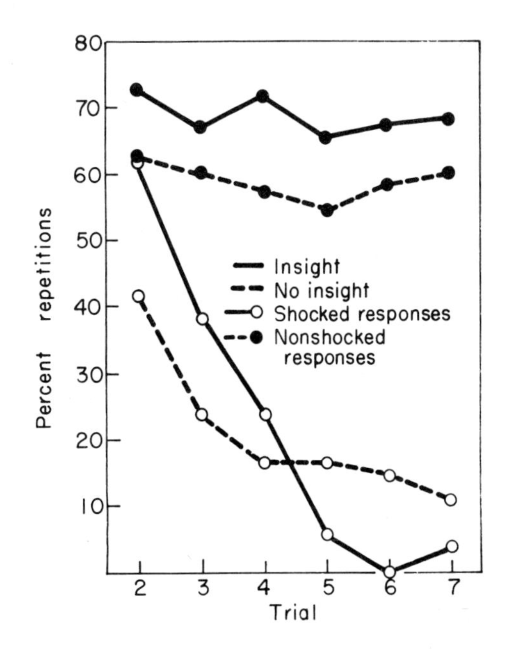

FIG. 10-1. Percentage of first-trial responses repeated on succeeding trials for all combinations of insight, no-insight, shock, and nonshocked groups. (Eriksen & Kuethe, 1956.)

of nonpunished associations which were repeated on successive trials by both the insight and the no-insight groups was relatively high (roughly between 60 and 70 percent), and it remained relatively constant throughout the experiment.

In contrast, look at the lower two curves, which represent the percent of punished associations which were repeated. There it can be seen that both the insight and no-insight groups learned quite successfully to avoid giving punished associations. It should be remembered that these punished associations were the first associations the subject had to a given word, and therefore would have been originally the strongest associative response in the subject's associational hierarchy. Therefore learning to avoid giving these relatively strong first associations should have been a fairly difficult task. Another factor which should have increased the difficulty of this learning is that the subjects thought that by responding quickly they could avoid shock. In spite of these facts, by the fifth trial both insight and no-insight groups were giving only a fraction of the original punished associations. Thus, clear avoidance conditioning occurred for both groups. Furthermore, the difference between the two groups in the rate of learning was not significant statistically. Both insight and no-insight groups learned at roughly the same rate.

We can see the qualitative difference in the learning between the insight and no-insight groups by inspecting the reaction-time data given in Fig. 10-2. This figure shows the time it took individuals to respond to the nonpunished words (noncritical stimili) and to the punished words (critical stimili) over successive trials. Careful inspection of this graph will reveal that for the no-insight group (the dashed lines) the time it took to associate to both punished and nonpunished words decreased steadily over trials. Now look at the curves of the insight group (solid lines). The times for the nonpunished words decreased with trials, just as they did for the punished and nonpunished words in the no-insight group, but for the punished words there is an *increase* in reaction time. The reaction times for punished words for this group only began to decrease on the fourth trial, and never did attain the speed of the other groups. This indicates that the subjects in the insight group deliberately withheld responses that had been punished while they attempted to think of new associations. When they were questioned after the experiment these subjects said that eventually they learned to think of nonpunished associations "automatically." This is reflected in the data of Fig. 10-2

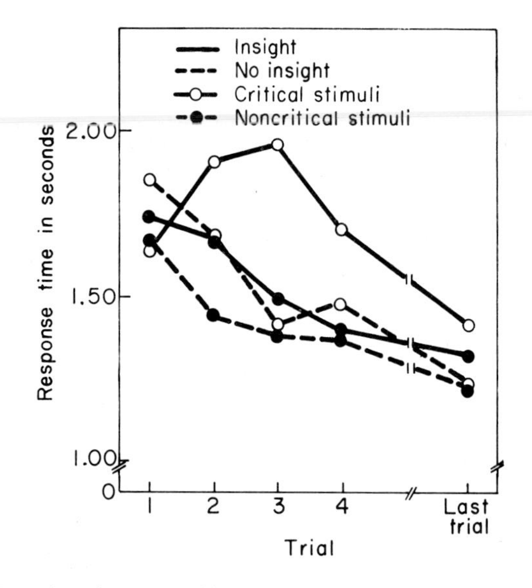

FIG. 10-2. Reaction times to critical and noncritical stimuli over successive trials. (Eriksen & Kuethe, 1956.)

since the reaction times for the last trials closely approach those for the nonpunished words.

A particularly interesting aspect of this experiment, because of its bearing on the repressive process, is that after the avoidance-learning phase had been completed the subjects were asked to chain-associate for 15 seconds to each of the stimulus words. They were also informed that there would be no more shocks in the experiment. In spite of the conscious knowledge that electric shocks would no longer occur it was found that the subjects gave significantly fewer first-trial associations to the previously punished words than to the nonpunished words. That is, the effects of punishment upon the subject's associations persisted.

Notice in particular in this experiment that the behavior of the insight group was equivalent to the conscious process of suppression and the behavior of the noninsight group to the process of repression. Punishment was avoided in both groups, but in the noninsight group avoidance occurred "automatically," without awareness and without an increase in their reaction times. On the other hand, the deliberate conscious act of avoidance by the insight group was reflected by an increase in their reaction times while they searched for other associations.

Further evidence that punishment has its effect on an unconscious, implicit, level is given by the fact that both groups of subjects gave fewer punished associations during the chain-association part of the experiment. Whether the learning occurs automatically and unconsciously, or is the result of deliberate suppression, the punished thoughts are eventually inhibited.

Denial

Normally, repression is inferred from the blocking of thought processes which may have anxiety-producing content and from the person's general style of coping with certain situations unimaginatively, or with a general constriction of ideation or creativity. Repressor-type personalities are also characterized by a certain naivete. They tend to cope with life situations in a generally unreflective way. There is also a tendency for them to have childish expressive reactions. For instance, such a person might respond to Rorschach ink blots by verbalizations such as, "Oh my, this is pretty," or "How ugly," or have phobic verbalizations, e.g., "That looks like a spider. Ugh! I hate bugs!". In responses to projective tests such as the Rorschach the repressor tends to emphasize color and simple form.

People judged by clinical psychologists to be repressors tend to be responsive to the more obvious and prominent aspects of their environment, rather than to detailed or complex aspects of it. Rapaport (1942) and Schafer (1948) report that one of the most commonly noticed traits of repressors is the difficulty they experience in being very specific or articulate in their recall or memory productions. For example, on the information part of an intelligence test if they are asked "Where is Brazil?" their response might be, "In a jungle somewhere—near Argentina."

This general lack of specificity in the repressor suggested to a group of researchers (Gardner, Jackson, & Messick, 1960) that there might be significant differences between repressors and other personality types in their reactions to certain perceptual situations. They hypothesized that repressors would tend to show a type of cognitive, or perceptual response known as *leveling.* Leveling refers to evidences of a low degree of articulation of the stimulus field. For example, people classified as levelers tend to have difficulty in extracting outlines of objects, pictures, or figures which are drawn or embedded within a larger context. They are people for whom gradual changes in a stimulus situation appear to go unrecognized. In one typical experiment subjects were shown a series of five squares projected on a screen. They were asked to estimate the size of the squares each time they were exposed. Unknown to the subject, the experimenter, over a series of trials, gradually removed the smallest square from the configuration, and added a square which was larger than the previous largest square. People classified as levelers tended not to notice the gradual changes. The evidence reported by Gardner *et al.*, indicated quite strongly that people diagnosed by clinicians to be repressors behave like levelers in experimental situations such as these.

All of these facts have relevance for personality defense. People classed as repressors tend to have cognitive dispositions to deal with threat and stress reactions in ways that are compatible with denial statements. They show a general tendency to be unable to cope with complexity; to be unreflective; to give relatively undifferentiated memory responses; and to easily assimilate gradual change without awareness: Given these habits of responding to life's situations it would be reasonable to expect that threatening stimili would be treated in a similar manner. Threat, unless it were considerable, would be expected to go unrecognized by the extreme repressor. Even given the fact that certain situations are recognized as threatening, it would be expected that individuals showing repression as their predominant method of handling anxiety might unreflectively respond to any evidence in the environment fostering denial.

Evidence is accumulating that this is the case. For people classified as repressors on the basis of the RS scale or other personality tests, fostering denial tends to short-circuit the effects of threat or reduce the level of threat. On the other hand, nonrepressors do not gain in terms of stress-reduction from the encouragement of denial. Let us consider one of an important series of experiments by Lazarus and others at the University of California. In order to study the conditions that determine the reduction of stress, Lazarus and Alfert (1964) asked 69 students to view a film dealing with a primitive ritual called "subincision." The film showed a young boy, 13 or 14 years old, being restrained by three or four older men. Agitation and stress are exhibited as 3 or 4 inches of the under side of his penis are cut to the depth of the urethra with a piece of flint.

Subjects were first given the RS scale, along with several other personality scales, and then randomly assigned to one of three conditions. In one experimental condition, called the "silent film condition," subjects watched the film, without any special introduction or commentary. However, as in all other conditions, continuous recordings were made of skin conductance and heart rate as they sat watching the film. In the second condition, the "denial commentary condition," the procedure was the same except that a short introduction and a running commentary during the film were presented. In this commentary the harmful threatening aspects of the events were denied. For example, the subjects were told that the operation was not harmful or painful. Also, the positive benefits of participation in the ritual were emphasized. It was described as a joyful occasion for the native boys, who look forward to gaining adult status through the ritual. The third condition was called the "denial orientation condition." It involved the reading of the denial passage before the showing of the film, without a running commentary while the film was going on.

The results indicated that the silent film elicited the highest degree of autonomic responsiveness from the subjects. Those subjects who were exposed to the denial commentary condition showed significantly lower levels of skin conductance and heart rate than those exposed to the same film, without the denial commentary, that is, the silent condition. The third group, those with the denial orientation, had lower skin conductance and heart-rate indexes than either of the other two groups. This latter finding probably occurred because the denial group was exposed to full denial information all at once, whereas the denial commentary group did not receive all of the information until well into the presentation of the film.

What is of particular interest is the results for those classified as repressors, or *deniers* on the various personality inventories. Lazarus and Alfert state: "high deniers refuse to admit disturbance verbally but reveal it autonomically, while low deniers are apt to say they are more disturbed while showing less autonomic reactivity. This is, of course, exactly what the concept of denial means . . . ." Figure 10-3 shows the difference between those classified as high deniers and those classified as being low in denial in their autonomic responsiveness over the course of the film. While not admitting it verbally, the high deniers showed autonomic responsiveness which was both significantly greater than that shown by the low deniers, and highly coordinated with threatening points in the film. On the other hand, the low-denial group exhibited lower levels of autonomic activity, and their autonomic activity remained somewhat stable throughout the film. Measures of heart rate reflected the same thing.

Figure 10-4 compares the skin-conductance recordings of high deniers and low deniers under the three experimental conditions. Figure 10-5 compares the high and low deniers in terms of self-reported anxiety under the three experimental conditions.

In Figure 10-4 it can readily be seen that distress reaction, measured in terms of skin conductance, was high in the silent film for deniers, greatly reduced under the denial commentary condition, and extremely low in the denial orientation condition. In other words, the effect of the denial statements was to short-circuit threat for these individuals. On the other hand, it can be seen that low deniers did not respond that way at all. In the first place, their autonomic response during the silent film was much less. In the second place, the denial statements, if anything, increased the autonomic stress reaction of these people. Presumably,

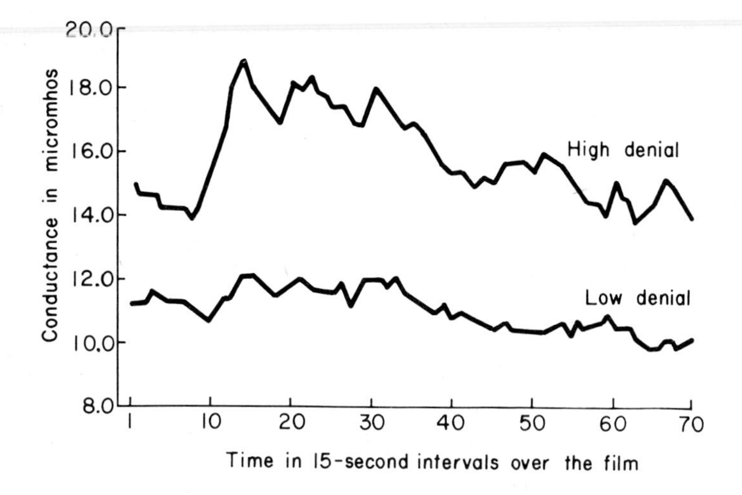

FIG. 10-3. The effects of the subincision film on skin conductance as a measure of autonomic responsiveness for high-denial and low-denial groups. (Lazarus & Alfert, 1964.)

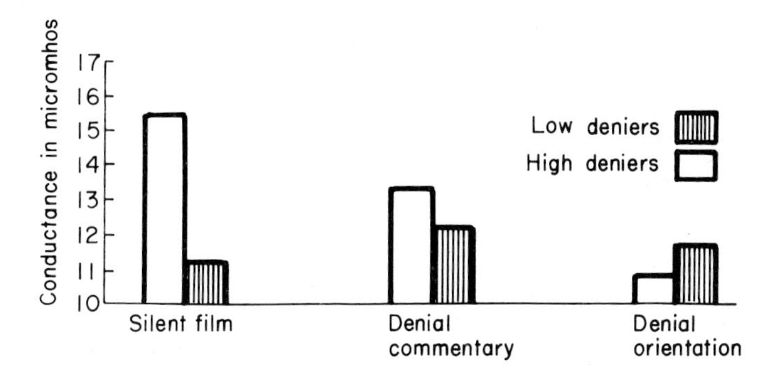

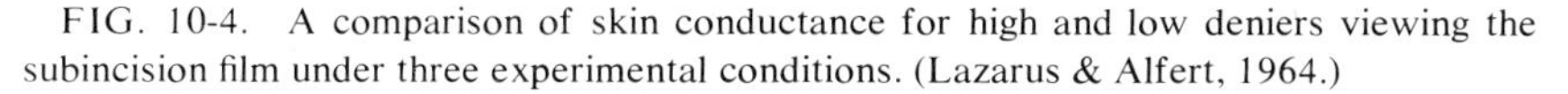
FIG. 10-4. A comparison of skin conductance for high and low deniers viewing the subincision film under three experimental conditions. (Lazarus & Alfert, 1964.)

since the low deniers do not have preexisting habits of unreflectiveness and avoidance, they cannot readily accept the denial statements.

The Lazarus and Alfert experiment clearly indicates that denial can operate as a defense against anxiety in certain personalities. Nor is the evidence restricted to just this experiment. In another study (Speisman, Lazarus, Mordkoff, & Dapison, 1964) business executives, found to be high in denial, showed reduction in stress under a denial condition, whereas students who were relatively low in denial did not. Exactly the same results have also been obtained when the stressor film was on the prevention of shop accidents, rather than the more esoteric subincision rites (Lazarus, Optom, Nomikos, & Rankin, 1965).

It should be stressed, as we have already noted in our discussions of hostility, sexuality, and dependency, that threatening or anxiety-producing stimuli frequently involve the person's own unacceptable impulses. Repression protects the individual from intolerable subjective danger and anxiety that would be generated if he were to permit himself to express certain impulses and feelings. Denial involves a further distortion —a complete disavowal of the impulse and feelings or ideas associated with it. Thus, the individual's conscious experience and his verbal communications to others may be in direct contrast to his real feelings and to his behavior in certain situations. Let us consider a fictional example:

Shirley's parents were divorced when Shirley was very young. Her mother never remarried and Shirley and her sister grew up without the benefit of a father. Her mother blamed their unhappy circumstances on their father. She frequently warned them that men were unreliable and in many ways undesirable. Now Shirley is married. She believes she loves her husband and that she would do anything for him. However, much of her behavior indicates the operation of unconscious hostility. Shirley is loath to prepare meals. She has her husband "fix himself a sandwich" much of the time. When they sit down at the dinner ta-

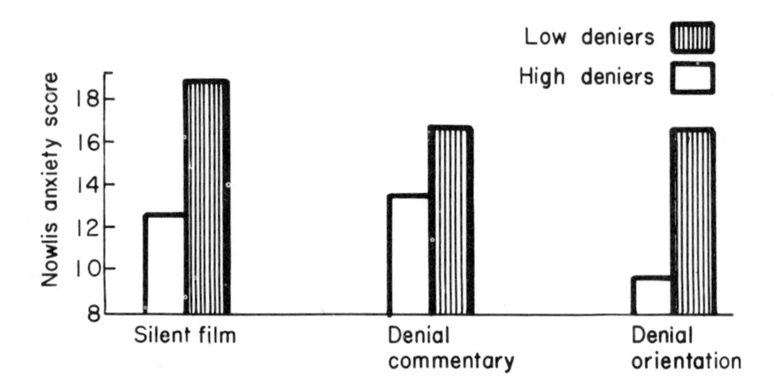

FIG. 10-5. Verbal reactions of high and low deniers viewing the subincision film under the three experimental conditions. (Lazarus & Alfert, 1964.)

ble she reads a magazine or a newspaper while they're having the meal. She spends much of the time with other women in the neighborhood, and little with him. On the occasions when they are alone, she quickly turns on the television set. There is actually little communication between them. When she leaves the house she seldom leaves a note, saying where she is going. Although she knows that Mr. A's business requires a certain amount of social life, she is decidedly unenthusiastic toward his business acquaintances and their wives. Her husband is quite unhappy about all this. Although her behavior undermines their relationship in many ways he cannot "prove" that she actually dislikes him. Shirley gets very upset whenever he indicates that real evidence of love is lacking and she emphatically denies it.

Reaction-Formation

The denial of unacceptable impulses or of external threat may be extreme. When the individual asserts the exact opposite of what he unconsciously feels the mechanism is called *reaction-formation.* Simple denial may have elements of reaction-formation. For instance, Shirley A, above, may not only deny any hostility toward her husband, but protest that she really loves him very much. Generally, however, the defense mechanism of reaction-formation refers to pervasive and stable attempts of the individual to defend himself against anxiety by thinking and behaving in a matter which is the exact opposite of his anxiety-producing feelings. Thus, the mother who actually harbors feelings of rejection and repulsion toward her child might defend herself against these impluses by showing elaborate displays of affection, extreme protectiveness, and anxious concern. The individual with an unusually violent temper who is quick to display irritability may learn, without his being aware of it, that his business and professional success is threatened by these traits. He may then repress these traits and go out of his way in his daily contacts to be pleasant and nice to people. The clue that such behavior is defensive in nature is primarily that it is "overdone." The mother is overaffectionate and overprotective; the businessman is so nice that his associates think of him as insincere. The following is an illustrative example:

> James E always thought of himself as "all boy." He never did anything "sissified." In high school he was very active in sports, especially weight-lifting, wrestling, and football. He detested poetry and English literature. In college he majored in engineering. His grades were not high but they were passable. A good deal of his extracurricular activities centered around his fraternity.
>
> His overconcern with being "a regular-fellow" probably helped his adjustment to army life. When he first arrived overseas, one night, seeking companionship, he went to the enlisted men's club on the base to have a few drinks. As he sat there another soldier sat down beside him and began chatting amiably. After a while the other soldier began to gently and firmly press his leg against his. It was unmistakeable. James recognized it immediately as a homosexual approach. He swung and knocked the soldier to the floor. Then he continued to hit him. When the military police stopped him he was kicking his victim violently. The charges were serious. The soldier had been almost fatally beaten. Furthermore, kicking with a GI boot constituted assault with a deadly weapon from a legal point of view. Eventually, however, the victim was transferred to another base and the military court dismissed the charges.

This kind of overreaction seems to indicate that reaction-formation, and repressive-type defenses generally, are maintained only at the cost of considerable energy expenditure. That is, when the impulses are repressed, denied, or actively opposed as in the case of reaction-formation, considerable tension is frequently created—tension which must be held in check more or less continuously.

Sometimes the defensive behavior itself provides a certain amount of gratification of the unacceptable impulse, and, therefore, a modicum of tension-reduction. For example, the overprotective mother, by making the child inordiantly dependent upon her may actually be behaving quite cruelly toward the child, insofar as she is crippling his chance for independence and autonomy in later life.

At other times the repressed impulse itself may remain unsatisfied, but a certain amount of reinforcement and tension-reduction may continuously result from the individual's defensive behavior. Thus, the extremely prudish or rigidly conventional person may be made so anxious by what he considers his own immoral tendencies that he manages a continual campaign to fight obscene literature, drinking, smoking, etc. Since the mores tend to support puritanical attitudes, he may not only successfuly deny his impulses, but also gain feelings of self-righteousness and self-esteem. In a different society, or even in a different subculture within his own society the irrationality of his attitudes and behavior would be much more obvious. (See p. 321.)

Tension-reduction may also accrue from the relief of hostile impulses under the guise of "decency." The extremely conventional person may have an especially punitive attitude toward social infringements. This mechanism, greatly exaggerated in fantasy, has been portrayed in a recent novel by Robert Sheckley, (1965) entitled *The Tenth Victim*. The setting of the novel is a society in the future. In the opening scene a litterbug is caught, red-handed, with a piece of a Hershey bar wrapper still in his hand. He is being led away to a public execution by impalement. The crowd gathers to watch . . . , etc.

Studies of Reaction-Formation. Mowrer (1940) conducted a classic experiment which he considered an analogue of reaction-formation. Rats were placed in an experimental box in which they received gradually increasing shocks through the electrified grid on the floor of the box. However, the shock could be terminated by pressing a panel at one end of the cage. The experimenter then arranged for some animals to receive shock whenever they pressed the panel. It was found that as the shock from the grid floor increased in intensity, some of the animals moved to the end of the cage opposite the panel. These animals appeared to behave as though they "wanted" to press the panel and terminate the shock, but instead retreated from it, in order to avoid further shock which might occur if they pressed the panel. Presumably, their move to the opposite end of the cage served to protect them from the anxiety-producing impulse.

In an interesting experimental demonstration of reaction-formation, Sarnoff (1960) randomly assigned 81 students to two experimental conditions that involved the manipulation of two levels of affectionate feelings toward others. Arousal of affection was induced by exposing subjects to modified excerpts of a play involving "the tender portrayal of spontaneous and unequivocal trust, warmth, and affection." For the high-arousal condition the excerpt was actually staged as a dramatic presentation. In the low-arousal condition subjects heard tape recordings of the same excerpt, but without the actors improvising emotional expression.

A "cynicism scale" was also administered to all subjects both before and after the arousal of affectionate feelings. That is, the scale was designed to measure the degree to which the subject had "a negative viewpoint concerning the inherent motivation of human beings."

Finally, a measure of reaction-formation was obtained for all subjects. Certain pictures taken from the Thematic Apperception Test were shown to the subjects, and they were asked to check their preferences for a number of statements which purported to summarize the interpersonal situations pictured in the test cards. The statements represented different defensive interpretations, so that it was possible to obtain an indication of the strength of the person's preference for any one of the defenses in relation to others.

Using a control group of 43 subjects not exposed to the affection-arousal manipulation, Sarnoff was able to show that those subjects found to have high reaction-formation against affection on the basis of their choice of statements about the TAT pictures were significantly more resistant to attitude change, as measured by the before–after scores on the cynicism scale, than those who were found to be low in reaction-formation against affection. This was true under the high-affection-arousal condition, but not under the low-affection-arousal condition. Of course, the general changes that occurred on the before–after cynicism rating after exposure to the affection-arousal conditions were in the direction of less cynicism.

Displacement

The mechanism of displacement was discussed at some length in the chapter on frustration (see p. 203). It involves the displacement of an impulse or a feeling which has been punished, and is, therefore, anxiety-producing, from the original object or person toward whom it was directed to other, neutral or less-threatening, objects or situations. Stimuli far removed from the original threatening situation come to elicit the punishable response. Displacement is a repression-type mechanism. The person is rarely aware of the reasons for his displaced feelings and actions. He is inclined to justify his behavior in terms of the present situation.

We noted in Chapter 6 that the mechanism of displacement is hardly as simple as was previously supposed. That is, the man who is reprimanded by his boss does not automatically come home and criticize his wife and perhaps kick his dog. For one thing, displaced hostility, and probably displaced sexual and affectionate behavior as well, usually depend upon some provocation on the part of the person upon whom it is displaced. However, the greater the amount of tension involved, the more likely it is that the individual will overreact to only slight provocation.

Another factor which adds to the complexity of the mechanism of displacement is that the displaced response can only rarely be attributed to simple stimulus generalization. That is, the similarity between the original object and the displaced object is liable to be symbolic or involve subtle contextual cues, rather than simple physical similarity.

When certain situations serve as positive cues for the repressed feelings or impulses, and when other cues in the situation signal what the person considers to be a lack of threat or danger, then the individual may be at a loss to control his impulses. A single example of displacement should suffice:

One morning on the way to school Beverly P., a girl of sixteen, saw a preschool child walking in the street and suddenly began beating on him. The child's parents called the police, and Beverly was eventually brought to the psychiatric clinic.

Interviews revealed that Beverly had always been considered a model child by her parents and teachers. That is, until about two years ago. Beverly always was a good student but her grades had suddenly dropped. Her teachers began to notice an adjustment problem.

Her parents appeared to be aloof and nondemonstrative people. Nonetheless it appeared that Beverly had always had a close affectionate exchange with her father. She had apparently competed with her older sister for her father's attention, and won. During the interviews her parents admitted that they had wanted a boy when Beverly was born, but they proudly pointed out that Beverly had a two-year-old brother who appeared to be the pride of the family. They were unaware of the deep-seated resentment which Beverly had toward the child, who had replaced her in their affections. They did note that they used to let her look after the child, but that she had unintentionally dropped him on occasion so that they had to prohibit her from caring for him—a job which she apparently enjoyed.

Sublimation

Sublimation involves the partial or symbolic gratification of unacceptable impulses by way of their being channeled into behavior which is socially acceptable. Classical psychoanalysts believed that a considerable amount of behavior represented sublimated primitive sexual and aggressive drives. It might be observed, for example, that when the surgeon enters the operating room a sudden change comes over him. He seems engrossed and completely absorbed. The obvious pleasurable concentration which he displays, the psychoanalyst might say, is the result of the sublimation of impulses which have deep unconscious roots—namely, murderous impulses.

This is a theory which is not easy to test. Most behavior is overdetermined. That is, it is a combined result of a number of sources of reinforcement. Nevertheless, people with repressive-type defenses might be expected to sublimate unacceptable impulses, particularly those of a sexual or hostile nature. Engaging in highly competitive sports, competitive business practices, playing games to win, etc., may be, at least in part, ways to work off a certain amount of unconscious hostility. Similarly, certain activities such as bowling, vigorous dancing, motoring, or horseback riding, may be associated with sexuality. Repressors would be expected to pursue these activities more vigorously than sensitizers. Furthermore, the degree to which the individual engages in such activities should fluctuate with tension-producing events in his life.

Identification

We saw in Chapter 8 that imitation learning or identification is related to certain sources of reinforcement, especially those involving social power. When either the acceptance of others or increased feelings of safety, security, and strength result from identification, the process can serve to reduce unconscious anxiety. Being a member of a group, a club, or a powerful organization, for example, tends to reduce anxiety by enhancing the individual's self-esteem.

There is some evidence (e.g., Silber & Baxter, 1963) that repressors tend to overestimate the amount of reinforcement which they receive in interpersonal contexts; that they respond more readily to approval, and when approval is withdrawn their behavior extinguishes relatively quickly. In contrast, sensitizers tend to resist being influenced by the approval of others. They appear to reject the demands of interpersonal situations. This implies that the repressor is more likely to behave in ways that are socially acceptable. He is more likely to spot cues in the situation related to interpersonal reinforcement, and consequently to imitate people around him whom he considers successful—that is, he is more likely to imitate, or identify with, reinforced models.

Rationalization

Rationalization simply refers to the fact that the individual disguises the true reasons for his behavior, or the fact that his behavior is irrational, by giving reasons which are plausible, logical, and socially acceptable. When the person intentionally distorts the reasons for his motives or behavior it is simply a case of lying, and not rationalization at all. Verbal bahavior is only classed as rationalization when it represents an unconscious defense against anxiety.

All of us tend to describe our motives in the most socially acceptable way—to ourselves and to others. Usually, merely giving a "logical" account of our behavior will suffice. As children we all receive a great deal of practice in doing this. Childhood training places an enormous stress on our being able to give a logical and sensible account of our behavior. Mothers commonly demand of 3-year-old children, "Why did you hurt your baby sister?" or "Why are you breaking your toy?" As the child grows he learns that plausible or sensible explanations are often accepted by adults. The relief in threat to his self-esteem strongly reinforces his learning how to rationalize.

Rationalization is common, of course, to both repressors and sensitizers, but there are differences in the types of rationalization used. Rationalization is an extremely ordinary, pervasive defensive phenomenon. It is often found in combination with other defensive patterns, which give it a particular quality. Two special kinds of rationalization are the so-called sour-grape and sweet-lemon varieties. The sour-grape type of rationalization is a reference to a story from *Aesop's Fables*, in which the fox who found he could not reach the grapes decided that they were sour anyway. This kind of rationalization is more typical of sensitizers. That is, they tend to admit failures and shortcomings—and then to rationalize them away. In sweet-lemon rationalization it is as if the individual denied that the lemon he received was sour. In fact, he claims that it is really sweet. It is immediately apparent that this type of rationalization contains the elements of denial and reaction-formation, typical of repressor defenses.

It will be recalled that the orienting passage in the denial condition of the experiment in which a film of the subincision rite was shown involved reaction-formation type comments, e.g., "the operation is not really painful. The boy actually enjoys it because it signifies his entry into adulthood." This is typical of the manner in which the repressor rationalizes away anxiety-producing thoughts in life situations—"The end justifies the means," "As long as you're successful society doesn't worry too much about it," etc. The repressor's tendency not to be able to cope with complex situations or complex explanations renders him an easy target for slogans, even though they might contain two incompatible ideas in the same sentence. Government spokesmen may justify a military budget with the slogan, "Armaments for Peace." A University administration may justify a "publish or perish" policy for its faculty by claiming that people who are actively engaged in research make better teachers. The man who purchases a costly new car stresses the special advantages—the added comfort, the tremendous horsepower, the supposed increased economy of a bigger and better engine, etc. The housewife claims that she saves money by buying luxury items on sale, overlooking the fact that she had not intended to, and would not have bought them at the ordinary price anyway.

Shortcomings, failures, or unpleasant motives may represent sour lemons which are really sweet. The sensitizer would argue that it would not be worth the time and effort to be highly successful in his occupation. The repressor goes a step further. Not only is success not worth having, but to be average, or normal, is a remarkably satisfactory state of affairs. The overprotective mother who refuses to have her boy go out on dates is "doing it for his own good"; so are the parents who prohibit their children from sharing in the adolescent culture. The parents who force their children into occupational choices they do not want or are not fitted for or who prohibit relationships which may end in what they consider an unacceptable marriage, "just know that he will thank us for it all later."

Producing Rationalization in the Laboratory

The conditions producing rationalization can be manipulated in the laboratory. The degree to which subjects rationalize their behavior varies with the level of self-esteem which has been experimentally induced. The mechanism is shown clearly in an experiment reported by Pepitone (1964). The object of the experiment was to create the perception in students that they had excessively and undeniably hurt another person. Because this is a violation of a cultural norm, it was expected that these subjects' evaluation of themselves would be lowered following the experience. The students were told that they would help diagnose the personal problems of a student who had sought help from the University Clinic. They were told that it had been found that the student's peers could often better identify with a troubled student than professional psychologists could.

The subject and two other students who actually were confederates of the experimenter were read a case history of Carter Fenwick. It was the case history of a problem student. The object of the case history was to get the student to express strong condemnation and criticism of the fictitious character, Carter Fenwick. Episodes were described in which Carter had used people to his own advantage; deliberately spread false stories about his roommate in order to get rid of him; stolen a sizable sum of money, then framed another student who was expelled from the University; headed a hate-monger group which directed its efforts against Negroes, Jews, and Catholics; got his roommate's girl friend pregnant and then rejected her, etc. By a series of skillful questions (e.g., How would you characterize this individual?, Do you like him as a person?) the typical subject was then led to express strong condemnation of Carter.

Then the second subject, actually a confederate, was asked the same questions, but he had been instructed to stress the fact that Carter was in trouble; that he was seeking help and needed to be supported and that he has been misunderstood. This was designed to bring the subject to think that his denunciation of Carter had been excessive.

Following this, the experimenter introduced the third subject as Carter Fenwick. The other two subjects, the real one and the first confederate, were told that he was present in order to learn how other people reacted to him. They were told that most subjects had been supportive and encouraging, and that was thought to be of therapeutic benefit.

Then Carter Fenwick got up and made a statement. Two different statements were made, for two different experimental conditions. The comments made by Carter in these two conditions are given below:

The Hurt Condition—Nobody has ever said such bad things about me . . . . People like you who don't give careful thought to what they say make me feel very depressed. I felt terribly let down by your comments; I feel worse now than when I came in; this therapy session hasn't helped me at all; in fact, it makes me want to give this whole thing up and drop out; on the whole, these group-therapy sessions are good—I've really benefited from those in the past, but today I think I've really been set back. Somehow I feel back where I started. Maybe it's because this group couldn't agree that I had possibilities.

The Help Condition—I don't feel at all bad because of your comments. People have told me things like that before, and, in fact, I'm beginning to appreciate comments such as yours for their thoughtfulness. Your remarks gave me a big lift—a sort of before–after picture. In general, I'd like to say that I derived much from these sessions, which have helped me a great deal. [Pepitone, 1964]

After these remarks Carter left the room—in the hurt condition on the verge of tears. There were other features of the experiment which need not concern us here. Following a questionnaire, and some self-evaluation ratings, the subjects were asked to write memos incorporating their viewpoints on the case. The memos were to be addressed to Carter, but it was stressed that he would not see them until a much later phase of the program.

The memos were classified into three categories: (1) self-criticism, (2) self-justification, and (3) rationalization. A few sample statements given by Pepitone for these categories are as follows: *Self-criticism*—"I should have thought more . . . . You've probably changed; you've got potential." *Self-justification*—"The traits you exhibited, I would never tolerate." "You are repulsive and you should be removed." *Rationalization*—"I would like to say that the passage I read was poor . . . . It did not give any good attributes of the person." "From the material presented, I do not feel that I can judge fairly . . . the material was incomplete."

Of the 37 subjects in the experiment, 19 were in the hurt condition and 18 in the help condition. Table I below shows the number of subjects who wrote each type of memo in both of these conditions.

TABLE I

CLASSIFICATION OF THE MEMORANDA WRITTEN UNDER HELP AND HURT CONDITIONS

Type of memo	*Hurt*	*Help*
Self-criticism	3	11
Self-justification	8	6
Rationalization	8	1
	19	18

Remember that the only difference between these two groups is that one group received an additional threat to its self-esteem. All of the subjects had been made to feel that their denunciation of Carter had been excessive, and that most people had been more understanding. However, one group, the help condition, was then led to believe that their behavior was not punishable, whereas the self-esteem of the other group, the hurt condition, was further attacked by making them seem blameworthy. Notice from Table I that most of the subjects in the hurt condition wrote defensive self-justifying and rationalizing notes. On the other hand, in the help condition where the subjects were under less pressure to defend their self-esteem, the majority of the subjects wrote self-critical notes. As people are made to feel more insecure, attempts at self-justification and defensive rationalization increase.

Psychosomatic Disorders

Evidence has gradually been accumulating that emotional factors resulting from certain conflict states in the lives of people with certain types of personality patterns can produce physical ailments and symptoms. Controlled studies of psychosomatic ailments are difficult to accomplish because the conditions producing them involve certain aspects of the individual's close personal life, e.g., of interpersonal relationships in the family, or of occupational setting. In many cases the gross facts are perfectly clear. For instance, the individual whose life situation stimulates a chronic, burning resentment which must be deeply repressed may develop a duodenal or gastric ulcer, for which no physical cause can be found. The hard-driving businessman may one day find that he has a case of arterial hypertension. Mr. Casper Milktoast may be dominated by his boss at the office and his wife at home—his daily half-hour commuting time may be the only time during his waking hours when unconscious resentment is not being mobilized, although he is not aware of it. As a result he develops an ulcerated stomach.

Psychosomatic medicine is still quite new. What facts have been pieced together find reluctant acceptance by the medical profession itself. The author was recently unable to obtain a copy of an issue of the *Journal of Psychosomatic Medicine* from the library of the medical school of a large University. He was told that it had been ordered, but "some doctor had discontinued its subscription." Patients, too, whose symptoms show obvious signs of psychological involvement resist any suggestions that their problems are not entirely physical. This is, of course, in keeping with denial and repression as a means for coping with emotional problems.

Space does not permit an outline here of the extensive, and rather heterogeneous, research in the area of psychosomatic medicine. One of the more promising series of studies has resulted in the finding that certain constellations of attitudes are significantly related to certain definite types of symptoms (e.g., Grace & Graham, 1952). For example, asthma has been found to be related to a constellation of attitudes involving the wish that certain unpleasant circumstances in the person's life would go away or that he would not have to have anything to do with them. Arterial hypertension appears to be related to an attitude of vigilance and preparedness to meet all possible threats. Duodenal ulcer appears to be related to revenge-seeking and so on.

The general relevance of psychological conditions to certain types of critical symptoms is nicely indicated by a study by Berle *et al.* (1951). They employed psychotherapy in the treatment of half of a group of 67 patients suffering from ulcerative colitis. The other half was given regular medical treatment—surgery, medication, special diet, etc. The results of this study are shown in Table II.

TABLE II
MEDICAL VS. PSYCHOTHERAPEUTIC TREATMENT OF COLITIS PATIENTS

Outcome	*Group receiving psychotherapy (percent)*	*Group not receiving psychotherapy (percent)*
Died	9	18
Were operated upon	3	29[a]
Became worse	0	6
Remained the same	23	21
Improved	39	18
Free of symptoms	26	15

[a]Two patients who died following their operation are included in this category, as well as in the one just above.

At the end of the study 65 percent of the psychotherapy group were reported improved or symptom-free, whereas only 33 percent of the group given the usual medical treatment were reported to be improved or free of symptoms. Notice also the difference in percentages of those who were operated upon in the two groups.

SENSITIZATION-TYPE DEFENSE MECHANISMS

Isolation and Intellectualization

Isolation as a defense against anxiety refers to the separation of feelings and emotions from ideas. If ideas, words, and things are deftly handled in such a way that they are not invested with feelings, then potentially unpleasant ideas are tolerable. When *intellectualization* becomes habitual, feelings and thoughts are viewed by the individual as two separate, distinct domains. Intellectualizing is incompatible with feeling. When the processes of isolation and intellectualization predominate, the individual remains oblivious to the relationship between his behavior and his feelings, and to the feelings and emotional nuances of others. He is likely to have little real feeling for others. Since he has little sense of the emotional significance of ideas and situations, he may behave in ways or entertain thoughts that would frighten repressors. Many ideas which would be intolerable to the repressor can be readily accessible to the individual who intellectualizes. By tolerating unpleasant ideas, ruminating, worrying on a conscious level, the sensitizer appears to be able to keep higher levels of anxiety under control. Little is known, to date, about the early learning experiences which determine the learning of this kind of coping device. It does appear that the reinforcement of attempts to approach and control threat by intellectualization comes from the learned expectancy that *future* threat can be avoided in this way. Emotionally, the sensitizer is more oriented toward the future than the repressor is. The two defenses have opposite effects. Repression tends to narrow awareness, whereas intellectualization tends to broaden it.

It will be recalled that repressors appear to have a cognitive disposition to avoid complexity, not pay attention to details, not to notice gradual changes which take place over time, and so forth. Persons whose major defense against anxiety is isolation and intellectualization appear to have just the opposite cognitive, or perceptual, disposition. They appear to be "scanners" (Gardner *et al.*, 1960). That is, they appear to pay attention to details of objects and events in the world around them. Consequently, they are more aware and more responsive to many objects and their properties. They are more sensitive to gradual changes in their environment than are repressors. In laboratory tests of the effects of gradual perceptual changes, these subjects tend to notice small differences and make accurate judgments. In their performance on projective tests or in ambiguous situations they do not react with expressive, childlike comments. Rather, they express a good deal of doubt and make qualifications. They may respond to highly specific aspects of stimuli, ruminate about similarities and differences, etc.

Let us return to the series of experiments by Lazarus and his associates employing stressor films on the subincision rite or workshop accidents. In one of these experiments (Speisman *et al.*, 1964) an intellectualization passage was employed to reduce threat. The subjects were presented with a scientific attitude toward the subincision ritual. They were invited to observe what took place in a detached manner, as an anthropologist might observe the interesting customs of primitive natives. There was no mention of feelings in the commentary.

The intellectualization commentary was somewhat effective for the group as a whole in reducing anxiety, as measured by the physiological indexes of skin conductance and heart rate. However, for students who appeared to use intellectualizing defenses on the basis of personality tests, this condition was highly effective; whereas the denial passage, which had been so effective with business executives was far less effective for these students. The striking differences between these two groups are shown in Fig. 10-6. Notice in this figure that whereas the intellectualization passage was highly effective in reducing the skin conductance of the student group, it was far less effective for the executive group. In contrast, the executive group, which on personality tests had been identified as repressors, benefited greatly from the denial, rather than from the intellectualization commentaries.

This experiment has been repeated with essentially the same results by Lazarus *et al.* (1965), using a film dealing with the prevention of shop accidents. In this experiment the effects of intellectualization defensive material were found to be most effective when it was presented as an orienting passage, as opposed to a running commentary. So that the reader may appreciate the type of verbal material which was effective as

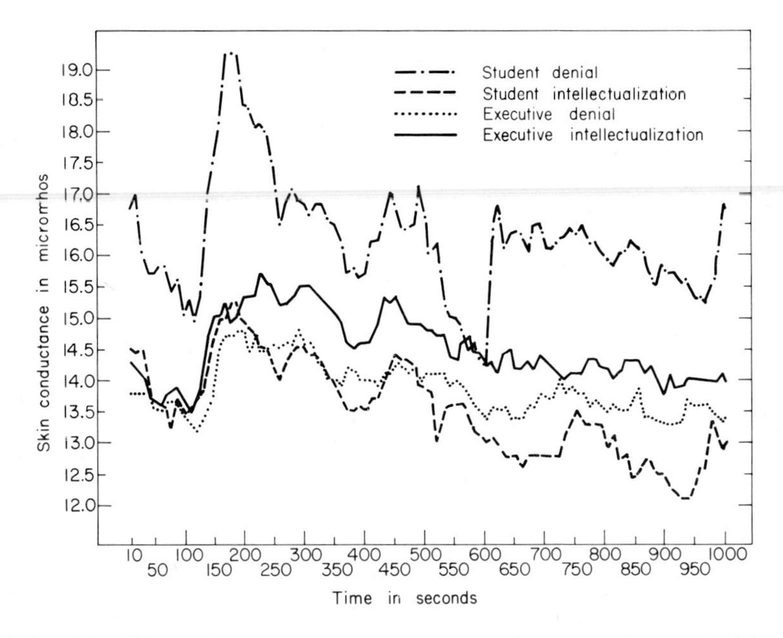

FIG. 10-6. Skin conductance measurements during the course of the subincision rite film for the subgroups of the Speisman *et al.* (1964) experiment, which used the denial passage and an intellectualization passage.

an intellectualization defensive passage in this experiment, the orienting passage presented to the students is given below:

> We have been conducting a series of investigations on the effects of motion picture films on physiological and psychological reactions. The short film you are going to see deals with an effort to train men in the use of safety devices. In doing so, two woodworking shop accidents are presented. In one a worker runs the tips of his fingers through a ripsaw, and in the other, a worker loses a finger. The events portrayed are interesting to analyze, and we may view them from the point of view of the socio- and psychological dynamics of the situation.
>
> A shop foreman is endeavoring to train the men in the use of safety devices. The pattern found here is typical of what might be observed in industry, which has been greatly concerned with accidents and their prevention. In the film you will see a foreman giving detailed explanations of the way in which the safety devices work. When the first accident occurs, he capitalizes on it, using it as an object lesson to the rest of the men, showing them what might have been done to prevent it, demonstrating the mechanical features of the guards on the machine, and explaining in detail how they work.
>
> In order to bring home the point, he remembers an experience of Armand, one of the older shop workers, an experience which has had a great influence in leading Armand to respect and use the proper safety devices. In using the past episode, he calls upon Armand, whom he knows will give him support in this matter, to tell his story, which is presented in the flashback. Armand's story reveals the value of direct experience with object lessons in influencing behavior.
>
> As you watch the film, you may reflect upon the psycho-dynamics revealed by the interchanges between the people, and be analytical about the devices used by the foreman to influence the men in the shop, that is, the psychological processes that are involved. This frame of reference is of course used by the psychological scientist as he observes, analyzes and understands these processes. [Lazarus *et al.*, 1965, pp. 624–625]

Projection

There is a tendency in all of us to view the world in terms of our own feelings and experiences. When we are happy we are likely to assume that others share the same experience. This is a basic principle of generalization which applies to all learning. We saw in the last chapter that people tend to evaluate their home town as they evaluate themselves. Defensive projection involves a somewhat similar process. In defensive projection impulses or perceptions which threaten the individual's self-esteem are disowned and externalized: "projected" to events and people in the individual's life.

Projection is a common result of intrapsychic conflict. When the person's realistic appraisal of himself is sharply at variance with the idealized image he has of himself, then the resulting negative self-evaluation, the self-dissatisfaction, may be disowned and projected onto others. Instead of experiencing conflict within himself, the individual tends to experience tension as if it were occurring between him and the outside world. The resulting projection can be active or passive (Horney, 1950).

If it is active, the individual tends to be irritable in his interpersonal relationships. He tends to blame others. He becomes a chronic fault-finder. Most of us have known such people. Their close relationships are characterized by continual dissatisfaction, irritability, fault-finding, and blame. Of course, under ordinary life circumstances it is often difficult to decide whether the individual's hostility and dissatisfaction are the result of his own negative self-evaluation, or whether the external circumstances or people actually warrant such reactions. Frequently, however, the fact that the person's reactions are chronic, disproportionate, and in sharp contrast to the reactions of others around him, and the fact that they vary in intensity with events in his life which threaten his self-esteem point to a defensive mechanism.

Passive projection involves the individual's experiencing others as being hostile toward him. Instead of the self-dissatisfaction being experienced, it is directed outward against people, events, institutions, etc. The person feels victimized by other people. He feels people exploit him, they treat him unfairly, and abuse him. Thus, when we ask Mr. Smith why he is depressed he might tell us of the problems he is having with his family, how they misunderstand him; how at the same time his job is going badly, his boss is unfair, his fellow workers align themselves with the boss for their own protection; and there is even more, his bank is constantly making mistakes, the company from whom he has ordered equipment has shown favoritism and placed him at the bottom of a waiting list, and the dentist has messed up his teeth.

Aside from generalized negative projection based on the individual's negative self-concept, specific, unacceptable impulses and feelings may be projected. The person who shows dishonesty in his dealings with others may be quick to point out how dishonest people are. He may accuse others of underhandedness on the basis of evidence with extremely low objective validity. The person who is convinced that others are unkind or hostile can justify his own cruelty and maintain his self-esteem at the same time. The sexy girl is likely to see others as oversexed.

The projective mechanism is often seen in trivial life situations. For instance, a person who is ignored, or feels ignored, may depreciate himself. Since this is intolerable to his self-esteem, however, depreciation of the person who ignored him (projection) may be strongly reinforced. In reality, perhaps the fact that he was ignored was unintentional.

The difference between projection and the mechanism of displacement is not always clear. *Cockroach projection,* for instance, is more likely to be a repressor-type mechanism. It can hardly be distinguished from displacement. Cockroach projection involves the projection of fear or anxiety onto the external world. The individual who must keep a relatively rigid watch over his own unacceptable impulses may be particularly prone to be anxious about seemingly trivial objects or events: he may be terrified of spiders, mice, grass-snakes, old houses, graveyards, etc. That is not to say that everyone who fears these things is projecting. There are many other ways in which such fears can be brought about.

Cockroach projection can be mediated by a variety of obscure associations. A large old house may symbolize mother. It is acceptable for young boys to be afraid of haunted houses, or for them to find a mysterious attraction in them. Young boys may experience an erection when they enter the forbidden domain of a supposedly haunted house. The connection between the awe and veiled mystery associated with certain impulses they may entertain toward the mother and their fascination with haunted houses is, of course, repressed. In this type of projection the object of the projected fear is not always trivial or unusual. The adult may fear strangers, foreigners, "communists in the government," etc., on little, or absolutely no, objective grounds.

In people suffering from severe emotional disorders, projected impulses and thoughts often clearly have the character of delusions. The person might believe that his thoughts and impulses are controlled from outer space, or by television. Having failed miserably in life situations, he may be convinced that others are trying to poison him, or that he is the victim of an international plot. Unacceptable homosexual impulses may be projected onto others. People in the street, hospital attendants, the doctor, etc., all secretly want to seduce him. The projective techniques which we have discussed from time to time are designed to represent ambiguous stimuli so that what the person "sees" in them is likely to reflect his own projected unacceptable impulses or conflicts. Generally speaking, the more concrete and unambiguous the external situation is, the less likelihood there is that the individual can project unacceptable impulses without himself or others being aware of it. Therefore, the less likelihood there is that projection will occur.

Laboratory Studies of Projection. The classic experiments on projection compared individuals' self-ratings on undesirable traits with the ratings of the individual by others on the same traits. For example, Sears (1936) asked 97 members of three college fraternities to rate themselves and the other members of their fraternity. A seven-point scale was used for the ratings of four traits: stinginess, obstinacy, disorderliness, and bashfulness. It was found that those students who were unaware of possessing one or another of these undesirable traits rated others as high in that particular trait to a greater extent than did the rest of the students. In a similar study, done in Vienna by Frenkel-Brunswik (1939), many of the subjects were found to rate themselves at the exact opposite pole form judges on certain undesirable traits. For example, the individual who claimed that he was "sincere under all conditions" was rated by judges as lacking in sincerity. The subjects of this experiment were graduate students.

Recent studies have been much more sophisticated, in that attempts have been made to manipulate projection directly in the laboratory. An experiment by Bramel (1962) has already become sort of a classic demonstration of projection. Male undergraduates were employed in Bramel's experiment. The purpose of the experiment was ostensibly to discover what kinds of people had insight into themselves. During the first session the subjects were given a number of personality tests. The tests were chosen to be varied and impressive. The subjects were asked to return about a week later. Two subjects were scheduled to appear at once. The experimenter first asked them in the presence of one another a series of questions about themselves and their attitudes toward current events. Thus, they were enabled to form an initial impression of each other.

The experimenter then privately communicated the personality test results of the previous session to each subject. The report, however, was falsified and totally unrelated to the individual's actual test performances. There were two conditions. Half of the subjects were given very favorable reports and the other half very unfavorable reports. Following this, the subjects were brought into a room where there was a slide projector and projection screen, and two instruments, shielded from each other. These were boxes with a dial and attached electrodes.

The subjects were first asked to make judgments of one another on an adjective check list, and then to fill out a self-concept rating scale. Then a lecture was given on the unconscious, and physiological nature of sexual arousal. Following that, the subject was asked to observe his galvanic response, indicated by the dial in front of him, while a series of photographs of men were projected on the screen. He was to record his own dial reading, and then estimate the dial reading of the other subject for the same photograph. The subjects were told that the dial readings indicated homosexual arousal to photographs. To avoid excessive threat, they were told that the dial would go off the scale if the homosexual tendency was very strong. It was also made very clear that the results would remain anonymous.

During this phase of the experiment, unknown to the subjects, the experimenter controlled the dial readings. They were programmed to be identical for the two subjects. Higher dial readings were given for those pictures where the men in the photographs wore less clothing. Bramel had predicted that those subjects to whom he had given favorable personality test results would rate their partners as having higher levels of homosexual arousal because the knowledge of this trait in themselves would contrast more sharply with their evaluation of themselves, and hence produce more tension (Bramel was testing a cognitive-dissonance theory, which need not concern us directly here).

That is exactly what happened. Subjects in the favorable condition attributed significantly higher levels of arousal to the other subject, than did those in the unfavorable condition. When the results were analyzed in terms of the subject's attitude toward the partner, the data suggested that defensive projection occurred especially when the subject's partner had been favorably evaluated by the subject.

It appears that in those for whom a positive self-evaluation was experimentally induced, the information that they were prone to homosexual arousal, a most undesirable personality trait, produced considerable tension. Accepting this information would threaten their self-esteem, and give them a confused picture of their worth. However, the threat could be reduced if their partners, whom they judged favorable, as they themselves had been judged, were "in the same boat." On the other hand, for subjects who had been unfavorably evaluated, the new threat apparently simply confirmed their existing self-evaluation. Festinger and Bramel (1962) have stressed the fact that in this experiment the information and its meaning was so unambiguous that it would have been difficult for subjects to deny it. If the situation had been more ambiguous it is possible that projection would have occurred toward those subjects whom they rated unfavorably.

Pepitone (1964) ran a very similar experiment, but he included female pictures as well. We will not discuss this experiment in detail here. However, he found "masculinity," or "virility" projected much more readily than homosexuality. That is, subjects who were supposedly aroused by female pictures projected more arousal to their partners, than those exposed to male pictures. He also found that subjects exposed to male pictures projected more vague pleasure than subjects exposed to female pictures, whereas those exposed to female pictures tended to project more shame and anger toward self. These are the very emotions that we would expect the subjects in these respective conditions to tend to *deny*. In this experiment whether the individual was high or low in self-esteem and how favorably he evaluated his partner appeared not to be significant factors. The need for future research in this area is indicated.

Obsessive-Compulsive Behavior

A compulsion is an irresistible urge to do or say something even though there is full conscious recognition that the act is irrational, unnecessary, or undesirable. Compulsions are frequently repetitive. If the person does not perform the act considerable anxiety occurs. Compulsive behavior may be highly ritualistic, with a great deal of symbolic content. It may be quite mild and limited to a few minor, recurring episodes in the individual's life; or it may be motivated by overwhelming anxiety and the resultant behavior may be so irrational and so frequent that it seriously disrupts any possibility for normal functioning. When compulsive behavior is pervasive and becomes a part of the person's style of life as a major defense against anxiety, the individual can be characterized as a *compulsive personality*.

The label *obsession* refers to repetitive thoughts and ideas which the individual has no control over. Obsessions also may be highly focal or they may be less situation-tied, dictating relentless striving for future goals. Probably all compulsive acts involve at least a shadow of obsessive thinking. Obsessive thoughts also tend to mobilize certain behavior patterns, although occasionally they occur on a purely ideational level.

Let us consider some examples of obsessive-compulsive behavior of varying degrees of severity and pervasiveness:

Ross likes to polish his automobile every so often—at least that's the way it appears to an objective observer. Actually Ross frequently becomes mildly depressed. When that happens he feels an absolute compulsion not only to clean and polish his car, but to do it perfectly. Afterwards he feels somewhat relieved—but not until he drives his car into town and buys a little present for himself—a book, a record, a small article of clothing—anything. If events of the day interfere with this ritual somehow Ross experiences considerable anxiety. It makes him feel just awful.

Just before *Henry* goes to the office in the morning he looks at himself in the bedroom mirror. It seems that he is always dissatisfied with whatever he's chosen to wear that day. Typically, he removes the jacket he is wearing and tries on another one—and another, and another. When he is satisfied with his jacket or suit he begins changing ties or shirts. The whole process may take fifteen minutes, forty minutes, or an hour or more. Henry knows that this behavior is absurd, but he can't help himself. Even when he is late for the office it appears important for him to change clothes until he feels "just right."

Professor Petty is a research psychologist. He can be found in his office week nights and almost any time on Saturdays and Sundays, although occasionally he works at home. His articles—and he writes many of them—always display a most thorough knowledge of the background literature. The reader is overwhelmed by references and cross-references. His publishers have told him that they will not publish his articles any more unless he stops including obscure references and learns to say things simply enough so that the readers can understand him. Professor Petty has tried to do just that but one still has to read his articles three times before it is clear that actually a relatively simple concept is being discussed. He tries to avoid this but it seems that he can not. Actually he always rewrites an article at least five times. Unfortunately, from the reader's point of view the first draft is always clearer than those that follow. Professor Petty's lectures are obscure and often rather dull. However, outside the classroom he may be found engaged in a conversation with an attentive student for very long periods of time. He talks on and on, qualifying, intellectualizing. His anecdotes often capture student interest. However, they are repetitive. It seems that he doesn't realize that he has essentially carried on the same conversation before with the same student. His gestures are also repetitive. During a conversation, for example, he will take a step toward the student, punctuating a point and then, still facing the person with whom he is conversing, he will walk backwards several steps, eventually to come forward again, and so forth.

Margaret came to the clinic terribly upset. She thought she was "losing her mind." Her life seemed filled with compulsive rituals and behavior which she could not understand. She had been able to live somewhat normally until recently, although most of her day was spent cleaning and recleaning the house and carrying out elaborate behavior designed to avoid people. For no apparent reason, for example, she would walk several blocks out of her way to avoid meeting the man next door on the street or in front of his house. Recently her irrational behavior has become intensified. She has begun to count her heartbeats. She feels that if she does not her heart will stop beating. She feels that this is "crazy," but she can't help herself. If she misses counting a beat or two she is overwhelmed with anxiety. She begins to tremble, shake and lose her breath.

Mary B also came to the clinic in a highly fearful state. When she was able to be calm enough to talk sensibly to the interviewer, the apparent reason became clear. Mary had been suffering from recurring thoughts that she would kill her three-year-old son. At first these thoughts occurred only when he happened to come into the kitchen, where there were sharp knives handy. Then at bedtime, the horrible thought of suffocating him with a pillow began to recur. Finally, whenever the boy was present, she was overwhelmed by an almost irresistible impulse to strangle him. Fortunately, a very primitive self-protective mechanism occurred at this point. Mary would have a fainting spell.

A great variety of behavior is compulsive in character. Our culture encourages compulsivity in certain ways: general cleanliness, having a neat appearance, being on time, doing our work with relative thoroughness, etc. Most of us become uneasy when we violate cultural norms in these respects. The difference between normal compulsivity and rigid compulsive behavior or pathological compulsivity appears, however, to be one of quantity rather than quality.

In terms of analyzing how particular compulsive acts or obsessive thoughts are learned, it is important to know the source of anxiety which the behavior serves to reduce. There are two broad classes of tension which sustain compulsive behavior: external threat and internal threat. External threat involves directly conditioned cues in the environment. For example, the compulsive cleanliness of the housewife is simply due to the fact that, as a child, she was taught to feel anxious in the presence of dirtiness or disorder, and rewarded for being neat, clean, and orderly.

Thus, any cues in the environment signaling the lack of neatness or cleanliness serve as discriminative stimuli for cleaning and straightening behavior. That is to say, compulsive behavior is frequently merely behavior under a stimulus control where the controlling conditions occur repetitively.

On the other hand, the source of threat may be internal. It may arise from unacceptable and punishable impulses and thoughts. It is in this sense that we generally think of obsessions and compulsions as unconscious personality defenses. External stimuli may also be important because they increase the level of distress. Thus Mary B's neighbor may elicit sexy feelings. Her compulsive ritual allows for anxiety-reduction. Making a positive, competing response only reduces anxiety, however, on a very temporary basis. It must be repeated again and again. When sexy thoughts threaten to break through into consciousness, thinking continuously of something else (counting heartbeats) may serve the same function. However, if the stress greatly increases, even that may not constitute an effective defense.

Compulsive personalities may display little open anxiety, but there are experiences in their backgrounds which have made them feel very insecure. Parents may have given the child little affection, and then only when he acted as a model child—living up to rigid standards of behavior. All spontaneity crushed, the child may have learned only to feel safe behaving in conformity with tested routines. The child may have learned to feel insecure in his personal relationships and to feel safe only by way of acting out certain rituals, or by clinging to routine performances. Intensive personality appraisal often reveals very elaborate chains of defensive mechanisms. For example, being afraid to express hostility, the child may substitute symbolic hostile activity, e.g., soiling himself, being messy. Apparently, an activity which is not in itself destructive, but which makes the parents angry or irritated may frighten the child almost as much as his original hostile feelings did. A compulsive preoccupation with cleanliness may result as a means of handling the now thoroughly disguised hostility (see Aldrich, 1966). The symbolic substitution of an act such as soiling for the more directly punishable aggressive impulse is probably unconscious, but the child appears to be sensitized to the guise. The ensuing "reaction-formation" is not quite like the more central and complete reaction-formation behavior of the repressor type of personality. Rather, there are overtones of continual anxiety, as evidenced by the compulsive preoccupation, and perhaps some recognition of conflict on a preconscious level.

Compensation

This mechanism involves the individual's increased efforts to find satisfaction in certain areas of his life in order to compensate for insecurities or frustrations experienced in other areas. If the bright child receives enough rewards from being studious and having serious purposes his increased esteem-income can lessen feelings of insecurity arising from his being physically weak and unathletic, his lack of close companions, etc. Compensation usually has a compulsive quality about it, because the achievement, recognition, or other rewards associated with compensating behavior not only enhance the individual's feeling of well-being, but serve to insulate him against threat as well.

This is particularly apparent when the compensatory behavior involves substitute satisfactions which are not socially acceptable, or are of only borderline acceptability. The young married woman who finds she is incapable of bearing children may compensate by aggressively pursuing a career. However, she may also find compensatory satisfactions in alcohol, promiscuity, or psychotic delusions of motherhood. Even though these latter behaviors may be self-defeating, they would also be pursued in a compulsive spirit.

Usually the compensating behavior is at least symbolically associated with the area of lack. Thus, the person whose need for dependency and affection is frustrated may compensate by excessive eating and consequently become obese. Early deprivation may result in an excessive tendency to hoard or in a materialistic orientation toward life. Oversleeping and severe restrictions of one's activities are also common defenses which provide compensation for a lack of positive reinforcements in the person's life.

Compensation is not a precisely delineated mechanism of defense. Like sublimination, it is difficult to demonstrate a direct relationship between sources of insecurity and the compensating behavior. Nevertheless, it is often relatively clear that a particular behavior which is a source of certain satisfaction becomes learned and engaged in excessively when the individual's security or self-esteem becomes threatened.

Attempts at compensation may be direct. That is, excessive strivings may occur in the very area in which failure or insecurity have been experienced. This is usually referred to as overcompensation. Thus, it occasionally happens that the student who just barely has the capacity to get through college puts forth enormous effort, resulting in an above-average performance. A person may compensate for threatening dependency needs by excessive strivings for independence and self-sufficiency. This latter kind of overcompensation is very similar to reaction-formation, except that it has an anxiety-driven quality, rather than a nonanxious excessive, insincere characteristic.

Overcompensation is relatively common. Nevertheless, the successes, rather than the failures of overcompensation are likely to be more stressed, partly because they are usually dramatic. Early psychoanalytic thinking, in particular Adlerian psychology, stressed the relationship between "organ inferiority" and overcompensatory styles of life. The theory was substantiated in the usual (for psychoanalysis), anecdotal manner. It was pointed out, for example, that Beethoven and Robert Franz were well-known musicians both of whom became deaf; that Demosthenes, the stutterer, became the greatest orator of Greece; that many famous painters suffered from visual defects, etc. Adler reported, for instance, that among art students, approximately 70 percent had been found to have some variety of optical anomaly (see Ansbacher and Ansbacher, 1956). Adler also postulated that negative self-evaluations, or feelings of inferiority on the part of the child result in his developing fictional or idealized goals as guarantees for future importance.

Horney (1950) has claimed that children develop an idealized image of themselves to compensate for feelings of insecurity and that individuals attempt to actualize their idealized images, which become greatly elaborated in adulthood. Horney was impressed by the great length to which many of her patients seemed willing to go in order to "prove" certain aspects of the idealized images they have of themselves. For example, she mentions two young girls who felt that they should never be afraid of anything. One of them became aware that she was afraid of burglers and forced herself to sleep in an empty house until her fear was gone. The other was afraid of swimming unless the water was clear because of the possibility of being bitten by a fish or a snake. This girl forced herself to swim across a shark-infested bay

Fantasies or Daydreaming

Fantasy or daydreaming is a much more common activity than most of us realize. In one study of daydreaming reported by Shaffer and Shoben (1956) a large number of graduate and undergraduate students were asked to report whether or not they had experienced a variety of different types of daydreams—whether they had ever experienced each type, and whether they had experienced them recently. The students were assured that disclosures would remain anonymous.

There was little difference between the graduate and undergraduate groups. Only two men and nine women (3 percent of the sample) said that they had experienced no recent daydreams. Almost half of the students reported daydreams. Men and women differed only in minor respects (e.g., there was a tendency for men to daydream more of physical strength and for women to prefer fantasies of physical attractiveness). Types of daydreams reported by large percentages of the students included vocational success, money and possessions, sexual daydreams, worry, mental feats, and physical feats. Many students also reported daydreams in what might be considered less culturally acceptable categories such as daydreams of death or destruction, martyrdom, and display.

Although these results must be interpreted with some caution, as is the case in any self-report type study, the evidence does indicate that the imaginary gaining of substitute satisfactions through daydreaming is a normal process (statistically). When the many other channels of borrowed fantasy such as television shows, movies, magazines, and paperbacks are considered, then fantasy as a means of gaining substitute satisfaction not present in reality must be considered a widespread phenomenon.

Fantasy behavior probably represents a first-line defense against anxiety. That is, excessive reading, television viewing, or daydreaming probably indicates that the individual is experiencing some threat. Daydreams of killings and bombings, of homosexual experiences, of repeated sexual or masturbation fantasies—these may point to the possibility of more serious maladjustment problems. The sensitizer is likely to do a great deal of ruminating and worrying. He may have active anxiety-generating thoughts which are qualitatively different from the imaginary satisfactions obtained by fantasying oneself a hero, an expert, or a great lover. The person who must depend upon a heavy fantasy life is certainly not obtaining many satisfactions in interpersonal relations and in his attempts to cope with his world. The sensitizer tends not to be sociable and to resist social approval. His fantasy productions provide considerable tension-reduction. They may be engaged in at the expense of real accomplishment and at the expense of time which could be spent learning how to experience reality-based rewards.

Depression

Depression is not a defense mechanism against anxiety, but it may have unconscious defensive features. The depressive reaction is commonly precipitated by external loss: occupational failure, financial loss, the desertion or death of loved persons, etc. However, individuals who have been taught to feel guilty, particularly about hostile feelings in themselves, may suffer mild depression in certain life circumstances, namely, whenever hostility or ambivalent feelings are mobilized. Depressive reactions following severe loss or death are normal, but when they are prolonged and disproportionate they indicate the existence of certain personality traits and mechanisms. It is usually found that the individual has actually had hostile or ambivalent feelings toward the loved object. He is unable to express hostility directly, however. A sort of reverse of the normal projective mechanism appears to operate. It is as if the individual substituted "I don't hate him, I hate myself" for the emotional reality "I hate him." Feelings of guilt, unworthiness, listlessness, loss of interest in living, well up in the individual—feelings of anger and hostility do not. Unaware of his hostility, he does not understand his feelings. Overwhelmed with guilt and self-punishment, he is likely to experience helplessness and loneliness, to weep on slight provocation and to show other symptoms such as an increase or a loss of appetite, inactivity, and avoidance of others. Essentially the same pattern may occur without an actual loss in the person's life. He may feel hostile, sexy, or ashamed of what he is doing, but only experience self-punishment in the form of the depressive reaction.

SUMMARY

Advances in experimental personality research have suggested that the classic mechanisms of defense can be meaningfully viewed within the framework of the person's overall anxiety-coping behavior. Individuals appear to differ in this respect along a continuum from coping with anxiety by avoidance and denial at one extreme to attempts to move toward anxiety and to control it at the other extreme. People for whom avoidance and denial tend to predominate as responses to anxiety are called *repressors*. Those who show a strong tendency to admit anxiety are called *sensitizers*. Research has indicated that repressors show two other broad tendencies. They tend to give socially desirable responses in testing situations and they tend toward emotional stability. Sensitizers, on the other hand, admit and report anxiety, failure, and inadequacies and they tend toward emotional instability.

Accordingly, in this chapter repression, denial, reaction-formation, sweet-lemon rationalization, and related classical mechanisms of defense are categorized and discussed as repression-type mechanisms. Similarly, projection, intellectualization, compartmentalization, compulsivity, sour-grapes rationalization, and related defenses are classified and discussed as sensitization-type mechanisms.

The repression of anxiety-producing thoughts associated with traumatic events is a dramatic instance of repression; more common is the repression of thoughts and feelings that have become associated with anxiety over a period of time. Under conditions in which it is strongly rewarded, repression may be learned as a generalized coping device for dealing with anxiety and anxiety-producing thoughts. An experimental analogue of repression was presented and discussed.

Repression as a coping device is also associated with denial mechanisms and with a tendency toward perceptual leveling. There is experimental evidence that people classified as repressors on the basis of their RS scale scores tend to deny experiencing anxiety even though physiological measures such as the GSR and heart rate indicate its presence.

When their impulses or feelings are associated with anxiety, people may assert the opposite of what they want or feel. This is called *reaction-formation.* Reaction-formation is typified by the so-called smother love of the essentially rejecting mother who goes out of her way to demonstrate affection and indulgence. Both animal and human experimental analogues of reaction-formation are available.

Feelings and impulses that are anxiety-producing may be displaced from the person or event that arouses them onto less threatening people or events. This involves an inhibition or a represssion of the impulse. The person erupts in the presence of the generalized stimulus only when there are specific cues signalling nonthreat and there is a certain similarity to the original stimulus person or event.

Partly because repression provides conditions so that the person is not able to see a connection between the two, partial or symbolic gratification of unacceptable impulses may be obtained by channelling them into behavior that is socially acceptable. Repressors seem to be more controlled by social rewards, and they appear to gain gratification through identification with symbols of safety and approval. People who exhibit the repressive pattern are likely to favor rationalizations with strong elements of denial, namely, the sweet-lemon type of rationalization, so-called because in this kind of rationalization the person tends to gloss over and deny failures, inadequacies, or other bad features of his experience and claim that they are actually advantageous or positive in some way. The lemon one receives is really sweet! Finally, psychosomatic disorders may be fostered by repessive-type defenses.

Whereas repressors are inclined to react emotionally, sensitizers appear to actually react less so. In experiments designed to measure both physiological and self-reported anxiety, sensitizers show less autonomic activity in response to threat than do repressors, but they report feelings of anxiety whereas repressors tend to deny them. Perhaps the major sensitization-type defense mechanism is isolation combined with intellectualization. That is, sensitizers tend to separate feelings and emotions from ideas. Potentially threatening thoughts appear to be tolerable for them because they are handled in such a way that they are not invested with feelings. Perceptually they tend to be scanners, and to notice small changes in their environment. They also appear to project or externalize their own feelings and conflicts onto others about them. Laboratory studies have demonstrated the reality of this phenomenon, although they have not to date attempted to demonstrate that it is primarily a sensitizer mechanism.

Attempts to reduce anxiety by order, neatness, repetitious behavior, etc., are assumed to be more typical of sensitizers than of repressors. Mild attempts to control anxiety in this manner may go unnoticed by others as unusual. They may even be adjustive. On the other hand, drastic attempts to control anxiety through obsessive thoughts and compulsive actions may incapacitate the person for normal living.

Another sensitizer-type mechanism for coping with anxiety involves increased efforts and strivings in certain areas of the person's life in order to make up for feelings of inferiority and inadequacy in some other area. This is called compensation. Sometimes increased efforts are made in the very area where inferiority is experienced. Such strivings invariably have a compulsive quality about them because they represent attempts to insulate the individual from threat.

In line with the sensitizer's tendency to ruminate, fantasy and daydreaming are assumed to be sensitizer anxiety-coping devises. Although these appear to be normal processes in a statistical sense, they probably represent first-line defenses against anxiety, much like excessive television viewing or excessive reading.

Although it is not traditionally classified as a mechanism of defense, under certain circumstances depression may serve as an anxiety-coping mechanism. It appears that it may occur in place of the more fear-increasing feelings of hostility and resentment.

Unfortunately, little is known at the present time about the antecedents of repressor and sensitizer patterns. Some research is available substantiating the reality of many of the mechanisms discussed in this chapter. Other than a few scattered observations, however, little information is available on the manner in which they are learned.

REFERENCES

Aldrich, C. K. *An introduction dynamic psychiatry.* New York: McGraw-Hill, 1966.

Altrocchi, J., Shrauger, S., & McLeod, Mary A. Attribution of hostility to self and others by expressors, sensitizers, and repressors *J. clin. Psychol.*, 1964, **20,** 233.

Altrocchi, J., Shrauger, S., & McLeod, Mary A. Qttribution of hostility to self and others by expressors, sensitizers, and repressors. *J. clin. Psychol.*, 1964, **20**, 233.

Ansbacher, H. L., & Ansbacher, Rowena R. *The individual psychology of Alfred Adler.* New York: Basic Books, 1956.

Berle, B. B. *et al.* Appraisal of the results of treatment in stress disorders. *Res. Publi. Ass. nerv. ment. Dis.*, 1951, **31**, 167–177.

Bramel, Dana. A dissonance theory approach to defensive projection. *J. abnorm. soc. Psychol.*, 1962, **64**, 121–129.

Bruner, J. S., & Postman, L. Tension and tension release as organizing factors in perception. *J. Pers.*, 1947, **15**, 300–308.

Byrne, D. The Repression-Sensitization Scale: Rationale, reliability and validity. *J. Pers.*, 1961, **29**, 334–349.

Byrne, D., Barry, J., & Nelson, D. Relation of the revised Repression-Sensitization Scale to measures of self-description. *Psychol. Rep.*, 1963, **13**, 323–334.

Byrne, D., & Sheffield, J. Response to sexually arousing stimuli as a function of repressing and sensitizing defenses. *J. abnorm. soc. Psychol.*, 1965, **70**, 114–118.

Byrne, D., Steinberg, M. A., & Schwartz, M. S. Relation between repression-sensitization and illness. *J. abnorm. soc. Psychol.*, 1968, 73, 154–155.

Carpenter, B., Wiener, M., & Carpenter, Janeth. Predictability of perceptual defense behavior. *J. abnorm. soc. Psychol.*, 1956, **52**, 380–383.

Dollard, J., & Miller, N. E. *Personality and psychotherapy.* New York: McGraw-Hill, 1950.

Edwards, A. L. Social desirability in personality tests construction. In B. N. Bass and I. A. Berg (Eds.), *Objective approaches to personality assessment.* Princeton, N. J.: Van Nostrand, 1959. Pp. 100–118.

Eriksen, C. W. Perceptual defense as a function of unacceptable needs. Unpublished doctoral dissertation, Stanford Univer. 1950.

Eriksen, C. W., & Kuethe, J. L. Avoidance conditioning of verbal behavior without awareness: A paradigm of repression. *J. abnorm. soc. Psychol.*, 1956, **53**, 203–209.

Festinger, L., & Bramel, C. The reactions of humans to cognitive dissonance. In E. J. Bachrach (Ed.), *Experimental foundations of medical psychology.* New York: Basic Books, 1962. Pp. 254–279.

Frenkel-Brunswik, Else. Mechanisms of self-deception. *J. soc. Psychol.*, 1939, **10**, 409–420.

Gardner, R. W., Jackson, D. N., & Messick, S. J. Personality organization in cognitive controls and intellectual abilities. *Psychol. Issues*, 1960, **2**, No. 4 (Monogr. 8).

Gordon, J. E. Interpersonal predictions of repressors and sensitizers. *J. Pers.*, 1957, **25**, 686–698.

Gossett, J. T. An experimental demonstration of Freudian repression proper. Unpublished doctoral dissertation, Univer. of Arkansas, 1964.

Grace, W., & Graham, D. T. Relationships of specific attitudes and emotions to certain bodily diseases. *Psychosom, Med.*, 1952, **14**, 243.

Horney, Karen. *Neurosis and human growth.* New York: Norton, 1950.

Joy, V. L. Repression-sensitization and interpersonal behavior. Paper read at Amer. Psychol. Assn., Philadelphia, August, 1963.

Lazarus, R. S., & Alfert, Elizabeth. Short-circuiting of threat by experimentally altering cognitive appraisal. *J. abnorm. soc. Psychol.*, 1964, **69**, No. 2, 195–205.

Lazarus, R. S., Eriksen, C. W., & Fonda, C. P. Personality dynamics and auditory perception recognition. *J. Pers.*, 1951, **19**, 471–482.

Lazarus, R. S., Optom, E. M., Jr., Nomikos, M. S., & Rankin, N. O. The principal of short-circuiting of threat: Further evidence. *J. Pers.*, 1965, **33**, 622–635.

Lomont, J. F. The repression-sensitization dimension in relation to anxiety responses. [Cited by Byrne in Maher (1964).] Unpublished manuscript, 1964.

Maher, B. A. (Ed.) *Progress in experimental personality research.* Vol. I. New York: Academic Press, 1964.

McDonald, R. L., Gyther, M. D., & Christakos, A. C. Relations between maternal anxiety and obstetric complications. *Psychosom. Med.*, 1963, **25**, 357–363.

Megargee, E. I. Under control and over control in assaultive and homocidal adolescents. Unpublished doctoral dissertation, Univer. of California, Berkeley, Calif., 1964.

Mowrer, O. H. An experimental analogue of "regression," with incidental observations on "reaction formation." *J. abnorm. soc. Psychol.*, 1940, **35**, 56–57.

O'Connell, W., & Peterson, Penny. Humor and repression. *J. existential Psychiat.*, 1964, **4**, 309–316.

Pepitone, A. *Attraction and hostility.* New York: Atheron Press, 1964.
Rapaport, D. Principles underlying projective techniques. *Charact. & Pers.*, 1942, **10**, 213–219.
Renaug, H., & Estess, F. Life histories interviews with 100 normal American males: "Pathogenesity" of childhood. *Amer. J. Orthopsychiat.*, 1961, **31**, 786–802.
Sarnoff, I. Reaction formation and cynicism. *J. Pers.*, 1960, **28**, 129–143.
Schafer, R. *The clinical application of psychological tests.* New York: Int. Univer. Press, 1948.
Sears, R. R. Experimental studies of projection. I. Attribution of traits. *J. soc. Psychol.*, 1936, **7**, 151–163.
Shaffer, L. F., & Shoben, E. J., Jr. *The psychology of adjustment.* Boston: Houghton, 1956.
Sheckley, R. *The 10th victim.* New York: Ballantine Books, 1965.
Silber, L. D. & Baxter, J. C. Reported via personal communication by Byrne, D. In B. A. Maher, (Ed.), *Progress in experimental personality research.* Vol. 1. Pp. 170–214. New York: Academic Press, 1964.
Speisman, J. C., Lazarus, R. S., Mordkoff, A. M., & Davison, L. A. Experimental reduction of stress based on ego-defense theory. *J. abnorm. soc. Psychol.*, 1964, **68**, 367–380.
Tempone, V. J. Differential thresholds of repressors and sensitizers as a function of a success and failure experience. Unpublished doctoral dissertation, Univer. of Texas, 1962.
Ullmann, L. P. An empirically derived MMPI scale which measures facilitation-inhibition of recognition of threatening stimuli. *J. clin. Psychol.*, 1962, **18**, 127–132.
Weinberg, N. H. Word association style, field dependence, and related personality variables. Unpublished manuscript, 1963.
Wolberg, L. R. *The technique of psychotherapy.* New York: Grune & Stratton, 1954.

Chapter 11

Complex Modes of Human Adjustment

INTRODUCTION

There has been a growing recognition that most people learn defensive adjustments as a part of their total pattern of coping with life in the modern world. All learning involves emotional factors. Feeling "good," confident, and secure results from successful learning. The reverse is also true: successful learning results from feelings of confidence and security. It is the emotional factors in learning which predispose the individual on the one hand to develop and adjust to the changing demands produced by his needs and by the environment and, on the other hand, to experience difficulties in adjustment, perceive threats on all sides, and attempt to avoid, or refuse to learn in response to life's demands.

The degree to which the individual has developed a capacity to cope with the many frustrations and problems associated with the normal demands of living is frequently referred to as the degree of *emotional maturity* he possesses. There are many facets of emotional maturity. Indeed, volumes have been written on the subject. The concept encompasses so many aspects of human adjustment that it is virtually useless in a scientific sense. On the other hand, the very lack of explicitness of the concept allows for the ordering and interpretation of a great variety of human adjustive behavior along a single dimension or continuum.

Teaching the foundations of psychology to beginning students is often frustrating because they tend to resist learning the basic principles. Instead, they want to jump immediately to the issues and problems of life.

In response, the author has often pointed out to his classes that the science of behavior is still in its infancy, although its potential and promise are great. Furthermore, it is only appropriate "to have life on the next page" when the basic principles and tools of interpretation have been learned and when the results of the scientific search for orderliness in behavior have been appreciated.

On the assumption that by this time the reader has acquired a fair interpretative background, in this chapter we are going to discuss certain aspects of gross human adjustment to life situations. Along with certain patterns of emotionally disturbed behavior, some consideration will be given to its antithesis—constructive responses to life problems.

The continuum of varying degrees of behavior reflecting emotional maturity is presented in Fig. 11-1. The boxes at the very top of this figure, connected by a horizontal straight line, represent a characterization of behaviors at either extreme of this dimension. For simplicity, other points along the continuum represented by the horizontal line are not presented. However, as we move from the upper right-hand box to the upper left-hand box defensive behavior patterns would decrease and optimally adjustive behavior would increase. Normal individuals in our culture, in a statistical sense, would be expected to occupy a position somewhere in the middle of the continuum.

The rest of the diagram traces the antecedent conditions of the behavior involved in either extreme of the continuum. "Basic Endowment" refers to the individual's inherited characteristics—physical, intellectual, and emotional. Given this, certain early learning experiences, primarily those centering around friendship, identification, assertiveness, and dependency qualitatively determine the individual's adjustment. The achievement of optimal learning in these areas is indicated by the square to the left.

Early learning which involves basic feelings of security in relation to others, along with healthy friction with others, allows the individual to learn tolerance, acceptance, and respect in conjunction with appropriate assertiveness in his interpersonal relationships. Good parental models result in the individual's gaining a sense of identity, security, and self-esteem. The existence of fortunate early conditions surrounding the learning of the expression of aggression and dependency allows him to become relatively expressive, confident, and autonomous. These learnings in turn predispose the individual to behave constructively in relation to basic life problems: relatedness to others; occupation; love and marriage (upper left of the diagram).

The squares to the right represent unfortunate conditions surrounding the early learning of relatedness to others, identification, assertiveness, and dependency. Here the individual learns anxiety in relation to others.

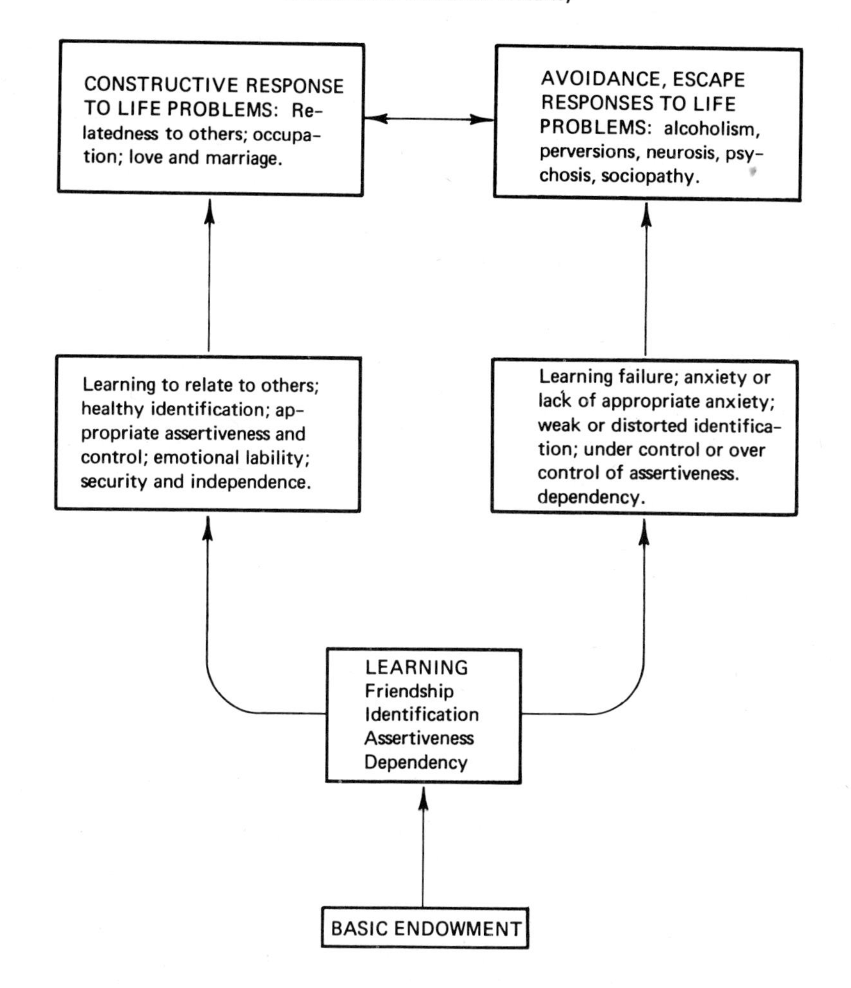

FIG. 11-1. A continuum of emotional maturity and the antecedent conditions related to its extremes.

Parental models either provide weak sources of reinforcement for the learning of identification, or reinforce inappropriate varieties of imitative behavior. It should be noted that when the individual does *not* learn to be anxious under conditions where anxiety is appropriate, for example, in response to impulses which violate important cultural sanctions, norms, and expectations, then his behavior and level of emotional maturity may be as maladjustive as those of the individual who has learned to be inappropriately anxious.

Poor learning in relation to the handling of aggressive impulses is evidenced by extreme apathetic, passive, or self-effacing trends on the one hand; and by a total lack of impulse control on the other. Analogously, poor learning relative to dependency may be evidenced by either overdependent behavior or anxiety-driven attempts to be independent (pseudodependency). These distortions of healthy early learning eventuate in emotional disorders which are reflected in avoidance and escape behavior toward life problems (upper right of the diagram).

The basic content of the present chapter will center around discussions of the disturbed behavior patterns indicated in the square in the upper right-hand corner of Fig. 11-1. Before turning our attention to disordered behavior, however, let us first discuss certain aspects of optimal adjustment as depicted briefly in the upper left-hand box of Fig. 11-1.

OPTIMAL ADJUSTMENT

Many criteria of optimal adjustment have been proposed from time to time. Psychologists have, however, accepted only the most general criteria for good adjustment. They have been rather reluctant to undertake a detailed study of people whose adjustive behavior is optimal or superior. Smith (1961) has pointed out that this reluctance has been largely due to our attempts to avoid making value judgments and a tendency to avoid issues relating to philosophical, moral, or ethical considerations. Part of this reluctance can simply be understood as psychology's eagerness to be scientific and objective—to practice science with a capital *S*. Somehow the investigation of personality disorders, on the other hand, has been viewed as appropriate. Perhaps this is because of the relative ease with which people can reach an agreement in forming the value judgments "bad" or "undesirable" when confronted with dramatic personality disorders such as hallucinations, delusions, anxiety states, extreme hostility, and severe impairment of the individual's ability to function.

Actually, given the vast amount of private and public support of programs designed to promote mental health, it would seem most reasonable for scientific researchers to attempt to cope with such problems as the incidence of positive mental health in the population, the antecedent conditions of emotional maturity, and criteria of optimal adjustment. Progress will undoubtedly be made in these directions in the future. Some psychologists have, over the years, been steadily engaged in the study of optimal personality functioning. It is somewhat unfortunate, in the author's opinion, that most of these investigators have not in fact been attuned to the nature of scientific interpretation. On the other hand, rigid, unrealistic conceptions of how scientific activities ought to be performed can only be regarded with amusement, if not with some chagrin.

Let us consider briefly some of the criteria which have been suggested for optimal adjustment. Certain criteria are almost universally accepted by clinicians. Common among these are the absence of inappropriate hostility or inappropriate anxiety; competence and effectiveness in coping with the environment; the ability to relate to others; the ability to love and do productive work; and the capacity for a realistic and accurate appraisal of one's environment. Generally, representatives of the psychoanalytic tradition were content with Freud's criteria—the ability to love and do productive work—although some of the psychoanalytic revisionists have added other criteria such as "social interest" (Adler), harmonious integration of conflicting elements in the personality (Jung), the ability to affirm one's own will (Rank). Rank viewed neurotics as persons with "strong wills" and creative urges, but who spent their lives trying to please others instead of asserting themselves.

More specific conceptualizations of optimal adjustment have been set down by such psychologists as Maslow, Allport, and Rogers. Maslow (1954) feels that the highest level of functioning is attained when the individual is able to commit himself to a cause outside of himself. The type of cause that Maslow has in mind is an abstraction—the scientist's search for order and regularity in nature, the artist's search for beauty, etc. As a by-product of this kind of active, relatively abstract commitment he sees the individual as "actualizing himself." He feels that this abstract kind of commitment does not occur and is therefore not reinforced (Maslow doesn't use this terminology) unless a number of conditions are met: The individual must feel safe, have a sense of belonging and esteem, and, of course, be free of physical privations.

Maslow made an extensive study of self-actualizing (SA) individuals. Some of the traits which he found consistently among these people were as follows:

1. *A more adequate perception of reality and more comfortable relations with reality than occur in average people.* The SA individual appears to be able to detect spuriousness or a lack of genuineness in interpersonal relations. They also appear to prefer coping even with unpleasant reality, rather than avoiding or defending themselves against it.
2. *A high degree of acceptance of themselves, of others, and of the realities of human nature.* They did not appear to be ashamed of being themselves, nor were they shocked or disturbed to find foibles and shortcomings in themselves or others.
3. *Spontaneity.* The SA people displayed spontaneity in their thinking, feelings, and behavior to a greater extent than average people.

4. *Problem-centeredness.* They seemed to focus on problems outside of themselves. They were not overly self-conscious. Since they were not problems to themselves, they could focus their attention on external problems.

5. *A need for privacy.* They appeared to be able to enjoy solitude, and even to seek it out on occasion, especially during periods of intense concentration on subjects of interest to them.

6. *A high degree of autonomy.* The SA people were able to stand some measure of rejection or unpopularity—they were able to pursue their interests and projects and maintain their integrity even when it hurt to do so.

7. *Close relationships with a few friends or loved ones.* The SA subjects were not necessarily very popular, but they did have the capacity to establish close, loving relationships with at least one or two other people.

8. *A strong ethical sense.* They were found to have a highly developed sense of ethics. Although their conceptions of right and wrong were not always wholly conventional, their behavior was always chosen with reference to its ethical meaning.

9. *Resistance to enculturation.* The SA subjects appeared able to detach themselves sufficiently from their culture to be able to adopt critical attitudes toward cultural inconsistencies or unfairness within their own society.

Allport's specification of the "mature personality" involves striking similarities to the characterization of Maslow's SA subjects. He has, for example, stated that the developed person is one who:

> . . . has a variety of autonomous interests: that is, he can lose himself in work, in contemplation, in recreation, and in loyalty to others . . . . Egocentricity is not the mark of a mature personality . . . . These goals represent an *extension of the self* which may be said to be a first requirement for maturity in personality. [Allport, 1937, p. 213]

A second requirement of personal maturity which Allport feels complements the first is spoken of as follows:

> We may call it *self-objectification*, that peculiar detachment of the mature person when he surveys his pretentions in relation to his abilities, his present objectives in relation to possible objectives for himself, his own equipment in comparison with the equipment of others, and his opinion of himself in relation to the opinion others hold of him. This capacity for self-objectification, is insight, and it is bound in subtle ways with a sense of humor, which as no one will deny, is, in one form or another an almost invariable possession of a cultivated and mature personality. [Allport, 1937, pp. 213–214]

Maslow's description of self-actualizing individuals and Allport's description of the mature personality are descriptions of rather unusual people. One can imagine that these people also have high intelligence quotients and that they are more "cultured," or perhaps better educated either in a formal or informal sense than the average person. It is tempting to draw an analogy between them and the RS dimension of personality. They seem like nonanxious sensitizers—even though that might appear to be a contradiction, because sensitizers tend to score high on self-report anxiety scales. Nevertheless, these people certainly do not try to escape, avoid, or deny perceptions which the average person might consider threatening. On the contrary, they seem to be able to see themselves and to be themselves, and at the same time to have a sense of security.

We have seen, then, that the descriptions of optimally adjusted people provided by Maslow and by Allport are very similar. It is possible that they were influenced by each other's writings. However, interestingly enough, if we attempt to derive a picture of the mature individual from observations of the functioning of people who have undergone successful psychotherapy we will again find a high degree of overlap in terms of descriptive characteristics. Although they represent very different schools of thought, and although the psychotherapeutic procedures are quite different, members of the neopsychoanalytic group (e.g., Horney, 1950), those who hold "eclectic" approaches to psychotherapy as it is commonly practiced, and members of the client-centered group have all made statements roughly similar to those given below. We will use as an example comments by Rogers (1961), whose particular approach to therapy is called client-centered. (Various approaches to psychotherapy will be discussed in the next chapter.) We will use Roger's statement to exemplify critical concepts of the optimally adjusted person because Rogers has been more articulate than others in describing the successful outcome of psychotherapy. Others appear to have focused their attention on the process itself, rather than descriptions of its result. As we review Roger's description of the end result of successful psychotherapy we will see at once a relationship between this description and those given above of the SA individuals and of the mature personality. Unfortunately, space does not permit our doing full justice to Roger's descriptions. However, we will outline a few of the more important characteristics as he sees them.

For one thing he seems to be impressed with the fact that successful psychotherapy involves the learning of an openness to experience. In this connection Rogers states:

In the first place, the process seems to involve an increasing openness to experience. This phrase has come to have more and more meaning for me. It is the polar opposite of defensiveness. Defensiveness I have described in the past as being the organism's response to experiences which are perceived or anticipated as threatening, as incongruent with the individual's existing picture of himself, or of himself in relationship to the world. These threatening experiences are temporarily rendered harmless by being distorted in awareness, or being denied to awareness. I quite literally cannot see, with accuracy, those experiences, feelings, reactions in myself which are significantly at variance with the picture of myself which I already possess. A large part of the process of therapy is the continuing discovery by the client that he is experiencing feelings and attitudes which heretofore he has not been able to be aware of, which he has not been able to "own" as being a part of himself. [Rogers, 1961, p. 187]

Elsewhere, Rogers adds to this:

As you might expect, this increasing ability to be open to experience makes him far more realistic in dealing with new people, new situations, new problems. [p. 115]

Two other characteristics which seem to be related to the client's increasing openness to experience are that he appears to be able to admit the therapist's positive feelings toward him to awareness, and that the client shows an increased liking for himself. Rogers comments on these:

It appears possible that one of the characteristics of deep or significant therapy is that the client discovers that it is not devastating to admit fully into his own experience the positive feeling which another, the therapist, holds toward him. Perhaps one of the reasons why this is so difficult is that essentially it involves the feeling that "I am worthy of being liked" [p. 86].

We have established the fact that in successful psychotherapy negative attitudes toward the self decrease and positive attitudes increase. We have measured the gradual increase in self-acceptance and have studied the correlated increase in acceptance of others. But as I examine these statements and compare them with our more recent cases, I feel they fall short of the truth. The client not only accepts himself—a phrase which may carry the connotation of a grudging and reluctant acceptance of the inevitable—he actually comes to *like* himself. This is not a bragging or self-assertive liking; it is rather a quiet pleasure in being one's self. [p. 87]

A second overall trend which Rogers discusses is the person's ability to be somewhat autonomous, to trust his own perceptions, feelings, and judgments. Rogers comments on this characteristic as follows:

Still another characteristic of the person who is living the process of the good life appears to be an increasing trust in his organism as a means of arriving at the most satisfying behavior in each existential situation . . . . As I try to understand the reason for this, I find myself following this line of thought. The person who is fully open to his experience would have access to all of the available data in the situation, on which to base his behavior; the social demands, his own complex and possibly conflicting needs, his memories of similar situations, his perception of the uniqueness of this situation, etc., etc. The data would be very complex indeed. But he could permit his total organism, his conscious participating, to consider each stimulus, need, and demand, its relative intensity and importance, and out of this complex weighing and balancing, discover that course of action which would come closest to satisfying all his needs in the situation. [pp. 189–190]

Let me return to the clients I know. As they become more open to all of their experiences, they find it increasingly possible to trust their reactions. If they "feel like" expressing anger they do so and find that this comes out satisfactorily, because they are equally alive to all of their other desires for affection, affiliation, and relationship. They are surprised at their own intuitive skill in finding behavioral solutions to complex and troubling human relationships. It is only afterward that they realize how surprisingly trustworthy their actions have been in bringing about satisfactory behavior. [p. 191]

Finally, in still another place Rogers extends this topic somewhat:

The individual increasingly comes to feel that this locus of evaluation lies within himself. Less and less does he look to others for approval or disapproval; for standards to live by; for decisions and choices. He recognizes that it rests within himself to choose; that the only question which matters is "Am I living in a way which is deeply satisfying to me, and which truly expresses me?" This I think is perhaps *the* most important question for the creative individual. [p. 119]

Unfortunately, in his writings Rogers does not appear to accept wholeheartedly the scientific, deterministic frame of reference. Yet later when we discuss his particular approach to psychotherapy, we will see that he clearly outlines the conditions under which the behavioral and subjective changes which he describes occur. It is probably not doing too much violence to Roger's observations to draw an analogy between his descriptions of successful psychotherapy and the behavior which is known to occur in the absence of anxiety conditioning. The reader will recall from Chapter 5 that laboratory demonstrations of anxiety conditioning have established beyond doubt two behavioral trends which result from the introduction of conditions producing anxiety: there is a decrease in positively reinforced behavior and an increase in behavior having an avoidance history (see pp. 143–145). The "spontaneous" behavior which occurs in the absence of anxiety conditioning constitutes behavior which has been positively reinforced in the animal's history.

There is no doubt that "free" human functioning is enormously more complex than the spontaneous behavior of lower organisms. At the same time, there is every indication that it is no more spontaneous in the sense of being "free." Apparently in the absence of anxiety the individual is responsive to many subtle sources of stimulation and reinforcement, the end result of which is a fully functioning, integrated person whose behavior might be described as optimally adjustive.

Now let us turn to behavioral variations at the opposite pole—disordered behavior.

DISORDERED BEHAVIOR

Resources for Understanding Disordered Behavior

In our discussions of neurotic behavior it will be seen that the neurotic is an individual who is miserable and unhappy, and who behaves, in spite of his being of normal or superior intelligence, "stupidly" in certain areas of his life. That is, much of his behavior is inappropriate, ineffectual, and unadaptive. Nevertheless, in spite of this, his behavior persists.

Mowrer (1948) has referred to this condition of behavior that seems to be self-perpetuating, and yet at the same time self-defeating, as the *neurotic paradox*. Others have noted the "masochistic" style of life of neurotic persons. Human behavior involves many more complex factors than the behavior of lower animals, but we have already seen from the results of laboratory experiments with animals how this type of behavior may be maintained by a variety of sources of reinforcement, although it appears to anyone who has not had a chance to observe the antecedent conditions that the behavior is either unreinforced or actually punished. Let us review some of the general principles relevant to what appears to be masochistic behavior. They fall into several classes.

The Partial Reinforcement Effect

First, it will be recalled that the removal of positive sources of reinforcement is frustrating to the organism. But it will also be recalled that behavior which has been learned on the basis of intermittent reinforcement, particularly variable-interval or variable-ratio schedules of reinforcement, is highly resistant to extinction. In some cases a single reinforcement occurring during the extinction process may bring the behavior back in full strength. Thus, under certain conditions behavior which appears to be frustrating or unreinforced may be maintained for long periods of time under very minimal reinforcement. Furthermore, when the possibilities for verbal and ideationally mediated sources of reinforcement enter the picture (as they do, of course, in human adjustment) and considering that such sources of reinforcement may be self-administered—then it can be seen that we are dealing with extremely complex circumstances.

Aversive Conditioned Stimuli

Many other basic principles also operate at the human level. Consider the experiments by Pavlov and by Farber which we discussed in Chapter 2. It will be recalled that Pavlov was able to use cauterization, pinpricks, and other painful stimuli as a CS for the conditioned salivary response; and that animals who were shocked just before reaching food in Farber's experiment showed a much greater resistance to extinction than animals who had not been shocked. In human situations many sources of reinforcement, subtle and otherwise, may be analogously preceded by noxious or punishing events. For example, children are often forgiven, or evidences of affection and love are reinstated *after* the child has been punished for a wrong-doing. A mother may behave punitively toward a child or continuously contribute to his learning a negative self-concept by, let us say, setting unrealistic standards for him. At the same time close observation of her behavior may reveal that her punitiveness is frequently followed by subtle positive sources of reinforcement. After shaming him she may behave in an overprotective manner.

When the incompatible responses elicited by competing cues of conflict situations are equal in strength, avoidance learning and responses characteristic of the activation syndrome are mobilized. However, as the strength of these cues is altered, under certain reinforcing contingencies the result is very persistent, stable behavior. Thus, in one of Masserman's experiments a cat was trained to accept a mild electric shock as a signal for feeding, and then taught to press a switch and administer the shock to itself in order to obtain food. The animal continued to obtain food in this way even when the shock intensity had been gradually increased to as much as 5000 volts. As Masserman himself has pointed out, this leaves little doubt as to why a neurotic might seek a series of punitive marital partners, when they represent the security he once had with an overdemanding, nagging, but attentive and protective parent.

Immediate vs. Delayed Reward

Turning now from situations in which punishment precedes reinforcing stimuli, let us consider those in which behavior is followed by punishing consequences. It will be recalled (see pp. 157–158) that a very important factor determining the effectiveness of punishment is the time interval between the occurrence of the behavior and the administration of the punishing stimulus. The longer the punishment is delayed, the less effective it is in inhibiting the response. Beyond a certain time interval it may have no effect whatsoever, except to attach fear to whatever behavior happens to be occurring when the punishment does take place. The effect of strong immediate reinforcement for behavior may thus easily offset the effect of equally strong punishment which is delayed. Mowrer and Ullmann (1945) have offered this hypothesis to account for the neurotic paradox.

This factor undoubtedly underlies certain aspects of neurotic self-defeating behavior. Of course, the reinforcement may be either positive or negative. The immediate pleasure associated with masturbation or other undesirable forms of achieving sexual orgasm may maintain the behavior in strength even though relatively strong feelings of guilt (anticipation of punishment) may follow the act. The relatively immediate rewards the individual receives for showing off, acting important, excessive ingratiation, etc., may maintain the behavior in strength in spite of the fact that these behaviors eventually lead to social punishment in the form of the alienation of others.

The reinforcing effect of negative reinforcers is frequently more immediate and stronger than positive reinforcers. Thus, variables which strengthen behavior by way of immediate anxiety-reduction frequently compete successfully with delayed forms of punishment. A great variety of neurotic symptoms, including all those behaviors we have described as mechanisms of defense, are maintained by anxiety-reduction. Long-term or cumulative effects of punishment for these responses have but little influence in altering them.

One might think that the human being's capacity for responding to verbal and other symbolic stimuli might greatly change the operation of these basic principles. To some extent this is, of course, true. However, verbal or symbolic reward and punishment is generally far less effective than we would like to believe. In order for cajoling, the "long-talk," reasoning, etc., to be effective, verbal stimuli must already have had a long history of conditioning in which their effectiveness as positive or negative conditioned reinforcers has been established. Trying to "talk a person out of" self-defeating behavior is singularly ineffective. In the first place the verbal cues may not arouse incompatible feeling states, or at least not in sufficient strength to be effective. Secondly, assuming the verbal cues have already been conditioned as discriminative stimuli for emotional or competing behavior, in order for them to be effective they must occur at the time the undesirable behavior occurs or, preferably, be attached to cues antedating the undesirable behavior.

Some work has been done in attempting to establish such associations for therapeutic reasons. For example, Ferster, Nurnberger, and Levitt (1962) attempted to develop techniques for teaching self-control of eating habits. They found that for the overeater, long-term or ultimate aversive consequences (UAC) of obesity were so delayed that they were ineffective compared to the immediate reinforcement associated with overeating. In order to overcome the time lapse, they sought techniques which would derive a conditioned stimulus from the UAC and apply it at the same time the disposition to eat was strong. The primary conditioned aversive stimulus employed was the person's own verbal behavior. They sought to establish an extensive repertoire of aversive verbal behavior by continually pairing the facts about the UAC with various kinds of eating performances, in an attempt to make the performance itself an aversive conditioned stimulus. Certain aspects of the findings of Ferster and his co-workers are worth quoting here for they have bearing on the control of self-defeating behavior in general:

> We found that simply recognizing the various aversive consequences did not give these subjects an active verbal repertoire which could be invoked immediately and whenever needed. To develop an active repertoire about the UAC, we arranged rehearsals, frequent repetitions, and written examinations. In general, the subjects were unaware of their inability to verbalize the relevant aversive consequences, and were surprised by the poor results of the early written and oral examinations. Verbal descriptions of aversive consequences the subjects had actually experienced were far more compelling than reports of future and statistically probable consequences, such as diabetes, heart disease, high blood pressure, or gall bladder disorder. In other words, descriptions of actual or imagined social rejection, sarcastic treatment, extreme personal sensitivity over excess weight, demeaning inferences concerning professional incompetence or carelessness, or critical references to bodily contours or proportions were much more potent. [Ferster *et al.*, 1962, p. 92]

The establishment of an aversive repertoire associated with the behavior they wished to control was only one of the techniques employed by these investigators. This and other experiments in behavior modification will be discussed in greater length in the following chapter.

Obviously, the effect of punishment upon behavior depends upon the amount of punishment as well as temporal factors. Behavior can be both rewarded and punished, but the punishment may be insufficient to offset the strength of positive reinforcement—the net effect being that the behavior occurs in spite of the fact that it is punished. In severe conflict situations, if the approach habit is very strong as the result of previous conditioning, increasing the painful or noxious stimulus beyond a certain limit may have no further effect. The conflict is resolved by a compulsive approach response, irrespective of the aversive stimulation. Little is known about the precise variables controlling behavior in these situations. Presumably the added increments of excitement and drive produced by the increased aversive stimulation merely serves to intensify approach behavior when the organism is very near the feared but desired goal. The reader may recall from Chapter 5 that several experiments have reported that when an animal is suddenly placed in the zone of conflict at a point near the feared goal it more often runs toward the goal than away from it.

Anxiety and Self-Defeating Behavior

Another source of what appears to be self-defeating behavior arises simply as a result of anxiety conditioning. Positively reinforced behavior becomes disrupted and inhibited. The individual may "really want to behave differently," but the inevitable operation of anxiety takes over.

Skinner has often said that the anxious individual is not interested in food, sex, or art. This is true. Many neurotics are simply not interested in things which are positively reinforcing for others. Depending upon the particulars of the individual case they may shun money, friendship, admiration—in some of the neurotic patterns which we will discuss it is as if the individual were starving in the midst of plenty. The person may be attractive, intelligent, accomplished, or moneyed, and yet be incapable of enjoying these resources, or of employing them effectively for his well-being.

Along with anxiety, escape and avoidance behavior produce excessive withdrawal and denial. If human contacts are not shunned, or their existence denied, inhibition prevents the person from enjoying the many sources of positive reinforcement associated with them. His ability to find substitute satisfactions is generally impaired. Again, at the human level the effects of anxiety are greatly complicated by the operation of verbal and symbolic processes; these latter processes underlie the self-concept, the self-ideal, and certain aspects of personality defense.

Another aspect of anxiety is the development of symptoms. It will be recalled from our discussions of conflict (see pp. 173–174) that cues associated with conflict situations function as conditioned aversive stimuli which evoke anxiety. When the individual cannot escape from the situation or resolve the conflict in one way or another, a wide variety of symptomatic behavior may be learned. These behaviors have partial anxiety-reducing value for the individual. The resemblance between the symptoms developed by experimental animals in conflict situations and those of human neurotics are sometimes striking. In Masserman's experiments (1943) (see pp. 159–160), in which he trained animals to manipulate an electric device which flashed a light, rang a bell, and then deposited a pellet of food in a food box, and then in following training, subjected the animals to a mild air blast or a mild shock when they began to manipulate the device, the animals developed behavioral patterns which were remarkably like those of human neurotics.

It is interesting to note that Masserman's anxious animals lost their position in the dominance hierarchy. This appears to be similar to the behavior of human neurotics, particularly of sensitizers. The sensitizers, manifestly anxious individuals, have also been found to be chosen less often as desirable partners in group situations and generally to have less leadership ability (e.g., Joy, 1963).

Inhibition and Learning Deficit

Frequently neurotic individuals are viewed by others as self-defeating simply because they do not show an ability to learn in important life situations. There has been some research relating an inability to learn in certain situations to specific personality variables. Learning deficits appear to result from inhibitory personality tendencies. Let us consider two examples of this type of finding.

Weiner and Ader (1965) placed 26 male medical and graduate students in the following operant-conditioning situation: the subject was seated in a small room at a table with a button conspicuously within his reach. Virtually no instructions were given. However, an electrode was strapped to his leg. Shock was administered at regular intervals, but the shock could be delayed for a fixed interval of time if the subject pressed the button. Two subjects walked out of the experiment thereby eliminating themselves from the experiment. Eleven were classed as nonlearners. Ten of these eleven subjects sat for an entire hour receiving repetitive shocks in from 5- to 30-second intervals, but did not press the button; the other subject pressed but he did not reach the criterion of learning previously decided upon by the experimenters. Thirteen subjects did reach the acquisition criterion, which was the avoidance of 80% or more of the shocks for a 5-minute period. Scores on various personality tests administered before and after the conditioning session indicated that the nonlearners were more likely to deny the existence of hostility, blame, or frustration, whereas the learners were more likely to externalize blame and become hostile in a frustrating situation. In line with this finding, several other studies have indicated that college underachievers tend to be people who have difficulty expressing hostility (e.g., Kirk, 1952; Shaw & Black, 1960).

Inhibition and a lack of learning have also been shown to be related in a study by Mosher (1965). He devised a test to differentiate between subjects with high and low guilt feelings about sex. He aroused 80 male subjects by showing them pictures of nude and seminude females and asking them to state various preferences. Half of the subjects were then assigned to a condition designed to create an expectancy of punishment for behavior related to sex, whereas the other half were given cues that sex-related behavior was not disapproved. Finally, the subjects were asked to identify taboo words.

Subjects with low levels of guilt were definitely influenced by the cues of acceptance or punishment, as indicated by significant changes in their recognition of taboo words. On the other hand, subjects with high guilt were found to be uninfluenced by the experimental manipulations. These subjects were insensitive to cues indicating either the presence or the absence of external punishment. The expectancy of punishment did not raise their recognition thresholds, nor did the expectancy of permissiveness lower them.

Thus, there appears to be a tendency for people who have developed inhibitions against the expression of impulses such as sex and hostility to be unresponsive to certain situational cues. Persons who are "wrapped up in themselves," in the sense of feeling guilty and hence attentive to their own response-produced cues, or in the sense of having to deny the existence of feelings of frustration and blame, may be generally less capable of focusing on external cues. When the person's attention is focused on himself, there is also an increased likelihood that situational or task-oriented behavior will be impaired (e.g., Beier, 1949). Even performance of very simple motor tasks is affected if the subject is told they measure some aspect of his personality (Neibuhr, 1955). His criterion of performance may remain the same, but his behavior is more diffuse and requires more effort; it is less efficient.

CLASSICAL PATTERNS OF NEUROTIC BEHAVIOR

Dependency, Aggression, and Sexuality

Emotional disturbances resulting in neurotic patterns of behavior in our culture are perhaps most frequently associated with dependency, aggression, and sexual expression. The learning of the expression and control of these behaviors has been discussed at length in Chapter 8. When the appropriate control and expression of these behaviors is not learned relatively early in the individual's development many possibilities for emotional disturbance and behavioral deviations exist. Generally, individuals may either never learn self-control and appropriate expression in these areas, or they may learn excessive and inappropriate inhibitions.

Excessive inhibitions frequently result in either inappropriate emotional and behavioral patterns, or in excessive anxiety. Very often quite stable inhibitory patterns relative to dependency, aggression, and sexuality are maintained throughout the individual's early adult years, until circumstances of his life combine to reawaken early fears and conflicts associated with them. Let us consider a typical example of emotional disturbance which may occur as a result of early fearful and conflictful experiences. The following case is an abbreviation of one reported by Wolberg (1954):

Mr. A came to the clinic shortly after the onset of a variety of emotional disturbances, including tension, anxiety, depression, and psychosomatic symptoms. Mr. A believed that his upset was caused by extreme pressures involved in the highly responsible job to which he had recently been promoted. Until recently Mr. A appeared to have been a well-organized individual who conscientiously cared for his family and performed his work duties, enjoyed his children, respected his wife, belonged to number of clubs and organizations, and seemed to get along well with his friends and associates. As the interview proceeded the psychotherapist learned much about the attitudes and expectancies of Mr. A. It appeared that, although Mr. A possessed a strong drive for achievement, his history also revealed a strong tendency to seek out another person who was the symbol of strength and omniscience. He tended to rely upon this other person as if he were a providing parent. Without being fully conscious of it, it appears that Mr. A assumed his demands and his needs would be automatically satisfied. Although he himself showed little spontaneity and initiative, he seemed to be comfortable and active when he was participating in projects for which his superviser bore the brunt of responsibility.

The therapist also discovered that his wife had been withdrawing her attention from him for some time before the onset of Mr. A's "nervous breakdown." Apparently she had become much more attentive to in-laws who, for financial reasons, were living with them on a temporary basis. It appeared that as Mr. A began to demand increasing attention she responded by giving him even less, until finally a sense of impending catastrophy, helplessness and fear of his own aggression began to well up in Mr. A. As these tensions increased, simply going to work became more and more of a burden. The work itself began to tax his capacities. Mr. A's breakdown was finally precipitated by the promotion to a more responsible position.

Other attitudes and trends in Mr. A's personality became apparent. For example, along with his general compliance to authority, he seemed to minimize his own assets and abilities, and, often naively, to overestimate the virtues of others. Part of his fear of being competitive and resourceful appeared to result from his anticipation of hostility from others, and the belief that he would surely fail and thus be ridiculed. The origins of these feelings became clearer as the interviews progressed. It appeared that Mr. A's mother had been strongly overprotective throughout his childhood and adolescence. Childhood attempts at assertiveness or independence had been very harshly dealt with until he had learned to be secure only by complying with his mother's demands. Other unconscious fears were revealed, such as fear of mutilation should he assume a clearly masculine role, passive wishes, etc. Finally, it became clear that Mr. A responded toward his wife much as he had responded toward his mother as a child. Partly because of her own impulses, and partly because Mr. A's behavior had subtly reinforced a parental role, she had responded, essentially, by mothering him. Her seeming withdrawal of affection subsequently mobilized intolerable anxiety. The assumption of new job responsibilities then produced more stress than Mr. A was able to cope with. [Adapted from Wolberg, 1954, p. 96ff.]

Case histories could be quoted indefinitely of people whose adult behavior is inappropriate, ineffectual, and frequently anxiety-ridden because of their earlier experiences of reward and punishment in the areas of sexuality, aggression, and dependency. There is the individual who finds his or her marital life poisoned by unconscious anxiety about sexual behavior; those who are so hostile in their relationships with others that they remain lonely and cut off; those who repress any desire for success or independence, and cling to others for support; those who work feverishly trying to be "successful" and desperately trying to avoid thoughts of dependency or affectional longings, etc.

As long as the individual can avoid stimuli in his life situation which activate early conflicts he may be capable of an acceptable adjustment. As long as Mr. A had an indulgent and protective wife and a routine job, everything went well. As long as the classic old maid or bachelor remains unmarried, severe stress is likely to be avoided. Anxiety may not be mobilized in the pseudoindependent person unless clear instances of failure occur, or circumstances combine to produce a situation in which he finds himself relatively helpless, etc.

All of the mechanisms of defense which we discussed in Chapter 10 serve to protect the individual from threat which many stimulus situations would represent. Although they distort reality and restrict the person's ability to function in various ways, they are also adaptative insofar as they prevent the reoccurrence of further tension. As environmental stress increases the individual's defenses must intensify so that larger segments of his environment become misinterpreted. Extreme loss of contact with reality is associated with psychotic reactions, which we will discuss later in this chapter. Certain varieties of psychotic behavior involve a complete breakdown of the repressive defenses and unrestrained direct expression of unacceptable impulses and drives.

Some Early Conditions

It will be recalled that in the normal process of socialization the child gradually loses his need for total reliance upon others and becomes increasingly capable and desirous of doing things for himself and of being relatively independent of others. To be sure, the normal development of independence involves ups and downs, advances and recesses, but the net result is that self-reliance and the ability to tap his own resources eventually predominate. In instances where the parents push too fast for achievement, the child may be left with a relatively severe problem of unresolved dependency, the awareness of which may be so painful that defensive measures are the inevitable results. On the other hand, if the child is overprotected, the normal learning of independence simply never has the chance to occur. In clinical literature mention of the "seductive parent" is not infrequent. This refers to the fact that sometimes, because of their own emotional difficulties, parents turn to the child for security and comfort which they are unable to attain in their adult relationships. In either case independent or assertive behavior on the part of the child, or later, the adolescent, constitutes a considerable threat to the parent.

Anxiety Reactions

In the case of Mr. A above, circumstances in his life combined to promote strong feelings of helplessness, impending disaster, tension, misery, restlessness, visceral responses, hopelessness, and physiological complaints. These are all varied aspects of a generalized anxiety reaction which was cued off by events which reawakened conflict and fearful anticipations learned in childhood. Of course, to an objective observer these acute manifestations of anxiety are inexplicable, because they seem so inappropriate to the circumstances. However, the elicitation of these emotional respondents has little or nothing to do with objective reality. These anxiety reactions are called *neurotic* because they are so inappropriate. Nothing has happened which would objectively justify such fearful reactions.

Neurotic anxiety reactions vary in intensity from those involving mild symptoms of restlessness and tension to more acute states in which the individual is overwhelmed by generalized fears, feelings of uncertainty and helplessness, visceral and somatic disturbances; to those involving absolute panic and acute episodes of extreme fright, misery, behavior disorganization, trembling, fainting, or the blotting out of memory.

Anxiety reactions may be relatively focal, in the sense that they are elicited by specific cues or environmental circumstances which the individual perceives as threatening. In this case removal of the specific cues, or a change in the individual's environment, may result in a sharp drop in anxiety and a return of emotional composure, feelings of security, and adequate behavioral organization and functioning.

Other anxiety reactions appear to be the result of a steady buildup of conflictful and anxiety-producing events in the individual's life so that, given the processes of generalization and incubation, so many cues in the person's life elicit anxiety that it becomes a chronic response. In these cases the individual may remain chronically anxious more or less independently of varied environmental situations. He may also experience more acute anxiety attacks over and above his chronically anxious responsiveness. The attacks are, of course, related to events which for him are especially threatening.

An anxiety reaction is well illustrated by the behavior of Thomas R, an eighteen-year-old high school senior who was referred to a counselor because he was failing in his studies, and had an attitude of apprehension and despair which was readily noticed by his teachers. Interviews showed that the boy's anxieties were not limited to any definite situation, but widely generalized. He was concerned about his academic standing, and especially about his father's reaction to it. Referring to his possible school failure, he said, "It will be the end for me." He felt an acute social incompetence, and said in a vague manner that he did not know much about the world, and that he had many things to learn. Thomas had little association with girls and appeared to be afraid of them, or rather of his inability to impress them as favorably as other boys. During the preceding year he had had a few dates, with a girl a little older than himself, on which he placed a high value—considering himself in love. The girl went away to college, and Thomas felt afraid of "losing" her. He was utterly unable to make decisions. The simplest problem caused him to seek advice or feel incompetent to face the difficulty.

In addition to his anxiety, Thomas had visceral symptoms centering around his heart. At times his heart beat very, very rapidly and his pulse pounded in his ears. Although several physicians examined him carefully and reported that he had no organic disorder, Thomas often rested in bed from early Saturday evening to until Sunday noon because of his supposed heart disease. The intensity of Thomas' anxiety was best revealed by notes he scribbled from time to time and gave to the counselor. He wrote, "I can never be at rest and am never satisfied. I fear of not being able to control my mental and physical actions. Something is always elusive. I am more afraid of life than the basest coward. Why can't I understand people? Why can I remember only my fears, the vacant mental situations and lonely places in my life? I seem to exist isolated. All the clean wholesome desires which make a man want to live seem to be crushed. Will I snap out of this, or will I never be a man?" [Shaffer & Shoben, 1956, pp. 277–278]

Basic to this type of case, in which the most conspicuous feature is obvious misery and continual manifestations of anxiety, are the many and varied conflictful stimuli which the individual does not have adequate resources to cope with. Invariably the underlying conflicts are found to be acute and prolonged. They invariably have their antecedents in parental attitudes and behavior which predispose the individual to lack skill in using his own resources to cope with his problems. In the case of Thomas R, a domineering father had never allowed him to be assertive or make an independent decision. His mother, too, had contributed by an extremely religious and moralistic attitude which added considerably to his generalized guilt and social isolation. Furthermore, his mother appeared to be a nervous or anxious individual herself so that some modeling effects may have occurred. Thus, as is typically the case, not only was there a history of unhappy experiences, but experiences which were of such a nature as to predispose him to further helplessness in the face of the many conflictful stimuli associated with adolescence—the problems of establishing identity, vocational choice, social and sexual adjustment, etc.

The chronically anxious adult personality also appears unable to rely on his own resources to resolve his problems, or to utilize defensive mechanisms against anxiety. Repression may function to keep the roots of his characterological conflicts out of awareness but a great deal of anxiety fails to be inhibited or avoided and there is considerable sensitization, rumination, and worry. Depression is common. Whatever defenses do develop are likely to be of the sensitizer type. For example, self-devaluation, feelings of inadequacy, and hostility are likely to be externalized. Others are seen as being hostile, incompetent, and blameworthy.

It is not uncommon for the adult to seek relief from his continual tensions by turning to alcohol, drugs, and sedatives for relief. Tranquilizers and sleeping pills are likely to be used in excess. In this connection it is interesting to note that significant relationships between sensitization as measured by RS scale scores and alcoholism, or alcoholic tendencies, have been reported (e.g., Joy, 1963). There are also data which strongly suggest a relationship between alcoholic tendencies and birth order. There is a significant tendency for alcoholics to be later-borns, rather than firstborn, or only children (Schachter, 1959).

The reader may recall that Byrne (1964) had reported that repressive scores indicate a home atmosphere characterized by permissiveness, acceptance, and confidence in assuming the role of the same-sexed parent. The clinical case history data relative to the background of anxiety neurosis tends to lend support to these findings. The remote causes of anxiety reactions appear to be childhood experiences which predispose the individual to make anxious responses to his conflicts. The parental traits which appear to significantly influence the development of anxiety neurosis include overdominant and belittling attitudes, rejection or nondemonstrativeness, and poor role or modeling definitions. These behaviors on the part of the parent may readily produce feelings of helplessness, a dependency–hostility conflict, and an attitude of inferiority, inadequacy, and anxiety on the part of the child. If it is true that the learning of relatively effective (repressor-type) defenses against anxiety requires a certain amount of outside support, then prolonged exposure to these insecure conditions would be expected to leave the individual more or less helpless to cope with his feelings.

Certain other tendencies of the anxiety neurotic deserve mention. The strong, sometimes profound, inhibition which the individual feels toward confiding in anyone or seeking advice appears to contribute to the maintenance of his anxious responses. Frequently such individuals have learned simply not to trust people. Since their backgrounds have been characterized by harsh or rejecting parents, or parents who did not inspire confidence, it is obvious how such inhibition gets learned. Moreover, people frequently try to help the anxiety neurotic by giving him advice, or trying to persuade him that his anxious reactions are unrealistic. This may actually constitute punishment, because he cannot control his feelings and behavior voluntarily. It may add further to his self-devaluation, isolation, and feeling that people do not understand him.

Rather severe repression of basic conflicting trends in the personality of the anxiety neurotic may be bolstered by his preoccupation with worrying, anxious concerns, and more or less frantic attempts to reduce his ever-present anxiety. There are many varieties of anxiety- reducing behaviors which may also function to avoid the recognition of real problems. Among these are obsessive-compulsive patterns of behavior, escape through excessive sleeping, reading, television viewing, and the like; fatigue and hypochondriacal reactions. Finally, other more or less distinctive behavioral patterns which belong under the general heading of anxiety reactions include phobic reactions and depressive reactions (discussed on pp. 317–318 and 367–368).

Dissociative Reactions

Once repression is learned, it becomes involuntary. The learning of habits of "not thinking" or of "not responding" as responses to threatening stimuli gives the appearance of being an all-or-none process. This is because when the response is first made reinforcement is immediate, and "automatic." When denial and repression become overlearned, generalized habits which are employed to disavow anxiety-laden aspects of the individual's personality, it is as if there were a splitting off or dissociation of certain aspects of the individual's experience and behavior. Stimuli which would be anxiety-producing or conflictful may include movements, sensations, memories, or whole patterns of personal habits. Cameron (1947) has referred to the excessive use of denial and repression as "over-exclusion." It appears as a predominate behavioral pattern in *hysterical personalities.*

Hysteria is an old psychiatric term used to describe motor and sensory disturbances and impairments such as tremors (oscillatory motor movements), paralysis, anesthesia, and memory loss (amnesia)—all of which appear to result from traumatic experiences which are fraught with emotion, rather than from any neural or anatomical defect. Hysterical reactions to traumatic events were thought to be exclusively feminine. In fact, "hysteria" comes from the Greek word hysteria, meaning uterus. Fainting is a common hysterical type of reaction. Less than 50 years ago it was still common for women to be able to faint on demand. The mechanism served as an effective and socially acceptable escape from unpleasant or threat-producing stimuli. Today we know that hysterical reactions may occur in either sex, although the fainting response in women appears to be passe, and never was socially acceptable for men.

"Hysterical personality" refers to a relatively ill-defined pattern of character traits. Most of them center around repressive patterns and basic characterological dependency needs. Thus, the hysterical personality has typically learned to repress or deny external unpleasantness. Minor repressive patterns have become habitual. The individual may forget names easily, occasionally not being able to remember the name of his or her friend or spouse. The mishearing of anxiety-producing communications; the inadvertant skipping over of unpleasant details in recalling events; a general sloppiness or lack of neatness in the individual's living habits; cockroach projection—these are all typical behaviors of the hysterical personality.

As an alternative to, or in conjunction with a nonchalant style of life the individual may be inclined to histrionics. Everyday events may be dramatized. He may also be over concerned with external appearances, and with justifying his behavior to others. In the same vein, there may be an inclination toward exhibitionism. Hysterical women frequently appear to be sexually attractive and seductive. Usually, however, such individuals are in reality sexually inhibited and frigid.

Two of the most often quoted general characteristics of hysterical personalities are social immaturity and external orientation. Social immaturity in this case refers to a general pattern of egotistical and selfish behavior. At bottom the individual seems to have a lack of concern for others, and a lack of ability to show genuine empathy for others. His interpersonal relationships are often solely on a superficial level. He may socialize with others a great deal, or even be the center of attention, but he actually confides little in other people. Since he talks little about himself, even close friends may know little or nothing about his values, ideas, and goals.

The individual's external orientation augments his social immaturity. External orientation implies that he is more dependent upon environmental or situational cues than he is on those provided by his own thoughts and feelings. Instead of his own internalized standards of conduct, the hysterical personality appears to be more dependent on momentary rewards and approval provided by others. He appears to lack thoughtful conditions of his own. These particular patterns of behavior have been discussed before under the headings of repressor-type defenses (see p. 349) and approval-dependency (see pp. 276–277). Of course, most of us exhibit these behaviors now and then, but only to a mild degree. The normal person is much more flexible. The classical descriptions of the hysterical personality apply to individuals who demonstrate repressive and approval-dependent patterns of behavior to an abnormal extent. Their excessive use of repressive patterns and striking lack of personal integration has also led to the more general label *dissociative personalities*.

Background Factors of the Hysterical Personality

Perhaps the most frequently reported background factors of hysterical-type personalities are histories of overindulgence or overdomination. The overprotected and overindulged child learns to expect to have his own way, irrespective of the wishes or rights of others. When demands are made upon him he may simply ignore them. This simple technique of coping with external threats and demands has a history of being rewarded for these individuals. It is easy to understand how the blotting out of any disturbing thoughts or unpleasantness may also be easily learned by such individuals. It is also clear that such people must be basically dependent personalities because there is little opportunity for independence or personal initiative and resourcefulness to be learned. Analogously, severely dominated children have little opportunity to learn habits of personal responsibility and independence. Repression may be strongly rewarded because cues associated with sexual, aggressive, or self-assertive impulses have become very frightening.

Severe hysterical-type reactions are much more frequent among persons of below average educational and cultural levels. Women may still faint or have more serious physical-conversion reactions (see below) in certain subcultures, such as the hill country of Tennessee. In such cultures sexual and aggressive impulses may be severely dealt with. The growing child may be exposed to fundamentalist-type religious fanaticism on the one hand and evidence of, let us say, decidedly amoral sexual behavior, including incest, on the other. Repressive and dissociative reactions may be strongly reinforced in such a tension-producing environment. Similarly, parental punitiveness and overdomination associated more with the lower socioeconomic levels may be predisposing factors in phenomena such as the mass screaming and swooning over popular entertainers by adolescent girls in our large cities.

Another common thread running through the case histories of hysterical personalities involves the learning of inconsistent or conflictful attachments to parental models. The individual may be torn between loyalty to one or the other parent. It is also common for hysterical persons to be afraid of the same parent or person whom they love. Difficulty in forming integrated values may also result from conflicting or inconsistent patterns of parental reward and punishment.

Amnesias and Fugues

Hysterical personalities may cope with neurotic conflict by blotting out, or dissociating, anxiety-producing thoughts and anxiety-producing aspects of their personalities from consciousness. Typically, the individual "acts out" his unacceptable wishes, but in order to do so, he must be amnesic for certain events and experiences. Most often the individual suffers amnesia for personal identity. That is, he "forgets" who he is and where he lives. Memories associated with symbols or stimuli that would give him a sense of the continuity of self-hood are dissociated from consciousness.

The most typical behavior exhibited in this state involves attempts to escape from an intolerable life situation. Instances in which the individual loses his sense of identity and runs away from familiar surroundings are technically referred to as *fugue states*. The individual may appear in a strange city, not knowing who he is or where he came from. Sometimes, in their confusion, such individuals report themselves to authorities, usually the local police. An occasional case is reported of an individual who has remained in a fugue state for years. Newspapers are sometimes able to document stories such as that of a man who, having abandoned his wife and family years before, is found living in a distant city, remarried and settled down, perhaps a respectable member of the community. Careful clinical appraisal reveals that he truly cannot remember his past identity and details of his past life, nor is he able to recognize real or symbolic stimuli associated with it.

We will not take the space to outline a detailed case of fugue reaction here. These cases are quite rare. A case involving a less severe dissociative reaction, although perhaps it is less dramatic, may be equally instructive:

> A man of twenty-nine had developed a high ideal of independence and manliness. He had been induced, however, to take work in his father-in-law's business, where he found himself dissatisfied and poorly paid. He was sometimes unable to meet family expenses, and was greatly humiliated to be extricated by his father-in-law on these occasions. One day, again in difficulties, he drove with his family to the town where his father-in-law lived, but could not bring himself to ask for the needed loan and turned the car homeward. He became so pre-occupied with the thought of finding a new job and making money, that by the time he reached home he no longer knew who he was nor recognized his wife and children in the car. Taken to the hospital, he spoke only of his new job. He falsified reality to the extent of interpreting everything in the hospital as though it were the operation of a business firm. Two days later he emerged spontaneously into his normal state, not remembering the amnesic episode. Shortly afterward he recalled the episode, including the suicidal despair that had filled him at the thought of asking his father-in-law for more help. [White, 1964, p. 268]

The fact that this individual interpreted everything in the hospital as though it were the operation of a business firm points to subtle sources of reward operative during the period of the dissociative reaction. Fantasy-reward value could only accrue to the individual's intense wish to get a job and maintain security and self-esteem if the realities of his present situation, including his identity, were obliterated from consciousness. Dollard and Miller (1950) have argued that the hysterical reaction, and the ensuing fear-reduction may be mediated by a thought or series of thoughts. It will be instructive for the reader to compare this case with the case of a soldier who suffered hysterical paralysis of his arm, given below under the heading Conversion Reactions.

Multiple Personalities

Another type of dissociative reaction involves the splitting off of the person's identity, and its replacement by an entirely different personality during the amnesic period. Cases have been reported in which the individual assumed two, three, or more different personalities and styles of life, each one being amnesic for the other. These are simply called cases of *multiple personalities*. They are extremely rare.

One of the most recently reported cases of multiple personality was that of Eve White:

> Eve White was a quiet, subdued, rather sober housewife—on the whole, a colorless individual. She came for psychiatric treatment because of severe headaches and "blackout periods." It was soon discovered that during her "blackouts" she behaved like an entirely different person, with many characteristics quite the opposite of those of her normal personality. She called herself "Eve Black" during these episodes. Her mannerisms, gestures, voice, and the basic idiom of her language changed. Eve Black was forward, impulsive, hedonistic and sexy. There was evidence that Eve Black was aware of Eve White, but Eve White wasn't directly aware of the existence of Eve Black.
>
> As therapy progressed a third distinct personality emerged: "Jane." Jane was more expressive and capable than Eve White, and more mature and less impulsive than Eve Black. By this time the therapists were able to "call out" whichever personality they wanted to talk to. Jane emerged only through Eve White. Eventually, with the continual breakdown of repressive blocks, Jane "stayed out" more and more. Finally, one day when the therapist called out Eve White, instead, an intense, emotion-laden episode occurred in which the patient appeared to re-live an early dramatic memory. Eve White was soon gone. Simultaneously the patient recalled a childhood experience of being forced at her grandmother's funeral to touch the face of the corpse. Thereafter, the patient appeared as an integrated personality which most resembled Jane. [Thigpen & Cleckley, 1954, p. 135–151]

This account is, of course, extremely brief. The change from one personality to another was a common occurrence outside of therapy, as well as during the therapeutic hours. The reader can imagine the disruptions and the confusion which characterized this person's life.

Elaborate consideration of a similar case led Prince (1920) to conclude that at least one of the personalities in his well-known case of Miss Beauchamp could be traced to certain aspects of her fantasy life as a child. The evidence also indicated that her various personalities represented clusters of wishes and strivings which were contradictory to one another. The conflict in such cases is apparently resolved by the splitting off, the dissociation of one entire system of behavior, values, and attitudes from another.

Conversion Reactions

There is a continuum of dissociative reactions which is reflected in bodily changes, or which have physiological manifestations. At the mild extreme of this continuum are simple motor and nervous mannerisms which are automatic and dissociative in nature. Among these nervous mannerisms are throat-clearing, finger-tapping, muscle-twitching, involuntary facial grimaces, coughing, sniffing, lip-smacking, and jerking. Oscillatory motor movements, more commonly known as tremors; the contraction of muscle groups (cramps); and the uncontrolled bodily reactions involving voice trembling, knee-shaking, and other anxiety-based respondents known as "stage fright" also belong in this group.

Stammering and stuttering are reasonably common reflexive reactions which are primarily found in individuals of the hysterical sort. Anyone who stutters is well acquainted with the uncontrollable, involuntary nature of the reaction. Stammering and stuttering appear to result from a learned fear of communicating verbally with others. A cluster of conditioned responses appears to be involved, including expectancies of criticism and rejection, and anxiety elicited by the stuttering itself.

In 80 percent of the cases considered stammering and stuttering are learned before the age of 6. The inception is typically in the 3 to 5 year age range. Many children have nonfluent speech. The figures vary, but normal adults have been found to exhibit various types of nonfluent speech on the average of at least once in every thousand utterances. People undergoing stuttering therapy are frequently surprised by this when they become trained to listen to the normal speech of others.

In the background of most stutterers an overanxious or domineering parent can be found. Parents who are strict, perfectionist, or anxious about the child's development of excellent speech, may easily condition the child to experience anxiety at the prospect of verbal interchanges. Aside from being rigid and strict, many parents of stutterers are oversolicitous and overprotective. Thus, once the damage has been done, so to speak, the child may be protected and given special consideration because of his speech defect. Thus, the child may be rewarded by being given attention by parents, teachers, and other adults which he would not otherwise have received. Of course, it is hardly being suggested here that a mere lack of special attention would significantly affect the cure of stuttering. Stuttering is highly resistant to treatment, especially for the late adolescent or adult. The cumulative reinforcement of a response that is learned on the basis of anxiety-reduction results in a highly ingrained repertoire of habits, which indeed, operate automatically. It is significant that even the severe stutterer may speak normally when anxiety-producing cues are absent. He may speak well alone, after the repetitive reading of material, or he may not stutter while singing. It is well known that alcohol tends to reduce anxiety. Stutterers frequently report normal fluency under the influence of alcohol.

The classical conversion reactions involve major motor and sensory systems. The label *conversion reaction* was originally used because it was believed that anxiety became converted into a physiological symptom, which served in turn to protect the individual from experiencing anxiety. Acute bodily motor and sensory disturbances which are psychologically, rather than physiologically, based are more or less distinguishable because, for one thing, the patient experiences decidedly less anxiety than persons with a comparable disturbance of organic or physical origin. Customarily the onset is acute. Furthermore, the symptoms tend to follow the patient's idea of neurology. Hysterical anesthesia, for example, may involve the entire hand or food (glove or stocking anesthesia), rather than follow the actual neural distributions. The particularly bodily function affected is also frequently found to be related symbolically to the person's unconscious conflicts. Serious conversion reactions are not extremely rare. When they do occur they can invariably be traced to an unconscious attempt to escape from an intolerable life situation of some sort. Typically, the individual has had a history of reacting to emotional conflict with ill-health, although this is not always the case.

It is interesting to note that certain borderline conversion reactions may occur, and perhaps are more frequent than we would ordinarily expect. Thus, the patient's recovery from real organic illness may be delayed. He may undergo a series of relapses, or develop new symptoms. The overprotectiveness, care, and special attention associated with the hospital environment may serve as very strong sources of gratification for unconscious dependency strivings in people who are prone to react to emotional conflict with hysterical-type symptoms. How many unnecessary operations are performed each year! There are few objective data; only the subjective impression of medical practitioners.

It has been suggested (Dollard & Miller, 1950) that the rewards associated with being physically incapacitated, cared for, escaping from extremely aversive stimulation, and so on, cannot explain the original learning of the symptom, because they occur after the symptom is firmly established. They point out that the original reinforcement for conversion reactions must occur while the symptom is being learned, and further, that it is well known that for reinforcement to be effective in strengthening a response it must occur soon after it. They suggest that the stimuli which mediate conversion reactions are verbal.

Let us take their description of a case of war neurosis, originally reported by Grinker and Spiegel (1945). This case illustrates the dynamics of conversion reactions in a relatively concise manner. The patient was an artillery officer who had been blown off the ground three times by the nearby explosion of shells. He found himself shaken, but otherwise all right. Some time later he found that he could not remove his right hand from his trouser pocket. He discovered that his right arm was almost completely paralyzed. He remained in the field, but eventually was sent to the hospital where, although the paralysis remained when he was in a normal state, under Pentothal narcosis it disappeared entirely. When he entered the hospital with his arm paralyzed there were no signs of anxiety. As the therapeutic interviews progressed and the condition began to clear up, symptoms of manifest anxiety appeared. Dollard and Miller discuss this case as follows:

> In the case of the soldier with the paralyzed arm the eventual hospitalization and escape from combat could not have been the original reinforcement because it occurred only after the symptom was firmly established.... Since the fear was not experienced as long as the symptom persisted but reappeared as soon as the symptom was interrupted, it is apparent that the symptom reduced fear. But a reduction in the strength of fear is known to act as reinforcement. Therefore, the symptom of partial paralysis seems to have been reinforced by the immediate reduction that it produced in the strength of fear.
>
> The symptom of paralysis produced this reduction in fear because the patient knew that it would prevent his return to combat. As soon as the patient noticed that it was difficult to move his hand, he probably said to himself something like "They won't let me fight with a paralyzed hand," and this thought produced an immediate reduction in fear. Though the fear reduction probably was mediated by a thought its reinforcing effect on the symptoms was direct and automatic. In other words, the patient did not say to himself anything like "since a paralyzed hand will keep me out of combat, I should try to have a paralyzed hand." In fact, when such a patient becomes convinced of the causal relationship between the escape from fear and symptom, a strong increase in guilt counteracts any reduction in fear. The reinforcement is removed and there is strong motivation to abandon the symptom. [Dollard and Miller, 1950, p. 166]

NEUROTIC PERSONALITY TRENDS

We have discussed the classical neurotic patterns: anxiety reactions, obssessive-compulsive reactions, phobias, neurotic depressive reactions, and dissociation. Many of the people who seek psychological help, especially from the private practitioner, do not exhibit these relatively clearly defined reactions. They do not have obvious manifest or focal fears. They may not exhibit hysterical-type symptoms. Their complaints are diffuse and varied. In fact, perhaps the only common denominator is that they are miserable, unhappy people.

These cases frequently show the very common trends which we have discussed; namely, emotional disturbances centering around dependence, sexuality, and handling of aggression. Shadings of the classical neurotic symptoms may be apparent too. Nevertheless, the most important sources of their difficulties do not seem to be readily apparent, unless one looks at their whole style of life. Let us examine certain observations and theories concerned with this.

Watson's Views

It might give us some perspective to consider some observations on personality which represent recurrent themes in the relatively short history of attempts to develop a science of human behavior. Watson published his classic book *Behaviorism* in 1924. He devoted his chapter on personality to an analysis of personality as consisting of organized habit systems. After arguing that personality is the outgrowth of the habits we form, he discussed some of the weaknesses of adult personalities. Bearing with Watson's simplified, and now archaic, use of language let us note some of his observations. Watson organized his discussion of personality weakness under four subheadings: inferiority, susceptibility to flattery, the constant strife to become kings and queens, and infant carry-overs. Let us consider a commentary or two taken from each of these headings. On the topic of inferiority, a sample of Watson's commentaries is as follows:

> Most of us have groups of reactions developed which do cover up, conceal and hide our inferiorities. Shyness is one form, silence is another, outbursts of temper are another, advanced stands on moral or social questions are other very common forms. The most selfish of individuals has a well organized verbal scheme which hides his selfishness from the uninitiated – the most "impure" of individuals often talks the loudest of purity....
>
> We likewise organize habit systems that serve the purpose of concealing our physical inferiorities. The little short man often talks loudly, dresses "loudly" wears high-heeled shoes, is "cocky" and forward. In order to be seen at all he must act in an unusual way. Women try to balance one thing off against another. Their faces may not be beautiful, but their forms are exquisite; their arms may be clumsy, but their legs are objects of admiration by discriminating artists. Nothing in their anatomy may be supreme – then they fall back upon style. When too fat to be stylish, they have wonderful cars to ride in, beautiful jewels, well appointed homes...somehow most human beings can't permanently face inferiority. [Watson, 1924, p. 228]

Under the general heading of "susceptibility to flattery," Watson turned attention more to the "vulnerability" of personality.

> I doubt very seriously whether any man or woman is invulnerable in any commandment, on any code of honesty, on any life-long settled conviction. I think the time was when invulnerability was a more nearly possible thing. Today conventions are so universally overstepped, religious mandates so often transgressed, business honesty and integrity so often a matter of legal decision, that all of us are vulnerable if and when approached long enough and hard enough and subtly enough on our weak side. . . . It often happens in business and in the professions that as long as a man ahead of you is helpful and useful to you, you are meticulous in giving him his due. He can do no wrong. You back him up, support him on every occasion. But when you reach him, when you begin to share the throne with him, you, without ever verbalizing it, somehow find your ear more attuned to his faults. A

strong visceral toning appears when you hear things not quite to his credit! Then again, when you pass him, you begin to wonder if your former rival cannot be replaced by a less expensive man. [p. 290]

On the constant strife to become kings and queens:

As a result of our training at the hands of our parents, of the books we read and of the biographies of those around us, every man deems it his inalienable right to become a king and every woman a queen. . . We try to carry over into everyday adult life the dominance we have put over on our parents in our childhood. The labor leader who says, "down with the capitalist and up with the labor," is just as anxious as any of us to be a king. The capitalist who says, "down with labor," is just as eager either to become king or else to stay a king. [p.292]

On infant carry-overs:

The weaknesses in our personalities which we have just considered are but examples of the general fact that we carry over many organized habit systems from our infancy and early youth into adult life. Most of these systems...are of the unverbalized types—the verbal correlates and substitutes are lacking. The individual cannot talk about them, he would deny that he has carried over his infantile behavior, and yet the appropriate situation brings out its expression. These carry-overs are the most serious handicaps to a healthy personality. [pp. 292-293]

It is apparent that Watson placed considerable significance on the struggle of human beings to gain some feeling of importance by fair or foul means, as it were. Watson was not alone in these opinions. His contemporary of the neopsychoanalytic camp, Alfred Adler, was beginning to elevate inferiority feelings and the consequent striving for superiority to the level of a major explanatory concept vis-a-vis human behavior.

Adler's Views

Adler was one of the first major personality theorists to call himself a scientist. He happened to be within the psychoanalytic tradition. Actually, his theorizing was highly speculative and couched in primarily nontestable terms. He eschewed an analysis of antecedent conditions and the deterministic frame of reference generally. He believed, for example, that the striving for superiority was innate. He conceived of the goal of superiority as a creative, self-actualizing process. What is important for us in Adler is not his philosophical flotsam and jetsam, but his observations. For, like so many personality theorists, he clearly outlined certain antecedent conditions and the relationships between them and behavior, and then proceeded to take but little account of them. Adler felt, that

The neurotic individual comes from a sphere of insecurity and in his childhood was under the pressure of his constitutional inferiority. In most cases this may be easily proven.

In other cases the patient merely behaves as if he were inferior (the inferiority is merely subjective). In all cases, however, his willing and thinking are built upon the foundation of a feeling of inferiority. [From Ansbacher & Ansbacher, 1956, p. 119]

He went on to outline the various patterns of development of abnormal inferiority feelings and the individual's neurotic ideas associated with them. He found, for example, that children with bodily defects tend to attempt to compensate for them, to be wrapped up in themselves as adults, and to be more selfish and inconsiderate than others. He thought that pampered children have as their goal of superiority to make their relationship with their mother a permanent one. Frightened by change, their feelings of inferiority increase as change is demanded. As adults they tend to consider other people enemies and to be unsuited for the complex adjustments required for marriage, occupational success, etc., because considering only their own welfare, they are not able to respond to the interest of others. He also thought that abnormal inferiority feelings develop in what he called hated or unwanted children. He believed they felt oppressed, and, in turn, hated others. He saw their goal of superiority to the suppression of other people, and so on.

An important contribution of Adler's was the observation of how much people tend to live in their imaginations. That is, he believed that man lived purely fictional ideas which had no counterpart in reality. The beliefs which the individual held about himself, about others, and about the world were seen by Adler as important determiners of behavior. Adler's interpreters (see Ansbacher and Ansbacher, 1956) use as illustrations of these fictitious ideas, "all men are created equal," "honesty is the best policy," the belief in a heaven for virtuous people and hell for sinners, etc. In any event, the ideas or concepts which the individual develops were seen by Adler to have an important influence on his behavior. Unfortunately, little emphasis was placed on learning and on the conditions under which the individual forms such concepts.

These observations and theories of Adler's were considerably changed, refined, and extended by later workers, particularly those within the psychoanalytic tradition. Perhaps the most elaborate and comprehensive theories of neurotic development to follow from the general notions of Adler were those of Karen Horney. It is also a body of theory which, in modified form, may be the most amenable to scientific test and subsequent reformulation.

Horney's Neurotic Personality of Our Time

We discussed Horney's concept of *basic anxiety* in relation to the early development of the self-concept (see Chap. 9, pp. 322–323). Instead of experiencing an atmosphere of warmth, feelings of inner security and inner freedom and the good will of others, the child who is exposed to early conditions producing anxiety in interpersonal relationships feels insecure, isolated, and helpless. He is forced to find ways to cope with his feelings—ways which relieve or diminish the painful insecurity and apprehensiveness which Horney called basic anxiety.

Horney thought that the way in which the child does this is determined both by his given temperament and by the contingencies of the environment. She argued that the major solutions are few in number. Namely, the child may try to shut others out and withdraw from them emotionally, he may rebel and fight others, or he may try to cling to the strongest parent or adult in his environment. Perhaps the most important aspect of these moves is that the child is driven to rebel, to keep aloof, or to cling to others without reference to his real feelings or the external realities of the situation. Thus, he may want to approach others and give or receive affection but because such spontaneous behavior in the past has led to rebuff and punishment these feelings undergo suppression and repression. Horney believed that this essential break with his spontaneous feelings consititutes a major source of conflict in the later adult personality. The individual becomes, in her terminology, "alienated from himself."

Horney felt that the child is driven to move in all three directions, but because they constitute fundamentally contradictory attitudes toward others, in time, one of the moves becomes predominant. This predominant move, presumably the one which has been most rewarded, usually becomes embellished with sets of related attitudes. For example, Horney stated that the predominantly complying child tends not only to be subordinate and dependent, but also tries to be unselfish and good. She felt that the integrating effect of the first solution was not as firm nor as comprehensive as later neurotic solutions.

> In one girl, or instance, compliant trends had become predominant. They showed in a blind adoration of certain authoritative figures, in tendencies to please and appease, in a timidity about expressing her own wishes, and sporadic attempts to sacrifice. At the age of eight she placed some of her toys in the street for some poor children to find, without telling anybody about it. At the age of eleven she tried in her childish way for a kind of mystic surrender in prayer. There were fantasies of being punished by teachers on whom she had a crush. But, up to the age of nineteen, she could also easily fall in with plans evolved by others to take revenge on some teacher; while mostly being like a little lamb, she did occasionally take the lead in rebellious activities at school and, when disappointed in the minister of her church, she switched from a seeming religious devotion to a temporary cynicism. [Horney, 1950, p.20]

Horney argued that since the individual lacks a basic sense of belonging and feeling basically isolated and hostile, he is pressured to develop more comprehensive neurotic solutions centering around a need to lift himself above others. Gradually he begins to embellish his particular solution by developing an idealized image of himself. He then begins to work very hard attempting to make this idealized image a reality. In full neurotic character development the self-idealization grows into a more comprehensive drive which involves elements of a need for perfection, neurotic ambition, and a need for vindictive triumph. The strength and degree of awareness of these needs are idiosyncratic for a given individual. They do, however, all have two general characteristics in common: their compulsive nature and the role which imagination plays in them.

The compulsive nature of these trends is apparent because, in the first place, the individual is driven "with an utter disregard for himself, for his best interests." To this point Horney mentioned an ambitious 10-year-old who thought that she would rather be blind than not become the first in her class. Secondly, these drives appear to be indiscriminate. The individual must be the center of attention, the most accomplished, etc.—whether or not in actuality he has the appropriate capacities and aptitudes. A third factor is that these trends are insatiable—actual success, for example, may hardly be experienced by the individual as success, or at any rate, bring little satisfaction to him. Lastly, the compulsive nature of these trends shows in the extreme anxiety or depressive reactions these individuals have whenever they are frustrated—reactions which are all out of proportion to the realities of the situation.

Imagination operates to distort reality in various ways for these people. The ideas of the individual has about himself may be frankly unrealistic. Horney stated, for example, that the neurotic may "turn his intentions to be honest or considerate into the fact of being honest or considerate. The bright ideas he has for a paper may make him a great scholar . . . . Knowing the 'right' moral values make him a virtuous person . . . ."

Horney further states:

> Imagination also operates in changing the neurotic's beliefs. He needs to believe that others are wonderful or vicious—and lo! there they are in a parade of benevolent or dangerous people. It also changes his feelings. He needs to feel invulnerable—and behold! his imagination has sufficient power to brush off pain and suffering. He needs to have deep feelings—confidence, sympathy, love, suffering: his feelings of sympathy, suffering, and the rest are magnified. [Horney, 1950, p. 34]

The neurotic is constantly faced with certain discrepancies between the idealized image of himself and the reality that he is essentially not particularly superior, but rather like everybody else. Other people treat him as though he were an ordinary person. His resolution of this conflict involves his concluding that there is something wrong with the world.

Horney stated that instead of tackling his illusions, he presents claims to the outside world. His claims vary with particular solution he has learned, and with the details of his idealized self. Thus, he may feel that he is always right, and entitled never to be criticized, questioned, or doubted. Another might feel entitled to being "understood," no matter how little he discloses himself to others, or how irritable, and petty his behavior might be. Others might feel that everyone should automatically love them, still others that they should be able to be exploitive without anyone complaining, and so on. These claims for special attention, consideration, and deference from others apparently may be widely generalized and directed toward institutions and toward life generally.

Of course, in his daily living these claims are continually frustrated. The ensuing anger may be expressed in fatigue or other psychosomatic complaints, by way of depression and self-pity, or it may be expressed directly. In any event, the individual is likely to suffer a chronic discontent.

Apparently, in the therapeutic situation patients show great resistance toward recognition of their neurotic claims. In the light of the fact that his claims are not acceded to in reality, and frequently lead to discontent, it would seem that the individual would be motivated to alter some of his beliefs. However, there is strong reward value associated with neurotic claims: they are his guarantee for *future* importance.

Another common trend which Horney observed in her patients was the operation of inner dictates by which the individual attempts to make himself over into his idealized image. We will see shortly that the childhood solutions of moving toward, against, or away from others have their parallels in adult solutions. The particular solution dictates to the individual what he should be like. These self-demands have many unrealistic characteristics: a disregard for feasibility, for the person's psychological condition, and for the conditions under which they can be fulfilled. For example, the individual may feel that he should be totally competent in his occupation, in which case the realities of his existing capabilities, of the limitations of time, of endurance, of circumstances beyond his control may be totally disregarded by his unreasonable ideas of what he should be or be capable of doing.

One of Horney's examples of the operation of the inner dictates involved a female patient whose solution was of the self-effacing variety, the adult counterpart of moving toward people. Her unconscious demands were that she should always be understanding, sympathetic, and helpful. The person felt that she should be like a priest, who symbolized these qualities. However, in reality she did not have any of the attitudes or qualities which enabled the priest to act as he did. Horney points out, "she could act charitably at times because she felt she should be charitable, but she did not feel charitable." Horney adds that as a matter of fact she did not feel much of anything for anyone.

It ought to be pointed out that these self-demands are distinguishable from genuine ideals. It appears that what the person actually is is less important than what is visible to others. Nevertheless, these "shoulds" and "oughts" appear to have a very strong coercive character. Some individuals try to mold themselves into behavioristic perfection, others suffer from continual feelings of guilt for not living up to their standards, whereas still others appear to continually resent their own (unrealistic) demands. Needless to say the individual's claims and his idealized standards are invested with a good deal of (false) pride.

Individuals with these patterns of insecurity and strivings for a feeling of importance exhibit a great deal of what might be called masochistic behavior. In some types of solutions, under stress, self-destructive trends may be quite obvious and objective. The individual may have frequent accidents, may drive, climb, or do many other things recklessly. The ultimate of self-destruction is, of course, suicide. Most of the masochistic tendencies, however, are more symbolic or covert in nature. Thus, because he does not live up to his idealized standards, the individual may torture himself with unreasoning self-reproaches. Self-discrediting attitudes are also common in certain types of solutions, directed especially against any strivings for improvement or achievement.

Part of the image the self-effacing person has of himself is that he should never be superior or better than others. Individuals with these self-disparaging trends constantly compare themselves unfavorably with others. Their negative concepts of themselves make them especially hypersensitive to criticism and rejection. Consequently, they tend to take too much abuse from others, perhaps without actually being aware of being exploited. Because of these continual negative self-evaluations, these individuals often experience a compulsive need for approval, attention, and the admiration of other people. In all of the types of neurotic solutions self-frustration is apparent. Affection and approval must be obtained at all costs and at the expense of the individual's actual feelings. The continual striving for importance, and the fulfillment of his inner dictates necessitate the inability of the person to relax and "be himself." The compulsive aloofness and withdrawal from others necessitate the individual's stifling impulses to participate with others, etc.

All of the tendencies we have discussed must of necessity result in considerable inner turmoil for the individual. We would expect, then, that defensive measures would be taken to relieve tension. Indeed, numerous defensive techniques are apparent in these patients. However, perhaps the most general, and one upon which Horney places great importance, is the individual's alienation from himself. Subjectively, it is the feeling the neurotic has of being removed from himself. Horney expresses this observation as follows:

> In fact, (this) analytic experience leads us straight into the core of the problem. For we must keep in mind that the patient does not talk about the weather or television: He talks about his most intimate personal life experiences. Yet they have lost their personal meaning. And, just as he may talk about himself without "being in it" so may he work, be with friends, take a walk, or sleep with a woman without being in it. *His relation to himself has become impersonal,* so has his relation to his whole life. If the word "depersonalization" did not already have a specific psychiatric meaning, it would be a good term for what alienation from the self essentially is: It is a depersonalizing, and therefore devitalizing, process. [Horney, 1950, p. 161]

Now, finally, let us see how all of the processes we have discussed—the search for importance, the shoulds, the claims, the self-disparagement, and alienation—operate in a particular case. As Horney points out, typing individuals is never satisfactory. Within the types one can distinguish subtypes. The degree of refinement of the classification is an arbitrary matter. Nevertheless, let us look at Horney's three major classifications of adult neurotics, each of which represents a particular extreme development. We must keep in mind that although people tend toward one major solution or another, and thus have characteristic similarities, they may differ widely both in the extent of their neurotic development and in many specific, unique aspects of their background and personalities.

Horney outlined two major conflicts, common to all neurotic personalities, and then outlined three major attempts to resolve these conflicts. The two sources of conflict involve, first, the conflict between the individual idealized image and all that it entails and his real, uncompulsive feelings, wishes, and cognitions. In Horney's special terminology this is called the *central inner conflict.* It is the conflict between the "real self" and the person's pride system. (Horney used the term "pride system" to denote the entire neurotic development—the claims, the shoulds, the idealized image, and the self-hate.) The second major neurotic conflict, and the one which is more immediately apparent, is that between elements within the pride system, namely, the idealized, grandiose claims on the one hand, and the operation of self-hate and self-contempt on the other.

How then, are these tensions reconciled in the adult? This can be accomplished in several ways: by compartmentalizing, by "streamlining," or by "resigning." According to Horney:

> Many patients . . . experience themselves successively as extremely self-effacing and as grandiose and expansive without feeling disquieted by this contradiction, because in their

minds the two selves are disconnected. . . . a more radical (solution) follows the pattern of *stream-lining,* which is typical of so many neurotic patients. This is the attempt to suppress permanently and rigidly one self and be exclusively the other. A third way of solving the conflict is by withdrawing interest from the inner battle and resigning from active psychic living. [p. 190]

Let us consider three of the major solutions which Horney outlines: the expansive solution, the self-effacing solution, and resignation.

The Expansive Solution. In the expansive solution, the individual experiences himself as a superior being. He tends to be expansive in his strivings and in his beliefs about what he can achieve. He may be more or less openly aggressive, ambitious, arrogant, and disdainful of others. Apparently, the feeling of superiority that goes with the expansive solution is not necessarily conscious. In any event, the individual places a high value on mastery of himself, of others, of difficult situations, and so on. Of course, the subjective conviction of superiority and mastery can only be maintained as long as any self-doubts, self-accusations, and evidences of failure are eliminated from consciousness. Above all, the individual must never be in a position where he feels helpless.

There is not one expansive solution; there are many of them. Three of these subtype expansive solutions are the narcissistic, the perfectionist, and the arrogant-vindictive type.

Horney's description of the narcissistic type is that of an individual who is optimistic, buoyant, enthusiastic, and tolerant. Consciously it seems that he identifies with his idealized self. However, he appears overly impressed with himself. Furthermore, he *must* impress others. Frequently such individuals are, at least on the surface, quite captivating and charming. On the other hand, closer inspection of their behavior may clearly reveal fraudulence, unscrupulousness, unreliability, and the like. As in all other solutions, difficulties occur in the individual's relations with others, and in his work. The narcissistic type tends to resent any indication that others look at him critically or are not willing to grant him special privileges. Above all, he must never be questioned seriously. Problems arise in his work because of his tendency to overrate his capabilities, plan unrealistically, become too diversified, and so on. Severe stress, for him, in the form of repeated rejection or repeated failures in enterprises may, according to Horney, result in depressions, psychotic episodes, suicide, or self-destructive urges.

Another expansive type of individual moves in the direction of perfectionism. As Horney says, these people seem to have the conviction of an infallible justice operating in life which gives them a feeling of mas-

The expansive solution. Possible traits: (a) he may consider himself a "superior being"; he may be (b) narcissistic; (c) perfectionistic; (d) arrogant; (e) vigilant; or (f) over-controlled. (Photograph by Tor Eigland.)

tery. That is, because they live up to standards of perfection, justice, fairness, and duty, they are entitled to be treated fairly. This gives them a feeling of mastery. It is not admiration, but respect that these people seek from others. A "polished friendliness" may disguise their feelings of moral and intellectual superiority, and their arrogant contempt for others. The whole adjustment of persons exhibiting this particular pattern is vulnerable because any failure or error which may occur in spite of strenuous efforts for perfection may expose them to destructive self-recriminations—that aspect of the pride system which has been streamlined or repressed. Also, any misfortune befalling such a person may shake his "deal" with life to the roots, overwhelming him with the paralyzing possibility of helplessness.

Still a third pattern of the expansive type of solution is that of arrogant vindictiveness. The primary motive of these individuals is vindictive triumph over others and a need to deny positive feeling. Generally, they impress the observer as arrogant, vigilant, and overcontrolled. However, aside from traits such as aloofness, detachment, pride in their honesty, and fairness, they may be militantly righteous, explosive, rude, and extremely competitive. Their unconscious claims revolve around the belief that they should be permitted to have an utter disregard for the needs and wishes of others. They move "against" others, and tend to justify this compulsive move by viewing the world as hostile. Although they may be unproductive in a creative sense, they tend to be indefatigable in their work.

When this type of personality is not highly gifted, people tend to simply reject him, although some might submit to his sadistic demands. On the other hand, when his capacities are considerable, through a combination of being rewarded by the submission of others and being rewarded simply for his leadership qualities, he may come to wield great power. The personal vindictive triumph of an Adolph Hitler, expressed in the mass murder of millions would, it would seem, make responsible people worry about cultural practices of admiration. Horney felt that people exhibiting the arrogant-vindictive pattern are those who have been treated with severe neglect, humiliation, hypocrisy derision, or brutality as children.

The Self-Effacing Solution. This is the adult counterpart of the early childhood clinging. The self-effacing person lives with feelings of guilt, inferiority, failure; feeling that he has no emotional rights, and appearing to want it that way. He cannot express hostility. He cannot be aggressive. He cannot make demands on others, nor reprimand others. He tends to be apologetic and self-minimizing. He scorns anything that is "selfish" or "presumptuous."

On the other hand, he longs to subordinate himself to others, to be dependent upon others, to be helped and protected. Thus, he has a strong need to be wanted, liked, approved, needed, and so on. He may try to make himself agreeable and useful. He may try to find people or causes for whom he can do things. His corresponding idealized image is loving, considerate, generous, helpful, and so on. Expansive trends are deeply repressed and may come to the fore only in his feeling of being abused by others, feeling secretly superior to others, or perhaps, getting back at others by making them feel guilty for exploiting or abusing his goodness. Such people feel entitled, of course, to understanding. When the circumstances do not provide the satisfactions of being liked, needed, and wanted, then the individual experiences an enormous increase in anxiety. Only at that point may there be an outpouring of accusations against others.

Love offers this type of individual the supreme fulfillment of his neurotic strivings, because it promises protection, support, surrender, self-riddance, harmony, and fulfillment for a longing of unity. Since these individuals fear rejection above all else, they are likely to be magnetically attracted to individuals who behave sadistically toward them, or even to detached individuals who, by implication, seem to reject them. Thus, the self-effacing person may develop a morbidly dependent relationship in which he is compulsively driven to "love" someone who treats him badly. Of course, obvious and serious morbid-dependency relationships are most likely to occur in people with severe self-effacing tendencies.

Horney seemed to feel that these people frequently "grew up in the shadow of someone"—perhaps a benevolent, but authoritarian, parent; perhaps a preferred sibling. The early background was such that it was precarious, in the sense that it was likely to arouse fears, yet affection was obtainable at the price of self-subordination.

The Resigned Type. The final distinguishable pattern which we shall discuss is that of resignation. In the resigned personality the emphasis is on detachment, nonparticipation, an aversion to effort, and an absence of serious strivings for achievement. These people tend to be underachievers. They tend to restrict their wishes and expectations. Their major unconscious claim appears to be that they are entitled "not to be bothered." They resent coercion or restrictions of any kind. They may even resist the simplest efforts of normal living. They may tire easily, suffer from chronic fatigue, be inefficient and inert, and yet present a picture of integrity and defiance.

These tendencies appear to be the result of the fact that their basic technique of handling the conflictful, compulsive tendencies of the pride

The self-effacing solution. Possible traits: (a) he has feelings of guilt, inferiority, and failure; (b) he feels he has no emotional rights and appears to want it that way; he is (c) apologetic, (d) self-minimizing, (e) submissive (with repressed hostility). (Photograph by Tor Eigland.)

system differs from that of either the expansive or self-effacing individuals. Rather than attempt to suppress self-effacing or expansive tendencies, and hence handle their conflicts by streamlining, these individuals maintain *both* kinds of conflicting, compulsive trends. The basic tension is handled by an attempt to immobilize (become detached from) the strongest elements of both. Thus the individual may idealize his self-sufficiency, he may be grandiose in his imagination, but he gives up the active pursuit of ambitious goals. Similarly, he may be sensitive and timid. He may be compliant too, but not through a dire need of approval—simply because he would like to avoid conflict. If there is anything that he strives for, rather than avoids, it is a freedom to do what he pleases without pressure from others.

It will be recalled that, according to Horney, the major, although not as apparent, conflict in neurotic character trends is the central inner conflict. That is, the conflict between the individual's spontaneous feelings and wishes and his compulsive, neurotic-based tendencies. The resigned neurotic appears to keep more of his real self intact, but he immobilizes this, too, by keeping it private. He abhors self-disclosure.

Horney thought the most important background factors in the resigned neurotic involved the existence of cramping influences against which open rebellion was not possible. An example might be a parent who was too wrapped up in himself to be sensitive to the child's needs, but made demands on the child for emotional support, either explicitly or implicitly. The child solves this conflict by suppressing attempts to fight as well. His taboo against being aggressive means that he can only cope with others passively, yet under the circumstances the expectations of others appear to acquire an exaggerated character. His "shoulds" may also be viewed by him as something to rebel against tacitly. Horney reported that the pattern of resignation often comes to the fore in adolescence and young adulthood. Previously the person may have been achieving, active, and outgoing. In the later period, however, the individual may have been crushed by circumstances, or simply gotten a thorough taste of his earlier learned, conflicting self-effacing and expansive neurotic tendencies. The solution, then, is to curtail them and live in a resigned style.

There are some reasonably distinct patterns within the general pattern of resignation, which represent subtypes. We will merely mention them here. There is persistent resignation, where inertia and resistance to change completely dominate the personality. There is "the rebellion group," where the pain of nonparticipation in living is finally rebelled against, either by fighting others, or by the individual's allowing himself

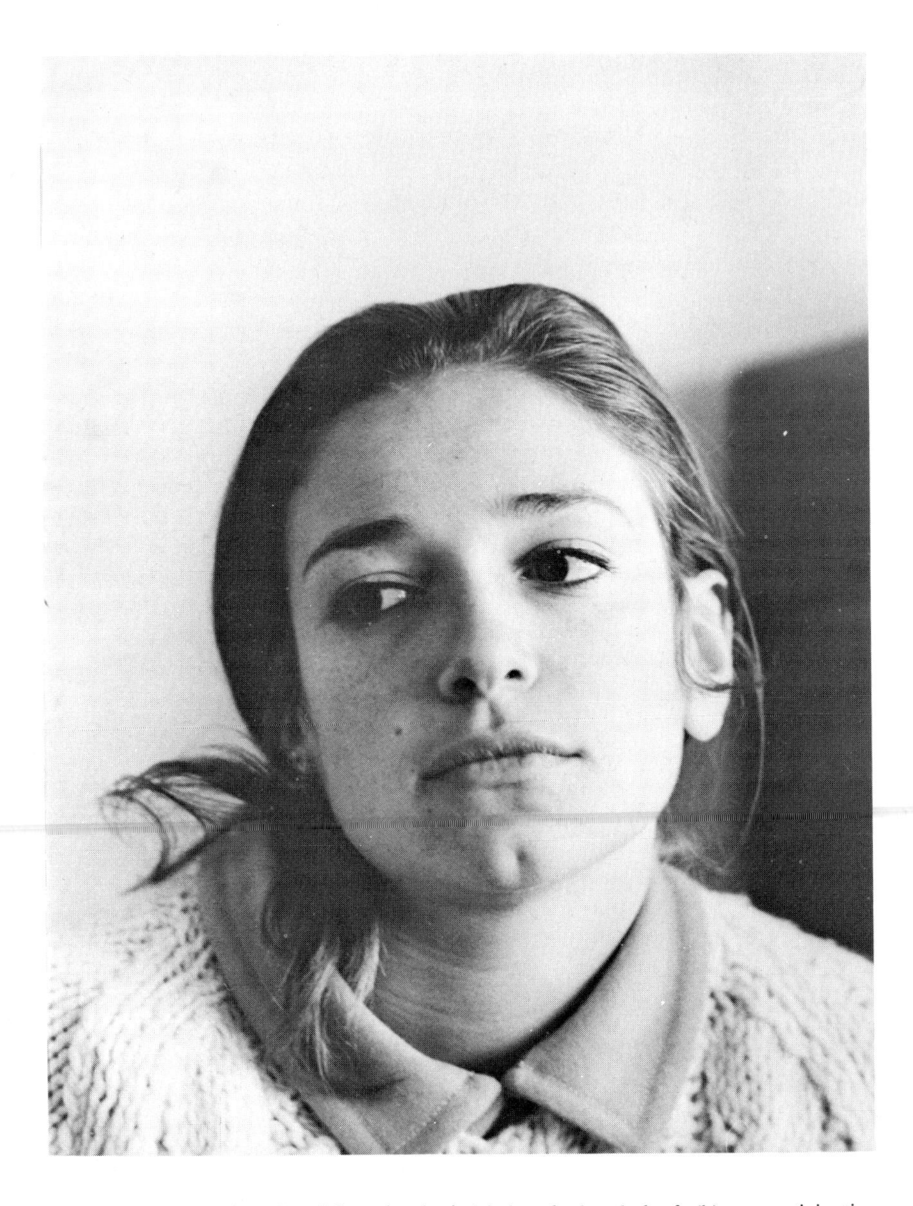

The resigned solution. Possible traits: he is (a) detached and aloof; (b) nonparticipating; (c) inefficient; (d) he may have no need for achievement—he "does not want to be bothered"; (e) despite lack of initiative and accomplishment, he tends to present a picture of integrity and defiance. (Photograph by Tor Eigland.)

some spontaneous interest and pursuits. There is a group of patterns which Horney refers to as "shallow living." Emphasis may be on prestige or opportunistic success. Finally, the individual may attempt to fit in with others and become a sort of well-adapted automaton.

Relevant Research. Horney's theory has some strong appeals for the behavioral scientist. It is parsimonious. An enormous complexity and variety of behavior is reduced to the operation of three major trends, or factors (moving toward, against, and away), and two intrapsychic conflicts (the central inner conflict and the conflict within the pride system). Moreover, the concepts are formulated in such a way as to allow for the possibility of defining them in operational terms. Antecedent conditions are also clearly suggested. Whether or not it will remain a theory and nothing more depends, in part, upon the development of research techniques which will enable the behavioral scientist to obtain reliable measures of the intrapsychic factors which Horney postulates, and upon which so much of the theory rests. It also depends upon the ingenuity and imagination of the researcher.

Let us look at some suggestive data. A rather interesting technique for measuring the unconscious self-evaluation of subjects was introduced by Wolff (1943). Without his subject's knowledge, Wolff obtained samples of their mirror-image handwriting, their voices, facial silhouettes, photographs of their hands, movies of them performing simple tasks, and so on. He was able to show that these materials could be disguised so that subjects were unable to recognize themselves. However, they sometimes became very excited when making judgments based on material derived from themselves; as opposed to making judgments on the basis of material which was the same in every way, save that it was derived from subjects other than the person himself.

Diller (1954) employed Wolff's measures to obtain subjects' unconscious self-evaluations, along with their conscious self-ratings following experimenter-induced success and failure experiences. He argued that there should be discrepancies between the individual's conscious and unconscious self-evaluations.

The results indicated, first, that success on the task led to a better opinion of the self, expressed by both conscious and unconscious self-ratings. For us, the results of induced failure are important, because the dynamics of the neurotic character trends which we have been discussing are based upon the individual's reactions to anxiety-producing circumstances. Diller found that failure did not change the subject's self-evaluations on the *conscious* self-report measures. However, self-evaluation was significantly lowered after failure when measured by the Wolff technique. This finding is schematized in Fig. 11-2. It is entirely in line with Horney's theory, which holds that threat to the individual results in an increase in *unconscious* self-evaluation.

But what about the fact that the conscious self-evaluations did not change following the failure experience? Horney's predictions would be more complicated here. Some individuals (expansive types) would be expected to *increase* their conscious self-evaluations, and thus compensate for the unconscious distress. Other individuals (self-effacing types) would be expected to lower their self-evaluations, fully conscious of inferiority feelings. Moreover, emotionally healthy individuals and resigned types might be expected not to change.

A new experiment would be required, specifically designed to test these predictions. However, Fig. 11-3, suggested by Professor G. Becker on the basis of available experimental personality research data, should help the reader to interpret the foregoing analysis. The straight solid line in the center of the diagram represents the result of the Diller study. That is, there was no change in the subject's conscious self-ratings following the experience of failure. The "before" and "after" points representing the subjects' mean ratings are connected with a solid straight line.

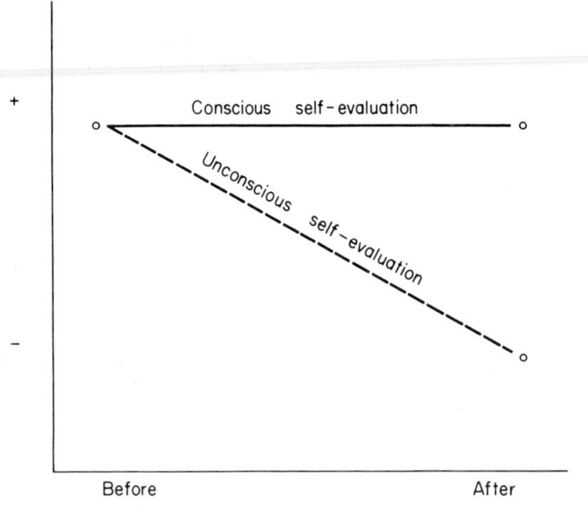

FIG. 11-2. A schematization of Diller's findings (1954) on self-evaluation. The plus and minus on the ordinate indicate experimentally induced success and failure experiences. The abscissa shows the time before or after induced success or failure.

FIG. 11-3. An interpretation of the fact that the conscious self-concept showed no change following induced failure in the Diller study. The opposing changes for repressors and sensitizers average to give an apparent "no change." (G. Becker, University of Manitoba, personal communication, 1968.)

Data from other experiments (e.g., Worchel & McCormick, 1963) indicate that repressors, people with low self-concept–self-ideal discrepancies, tend to enhance their self-evaluations after being exposed to threat, whereas sensitizers, individuals with high self-concept–self-ideal discrepancies, tend to decrease their evaluation. These respective hypothetical data points are shown in Fig. 11-3 labeled "repressors" and "sensitizers." The two dotted lines indicate the corresponding "before" and "after" trends for these groups.

Notice that if the data points representing the repression mean and the sensitization mean were averaged the data point for "no change" would be obtained. In a randomly selected group of subjects, individuals with repressor-type tendencies and those with sensitizer-type tendencies might balance each other off so that the resulting average ratings for the entire group would be the middle ("no change") value. Thus, Diller's results might have been obtained simply because the results were not computed separately for these two subgroups.

Finally, notice the solid arrows leading from the repressor and sensitizer data points at the top and bottom of Fig. 11-3. Recall all that we have said about repressors and sensitizers in Chapter 10 and elsewhere. They correspond approximately to Horney's expansive and self-effacing types. In the former, streamlining occurs by way of the suppression of self-effacing trends and the enhancement of conscious self-idealization (labeled Streamlining I in Fig. 11-3); in the latter, expansive trends are suppressed and self-effacing trends are consciously enhanced (Streamlining II). Needless to say, only future research can supply data which are more than merely suggestive.

SUMMARY

Given the individual's basic biological endowment, early learning experiences involving friendship, identification, assertiveness, and dependency qualitatively determine his life-adjustment. The existence of fortunate early conditions of learning in these areas allow him to become relatively expressive, confident, and autonomous. Thus, he is capable of finding constructive solutions to the basic problems of adult life: relatedness to others, occupation, love, and marriage. On the other hand, poor early conditions for learning in these areas result in the learning of anxiety, weak or distorted identification, undercontrol or overcontrol of assertiveness, and dependency. These foster emotional disturbances and an inadequate adjustment to the demands of life.

Some psychologists have given thought to the optimally adjusted personality. One such type of person, termed by Maslow the *self-actualizing personality*, possesses such qualities as spontaneity, acceptance of self and others, accurate perception of reality, autonomy, and intimacy with a select number of friends.

Rogers describes a kind of optimally adjusted person who emerges from successful client-centered psychotherapy. Such a person has an openness to experience, especially the experience of his own feelings and attitudes. As a result he is more realistic in dealing with new people, situations, and problems. He also is willing to admit the positive feelings that others have toward him into awareness, and he shows an increased liking for himself. Finally, he is able to be somewhat autonomous and to trust his own perceptions, feelings, and judgment.

In contrast to this, the emotionally disturbed individual develops behavior that is symptomatic of his disturbance. He is miserable, unhappy, and "stupid." His behavior appears to be self-defeating and self-perpetuating. Some of the principles relevant to the learning and maintenance of this kind of behavior have been reviewed, including a review of certain aspects of partial reinforcement and its effect, the pairing of aversive stimulation with reward, immediate vs. delayed reinforcement, discriminative stimuli, anxiety, and inhibition. Classical patterns of neurotic behavior involving dependency, aggression, sexuality, anxiety reactions, and dissociative reactions were also discussed.

Neurotic personality trends may be viewed within the framework of certain observations and theories, particularly those of John B. Watson, Alfred Adler, and Karen Horney. Particular attention has been paid to Horney's theory of neurotic personality development. It is suggested that Horney's notions of neurotic character structures and the intrapsychic conflicts related to them are amenable to the techniques of experimental personality research, and that very likely they are related to established personality dimensions, such as the repression-sensitization dimension, discussed in Chapter 10.

REFERENCES

Allport, G. W. *Personality: A psychological interpretation.* New York: Holt, 1937.

Ansbacher, H. L. & Ansbacher, Rowena R. *The individual psychology of Alfred Adler.* New York: Basic Books, 1956.

Beier, E. G. The effect of induced anxiety on some aspects of intellectual functioning. Unpublished doctoral dissertation, Columbia Univer., 1949.

Byrne, D. Repression-sensitization as a dimension of personality. In B. A. Maher (Ed.), *Progress in experimental personality research.* New York: Academic Press, 1964. Pp. 179–215.

Cameron, N. *The psychology of behavior disorders.* Boston: Houghton, 1947.

Diller, L. Conscious and unconscious self-attitudes after success and failure. *J. Pers.,* 1954, **23**, 1–12.

Dollard, J., & Miller, N. E. *Personality and psychotherapy.* New York: McGraw-Hill, 1950.

Ferster, P. B., Nurnberger, J. I., & Levitt, E. B. The control of eating. *J. Math.,* 1962, **1**, 87–109.

Grinker, R. R., & Spiegel, J. P. *Men under stress.* New York: McGraw-Hill, 1945.

Horney, Karen. *Neurosis and human growth.* New York: Norton, 1950.

Joy, V. L. Repression-sensitization and interpersonal behavior. Paper read at Amer. Psychol. Assoc., Philadelphia, August, 1963.

Kirk, Barbara A. Test versus academic performance in malfunctioning students. *J. consult. Psychol.,* 1952, **16**, 211–216.

Maslow, A. H. *Motivation and personality.* New York: Hafner, 1954.

Masserman, J. H. *Behavior and neurosis.* Chicago: Univer. of Chicago Press, 1943.

Mosher, D. Interaction of fear and guilt in inhibiting unacceptable behavior. *J. consult. Psychol., 1965,* **29**, 161–167.

Mowrer, O. H. Learning theory and the neurotic paradox. *Amer. J. Orthopsychiat.,* 1948, **18**, 571–610.

Mowrer, O. H. *Learning theory and the symbolic processes.* New York: Wiley, 1960.

Mowrer, O. H., & Ullman, A. D. Time as a determinant of intergrated learning. *Psychol. Rev.,* 1945, **52**, 61–90.

Neibhur, H. R. Muscle action potential patterns as a function of practice and task continuing. Unpublished doctoral dissertation, Univer. of Pennsylvania, 1955.

Prince, M. *The dissociation of a personality.* New York: Longmans, Green, 1905.

Rogers, C. R. *On becoming a person: A therapist's view on psychotherapy.* Boston: Houghton, 1961.

Schachter, S. *The psychology of affiliation.* Stanford, Calif.: Stanford Univer. Press, 1959.

Shaffer, L. F., & Shoben, E. J., Jr. *The psychology of adjustment.* Boston: Houghton, 1956.

Shaw, M. C., & Black, M. D. The reaction to frustration of bright high school underachievers. *Calif. J. educ. Res.*, 1960, **11**, 120–124.

Smith, M. B. "Mental health" reconsidered: A special case of the problem of values in psychology. *Amer. Psychologist,* 1961, **16**, 299–306.

Thigpen, C. H., & Cleckley, H. A case of multiple personality. *J. abnorm. soc. Psychol.*, 1954, **49**, 135–151.

Watson, J. B. *Behaviorism.* Chicago: Univer. of Chicago Press, 1924.

Weiner, I. B. & Ader, R. Direction of aggression and adaptation to free operant avoidance conditioning. *J. pers. soc. Psychol.*, 1965, **2**, No. 3, 426–429.

White, R. W. *The abnormal personality.* (3rd ed.) New York: Ronald Press, 1964.

Wolberg, L. R. *The technique of psychotherapy.* New York: Grune & Stratton, 1954.

Wolff, W. *The expression of personality: Experimental depth psychology.* New York: Hafner, 1943.

Worchel, P. & McCormick, B. L. Self-concept and dissonance reduction. *J. Pers.*, 1963, **31**, No. 4, 588–599.

Chapter 12

Approaches to the Modification of Behavior

INTRODUCTION

All cultures attempt to shape the behavior of their members along culturally desirable lines. The sole aim of certain formal institutions of society is the modification of behavior. These include our educational institutions, prisons, and mental hospitals. In a less formal, but in a no less organized way, public relations firms and certain other interest groups attempt to change such diverse behaviors as buying choice, consumer spending, and voting.

The great technological advances and the unique organizational problems of the modern world require personality and behavioral change and have fostered the growth of techniques to bring them about. Rapid cultural change itself has created adjustive difficulties for the individual. Within the last 50 years the professions of psychiatry, clinical psychology, psychoanalysis, and psychiatric social work have evolved to deal with personal maladjustment. The common aim of these professions is the modification of the individual's motivation and behavior.

This is the topic of the present chapter. Certain theories, conceptual schemes, and goals underlying present-day psychotherapy will be discussed. Special emphasis will be placed upon the relationship between the techniques of psychotherapy and established principles of learning. Directly relevant to this are the more recent formulations of psychotherapy based upon behavioral science, involving such approaches as

Rotter's social-learning theory, Phillips's assertion therapy, behavior therapy involving the operant conditioning of adaptive responses and desensitization, and cybernetic concepts applied to behavior change. In the latter part of the chapter certain aspects of psychotic behavior will be discussed. Emphasis will be upon the problem of its etiology and special methods of treatment it requires.

PSYCHOTHERAPY

Definition and Criteria

We may ask what is the meaning of the term "psychotherapy," and by what criteria is it decided whether an individual is in need of it? Perhaps the best way to convey the most accepted meaning of the term is to present authoritative definitions of it. The official definition of psychotherapy provided by the American Psychological Association is "a process involving personal relationships between a therapist and one or more patients or clients by which the former employs psychological methods based on systematic knowledge of the human personality in attempting to improve the mental health of the latter" (Raimy, 1950). Wolberg (1954), representing the view of the medical profession, defines psychotherapy as "a form of treatment for problems of an emotional nature in which a trained person deliberately establishes a professional relationship with a patient with the object of removing, modifying, or retarding existing symptoms, of mediating disturbed patterns of behavior and of promoting positive personality growth and development."

Whereas Wolberg's definition uses the term "treatment" and the American Psychological Association employs the term "process," both stress the improvement of mental health resulting from a relationship between a professionally trained person and a person whose emotional or mental health is being improved. Both of these definitions also imply that the therapist has systematic knowledge or conceptual tools which enable him to achieve these goals, and also that there are some criteria for the presence of mental health and of emotional disturbance.

A definition which is a little more explicit has been put forth by Eysenck (1961). It provides the following criteria:

1. There is an interpersonal relationship of a prolonged kind between two or more people.
2. One of the participants has had special experience and/or has received special training in the handling of human relationships.
3. One or more of the participants have entered into the relationship because of a felt dissatisfaction with their emotional and/or interpersonal adjustment.
4. The methods used are of psychological nature, i.e., involve such mechanisms as explanation, suggestion, persuasion, etc.

5. The procedure of the therapist is based upon some formal theory regarding mental disorders in general, and the specific disorder of the patient in particular.

6. The aim of the process is the amelioration of the difficulties which caused the patient to seek the help of the therapist. [Eysenck, 1961, p. 698]

This definition indicates that the client enters the therapeutic relationship because of self-dissatisfaction involving his emotional adjustment, his interpersonal adjustment, or both. The fact is that people do come for psychotherapy or are sometimes advised to seek professional help for no other reason than that they are unhappy. However, it is usually only the very sophisticated person who seeks help merely because he is experiencing a lack of positive satisfaction in his life. Actually, in terms of a conflict analysis of neuroses this is also likely to be the very person for whom change would be the most difficult and for whom the motivation for change would be minimal (see pp. 163–168). That is, he would be a person with strong avoidances, for whom the felt conflict is not very intense. Such a person is also likely to have repressor-type defenses against anxiety (see Chapter 10).

When the avoidance gradient is weaker, so that the person is not kept far from the feared goal, the experienced tension and misery is likely to be greater because under these conditions the conflict is more intense. It is the individual with unresolved conflicts and experienced misery and tension who is likely to seek therapy. The defensive pattern of these patients is likely to involve sensitization. In general, the greater the subjective feeling of misery, the greater is the positive motivation for seeking therapy.

A second criterion of neuroses and hence for therapy is whether or not the individual is incapacitated in his daily living. Many of the classic patterns of neurotic behavior seriously interfere with the individual's ability to function normally. This is clearly the case with severe rituals, obsessions, compulsions, or hysterical symptoms like amnesia and conversion reactions. Disturbances of the person's ability to handle aggressive, dependent, or sexual impulses often interfere with his ability to function effectively. Neurotic patterns almost always interfere with the person's relationships with others and with his ability to do productive work. Very often the symptoms involve a vicious circle as, for example, when a person anticipates rejection from others and hence reacts towards them in a distant or hostile manner. This increases the likelihood of his being reacted to in kind. His (now) more or less reality-based feelings of rejection increase his need for others, which intensifies his defenses of detachment and hostility, and so on.

Although the neurotic may be distressed, and although his sense of mastery may be shattered, there is no loss of contact with reality and no gross personality disorganization. Psychotic disorders (see pp. 465–466), on the other hand, do involve a loss of contact with reality, and/or serious personality and behavioral disorganization. The individual may not respond normally to the simplest kind of situational or social demands or he may be overreactive, hold bizzare beliefs, talk in a private language, hallucinate, show inappropriate affect, and/or a wide range of other untoward symptoms. The need for therapeutic intervention in such cases is clearly indicated.

Models of Psychopathology

In all cases where individuals are described as neurotic or as psychotic the judgement is based upon observed behavior. The individual is observed doing eccentric things, talking foolishly, behaving in antisocial ways or being shy, withdrawn, etc. Even when the subjective feeling of distress is the criterion, objective observation of the person in his daily living typically reveals inadequate or inappropriate response patterns of various kinds. His behavior may indicate that he depends too much on others, becomes easily upset, or makes any of a great variety of other maladaptive responses. The scientific study of behavior encompasses a body of principles that can be used to account for maladaptive behavior equally as well as for adaptive behavior. However, largely for historical reasons, modern clinical practice is strongly affected by concepts taken from other frames of reference. Sometimes these are called models of psychopathology. The most influential of these conceptual models are, first, the *disease model,* sometimes called the *medical model,* and second, the dynamic model. Still a third is the *moral model.*

The Moral Model. The moral model is not very common today, although there are some relatively recent proponents of it. Early attempts to explain deviant or abnormal behavior interpreted it in terms of supernatural forces. Deviant individuals were thought to be possessed by good or by evil spirits, gods and demons, which made them behave in the way they did. Although occasionally bizzare behavior was worshipped and glorified, more often it was regarded as the machination of evil spirits and severely dealt with. Victims were put to death or cruelly tortured in order to exorcise the evil spirits. After the Middle Ages there was some lessening of the influence of these ideas so that deviant individuals were merely segregated from society in detention houses. The work of Philippe Pinel (1745–1826), a French physician, is often cited as one of the first attempts to provide humanitarian treatment for the insane, and a major inroad in the shift from the moral model to the medical model.

The Medical Model. The medical model held, by analogy with the concept of disease in medicine, that the insane were actually "mentally ill." The medical model interpreted the mental processes thought to be involved in abnormal behavior in terms of analogies from physical disease processes and from the medical treatment of them.

This model is still prevalent today, although it is often more implicit than explicit. Indeed, the clear statement of the concepts of the medical model makes it seem anachronistic to us. The concepts involved are merely analogies, and further, they do not relate to actual behavior, nor make use of the established principles of behavior. The concepts of the medical model have been succinctly stated by Professor B. Maher, a well-known experimental psychologist, as follows:

> Such (deviant) behavior is termed *pathological* and is classified on the basis of *symptoms*, the classification being called *diagnosis*. Processes designed to change behavior are called *therapies* and are applied to *patients* in mental *hospitals*. If deviant behavior ceases, the patient is described as *cured*. [Maher, 1966, p. 22]

This model has gained widespread acceptance over the years. Many positive contributions have stemmed from it, not the least of which have been the humanitarian treatment of deviant individuals and large-scale spending, both public and private, for prevention, treatment, and research in the area of "mental illness." However, the very notion of deviant behavior as disease has not encouraged its interpretation in terms of the science of psychology. It has also led to a system of diagnosis and symptom classification which many experts feel is not only unreliable, but also has little relation to subsequent improvement.

Several studies (e.g., Mehlman, 1952; Raines & Rohrer, 1955; Schmidt & Fonda, 1956) have concluded that the way a patient is diagnosed depends to a large extent upon the psychiatrist doing the diagnosing. Typical data, from a study by Schmidt and Fonda (1956), are given in Table I. The percent of agreement between pairs of psychiatrists diagnosing the same patients from state hospital populations is shown.

It can be seen in Table 12-1 that for general diagnosis the agreement among the categories is fairly good, although it declines steadily in the order: organic, psychotic, characterological. The reliability among specific diagnoses, however, leaves much to be desired. Agreements are somewhat higher in the organic category than in the other two general categories, but even there, in classifying what would seem to be relatively easily recognizable conditions, such as the acute brain syndrome or mental deficiency, the percent agreements are not particularly encouraging (68 and 42 percent, respectively). When the diagnosis involves patterns of maladjustment which are much more common, such as neu-

TABLE I
PERCENT AGREEMENT IN DIAGNOSIS BETWEEN PAIRS OF PSYCHIATRISTS

General diagnosis		*Specific diagnosis*	
Organic	92%	Acute brain syndrome	68%
		Chronic brain syndrome	80%
		Mental deficiency	42%
Psychotic	80%	Involutional	57%
		Affective	35%
		Schizophrenia	51%
Characterological	71%	Neuroses	16%
		Personality pattern	8%
		Personality trait	6%
		Sociopathic	58%

rosis or personality patterns or the personality-trait diagnosis, the percent of agreement dropped to completely unacceptable levels (16, 8, and 6 percent, respectively).

Another aspect of the present classification system is that there appears to be little relationship between the diagnostic categories and the prognosis or therapeutic outcome. This in itself reflects a number of problems, including the irrelevance of certain features of the behavior upon which the original diagnoses were made, the lack of proven effectiveness of many commonly used therapeutic techniques, and the occurrence of "spontaneous recovery"—improvement without apparent cause. These will be discussed later in this chapter.

The disease concept itself, at the basis of the medical model, is difficult to defend. A small percentage of mental patients suffer from demonstrable organically based abnormalities related to damaged brain tissue or hormonal imbalance. Otherwise there is little evidence for organic pathology as a causal factor throughout most psychiatric classifications. Schizophrenia, for example, affects the largest single group of patients in mental hospitals, yet if its cause is organic, it remains to be demonstrated—and this is true in spite of the fact that research in this area has been extensive and has been given overwhelming support for the last 20 to 30 years. (See pp. 465–466.)

The Dynamic Model and Psychoanalysis. In the dynamic model, analogies are drawn from molecular or physiological homeostatic processes and applied to the adjustive strivings of the individual. At the physiological level bodily processes tend to maintain a state of equilibrium, or a "dynamic balance of forces." This is called a *homeostatic balance*, or the maintenance of *homeostasis*. Thus, body temperature is kept constant at approximately 98.6°F. A change in body temperature of only 1 degree stimulates a variety of bodily processes which function to oppose the change. If the deviation is in a positive direction, that is, if body temperature rises, there is an increase in sweat-gland activity, dilation of the blood vessels of the skin, and initiation of other bodily processes which function to reduce metabolic activity. Should external or internal conditions operate to produce a change in the opposite direction, other processes are mobilized. Thus, when the body temperature falls below 98.6°F, there is a constriction of the small blood vessels of the skin, muscular contraction (shivering) which increases heat production, etc.

These homeostatic processes can be observed on the molecular level, as for example, the maintenance of osmotic pressure across the cell membrane. They can also be observed on the molar level, as for example, an increase in the activity level, and hence in the probability of finding food, in response to those bodily processes that signal a state of hunger to the organism.

Analogously, the psychological reactions in people which make them respond to pressures in the environment in particular ways can be viewed as dynamic processes of adjustment which function to maintain homeostasis or a *steady state*. By analogy, the neurotic patterns which were discussed in the last chapter, according to this view, represent attempts on the part of the individual to maintain a steady state of self-esteem and feelings of security in the face of events which happen in his experience to threaten them. Thus, neurotic and defensive behaviors represent an attempt to maintain a dynamic balance of forces. As a single example consider the self-effacing individual who is unable to show appropriate assertiveness. Typically, upon analysis it is found that such a person feels that the expression of his true feelings toward significant others would lead to punishment, loss of love, abandonment, etc. Accordingly, these self-effacing patterns of behavior are viewed as dynamic, or homeostatic, attempts to maintain the approval and love of others.

The dynamic model was employed by the early psychoanalysts, Freud and his followers, in a speculative account of complex human adjustment. Analogous to physical energy, Freud conceptualized the operation of *psychic energy*. The life-force or reservoir of this energy, which he conceived as having a broadly sexual character, he termed the *libido*. Libidinal energy was thought of as following the first and second laws of thermodynamics. That is, there was thought to be only a given quantity of it available to the personality and it could not be diminished, although it could be diverted from one place to another. Thus, when *x* amount of libidinal energy is invested in one place, that particular amount of energy is not available for investment elsewhere and the total amount of otherwise investable energy is reduced by that amount.

Freud theorized that the libido became attached to certain objects, and when it did, these objects functioned to arouse libidinal energy. He called the investment of libidinal energy in this manner *cathexis*. The objects of cathexes may be ideas, persons, or aspects of the individual's own body. The stages of psychosexual development during which the libido becomes attached to successive erogenous zones of the body, and by extention, to things and events in the world that symbolize them were discussed previously (pp. 279–281). Briefly, the earliest cathexis was postulated to be diffuse, and to involve the entire body. Later, the focus of pleasure and psychic energy involves successively, as development proceeds, the oral erogenous zone, the anal erogenous zone, and the genitalia, followed by an inhibition period and then phallic attachment, and finally, a partner of the opposite sex. Whether or not fixation occurs at any given stage depends upon the amount of libidinal investment, the counterforces opposing it, and the amount of frustration which the person encounters in progressing to a later stage.

Freud also invented the *mental apparatus* in order to account for his observations. The mental apparatus consisted of the operation of three systems of personality, namely, the *id*, the *ego*, and the *superego*. All libidinal energy was postulated to be id energy. In Freud's highly figurative and literary language, the id was conceived as being a "seething caldron of psychic excitation." It was conceived of as having no sense of time, no morality: only the instinctual libidinal energy seeking satisfaction. The id, as a system of personality is particularly represented by the expression, "I want."

The instinctual energies of the id were thought to seek relief according to the *pleasure principle*, which means simply that the organism seeks to obtain pleasure and to avoid pain. However, in the socialization of the individual many pleasurable goals are not allowed and pursuance of them leads to punishment and frustration. Society punishes inappropriate sexual attachments and behavior, aggression which it considers to be inappropriate, etc. By thus imposing frustrations on the individual, society was viewed by Freud as fostering behavior according to a second principle: the *reality principle*. This is the principle that the individual tends to forego immediate pleasure or to put up with some frustration in order to obtain greater pleasure or security in the future. Other parts of the mental apparatus allow him to do this because in their development they were thought to "trap" some of the energy from the id and to use it to counteract id impulses and to satisfy their own functioning.

The ego is the executive part of personality. It perceives reality, organizes, plans, mediates between the demands of the id, of external reality, and of the third system of personality, the superego. The superego is analogous to what we ordinarily think of as the person's conscience. It involves the operation of the moral teachings of society, incorporated by way of early socialization experiences involving primarily the parents, into the individual's personality. It substitutes moralistic goals for real ones. In doing so it also traps id energies and uses them to counteract the strivings of the id, and, sometimes, of the ego.

In order to complete this brief account of the mental apparatus it ought to be noted that the libidinal energy and its aims were viewed by Freud as being largely unconscious, that is, outside of the individual's normal awareness. In keeping with the literary, metaphorical language in which Freud couched the other aspects of his theory, he used the expression *the unconscious,* thus setting up another entity in the mind reminiscent of the way that faculty psychology of the late nineteenth century talked about "the will." Thus, the psychoanalysts apparently take the unconscious as representing a real force, much as the "will" was taken to represent a real force in an earlier era.

The id, ego, and superego were thought of as varying along a conscious–unconscious dimension. Originally Freud conceived of only the id as unconscious. Later he came to view the ego and the superego both as partly conscious. The person's awareness of his motives and activities was viewed as representing only a very small reflection of the dynamic processes actually controlling his behavior. Certain material was also considered to be *preconscious.* It could be brought to conscious awareness relatively easily.

Part of the reason for our considering these elaborate and scientifically irrelevant concepts is that many practitioners of psychotherapy have viewed the process as one in which the central problem is to make the unconscious conscious, and in doing so release libidinal energies for constructive, rather than for destructive or self-destructive uses. Also, after the elaborate trappings of the theory are summarily understood it is appropriate to look at the broader aspects of the theory.

Thus, it can be seen that the psychoanalytic theory is related to the homeostatic view of adjustment and that it is a dynamic theory. It follows the dynamic model of adjustment. The dynamic principles governing the individual's behavior are the pleasure principle and the reality principle. The individual seeks pleasure and avoids pain. If sufficient pleasure in terms of realities that will satisfy the id "wants" are not available, then there is a buildup of undischarged libidinal energy. Should the tension become too great under continued frustration, the person will attempt drastic and highly irrational means of providing an outlet. Violent crimes or acts of passion could result. More often, however, when these tensions threaten to break through into consciousness or to erupt in behavior, defensive mechanisms are used to prevent disorganization and a breakdown of the individual's ability to function. If these defenses do not prove effective, personality disorganization and the psychological destruction of the individual occurs. Thus, psychoanalysis strongly postulates that man's drives are primitive, not ideal, and that rational and idealistic motivation are the result of social and parental indoctrination. When society makes more demands than the individual can tolerate, human energies are bottled up and turned inward, as it were, to the destruction of man.

The individual can tolerate a good deal because of the operation of the reality principle. When the ego and superego have taken enough energy from the id, through the necessity of the ego to cope with the external world and the demands presented by it, then it makes it possible for the infant to learn to obtain future satisfaction by inhibiting immediate pleasure-seeking. Hence the individual learns to tolerate tension and frustration. In terms of the equilibrium of energy distribution, the reality principle makes it possible for the tension which would otherwise occur, and possibly cause a breakdown in the system, to redistribute itself, to become more diffuse, or to enter a reservoir, and hence the dangerous level of pressure is avoided. Eventually, however, with continued lack of satisfaction, libidinal energy again builds up so that if an outlet is not found the pressure becomes too great for the individual to withstand.

Psychoanalytic Therapy

Psychoanalysis holds that anxiety and the repression of it are the most important processes in maladjustment. Although repression has been reinterpreted in terms of modern learning theory (see pp. 335–339), the psychoanalytic notion is that repression is a dynamic process, that repressed thoughts and feelings are continually seeking expression, and that energy must be used to keep them from becoming conscious. Accordingly, although he is not aware of them, repressed thoughts and feelings control much of the individual's behavior. Therefore, the object of psychoanalytic psychotherapy is to remove repression and to make the unconscious conscious.

In order to do this Freud devised the technique of *free association.* On the face of it free association seems like a procedure which is so simple that it is trivial. It simply requires of the patient that he say whatever comes to mind in a continuous, spontaneous chain of associations. The patient is sometimes said to have been given the "compulsion to utter." He is requested to say whatever comes to mind without censoring his thoughts or feelings.

Actually, free association is very difficult to learn. Ordinarily we are taught to think that our thoughts are rational, orderly, and organized around reality considerations. When the person is required to say out loud whatever comes to mind the lack of organization of his associations and the emotional intrusions disturb him. Within a very short time thoughts come up which for one reason or another are censorable, thoughts which, if shared with another person, are anxiety-producing. These cause a block or a break in the chain of association. The psychoanalyst knows that there is always some content present in a mind that is free of fear and the associational cues and themes leading up to the blocking are used as indicators of areas of unconscious anxiety.

Another aspect of psychoanalytic therapy involves the notion of *transference.* Transference is analogous to the concept of generalization in learning theory. According to the theory of transference, as the patient–therapist relationship continues, certain thoughts and feelings of the patient's become mobilized and directed toward the therapist; thoughts and feelings which are not reality-based. They are not reality-based because in reality the therapist has done nothing to earn them. Analysis of them often reveals that they resemble attitudes from earlier periods in the patient's life, especially attitudes which had existed in the patient's relationship with parents and other significant people.

Sometimes the analysis of the transferred attitudes helps the patient to learn a discrimination. He may learn to stop reacting to people with whom he is presently intimately associated (e.g., his wife) as he had toward parents or others intimately associated with him in his past. When the two are no longer psychologically equated, he can better respond to people in the present as they really are.

Modern learning theory's interpretation of the psychoanalytic process is basically that the therapist assumes the role of a nonpunishing audience (e.g., Dollard & Miller, 1950). The therapist listens to the patient's free associations without judging or condemning. He offers interpretations in such a way that the patient can gain insight into his unconscious motivation and yet not feel threatened by having it revealed. As we have seen, however, the extinction of learned fears is likely to be a very slow process.

Psychoanalysis deals with the "depth" aspects of personality. Free association, transference, the analysis of dreams (dreams are also thought to indicate unconscious motivation), getting the patient to explore early experiences and to regress to them within the analysis—all of these techniques are designed to encourage the patient to develop insight into his deeper motivations and to utilize his psychic energies in more mature ways. The goal is the reconstruction of the patient's personality. Psychoanalysis is a costly venture. It may take from 2 to 5 years or longer. Furthermore, demonstration of the success of the technique is largely lacking.

Client-Centered Therapy

Traditional psychoanalytic therapy is depth-oriented insofar as an attempt is made to get at unconscious motivations by means of lifting repression. Also, the therapist interprets the material provided by the patient's behavior, free associations, dreams, etc., according to an elaborate formalized theory of human motivation and personality.

The approach known as *client-centered therapy* developed by Carl Rogers (1961) and his associates, in contrast, does not require the patient (called a *client*) to free-associate or report dreams. Neither does the therapist attempt to interpret the client's actions or make use of transference in order to give the him insight into his unconscious motivation. Instead, this approach emphasizes techniques which help the person to help himself cope with his problems and emotional disturbances. This is done primarily by establishing a relationship with him which is permissive, warm, accepting, and understanding. In such an atmosphere, in which the elements of evaluation, impersonality, and other threats which the person so often experiences in his daily living are not present, he is enabled to examine his thoughts and feelings.

Rogers believes that the creation of a therapeutic atmosphere is so important in producing personality change that he has developed special phrases emphasizing this feature of therapy. He talks, for example, of *empathic understanding* and of *unconditional positive regard* on the part of the therapist, whom he calls a *counselor*. A significant portion of his theoretical and research efforts has been directed toward discovering the conditions under which these qualities make their appearance in the therapist and are perceived by the client.

A second major aspect of client-centered therapy involves techniques for changing the individual's perceptions of himself and of his behavior in the direction of increased awareness, especially awareness of his own latent capacity to understand the events of his life which have led to unhappiness and pain and to discover ways to overcome them. The client learns these things indirectly, almost as a by-product of his discovery of himself as a person. As the client discusses his unhappiness, disturbances, the events of his life, etc., the therapist interprets them only insofar as he clarifies and reflects back to the client the thoughts, and especially the feelings, expressed by him. As a result the client gradually learns to admit into his awareness his own motives, thoughts, and feelings. In doing so his accepted experience, those aspects of his experience that he is willing to perceive as being related to himself, becomes increasingly congruent with his actual experience. When this happens, distortions and defenses become no longer necessary. That this should be the case is not really surprising. It was seen in Chapter 8 that warmth and acceptance and the experiencing of a positive attitude is one of the primary conditions under which children will identify with adults and imitate them (see pp. 260–263).

Rogers has emphasized that when client-centered therapy is successful the individual shows an increasing spontaneity, openness to experience, self-respect, and trust in his own capacities and resources. He describes the client as becoming a more "fully functioning person." Apparently as a result of his experience with clients, Rogers has postulated that self-actualizing or positive growth tendencies are present, in fact, are probably biologically determined, but that through the socialization process personality defense and perceptual distortions are learned which inhibit the natural expression of these forces. When the special conditions described above—lack of threat, and the fostering of perceptions that are congruent with experience—are present then the natural positive growth tendencies come to the fore.

Spontaneous Cures

The chance combination of life circumstances sometimes produces spontaneous cures. Sometimes this is because the person's defenses become inadvertently strengthened, anxiety-producing stimuli are removed or diminished, or chance opportunities to establish new interpersonal relationships provide an opportunity for change. Thus the "successful" individual may be so because of achievement strivings that are neurotic in the sense that they are based upon an attempt to disguise and counter strong dependency needs. The fact of his success, however, may satisfy other motives and serve to strengthen his previously threatened defenses.

At times the conditions producing anxiety may change so that the individual may experience a sharp reduction in his symptoms. Thus, the man who is insecure in his job because of his fear of responsibility may get demoted to a less-demanding position or find himself engaged in a job that does not mobilize these conflicts. He may subsequently experience considerable relief and a lessening of symptoms. As the neurotic individual finds opportunities to interact with people, previously distorted or unacceptable aspects of his personality may enter his awareness for the first time, and once experienced, he may change them. Similarly, repressive defenses may be strengthened through the individual's life experiences or activities. The unmarried inhibited woman approaching middle age may decide to seek a husband and in the process of doing so may meet with such unpleasant experiences that a strengthening of repressive defenses may occur. Again, a strengthening of repressive defenses may occur when a person finds that engaging in sports, hobbies, or social activities occupies his time and keeps him from thinking about himself.

These so-called spontaneous cures provide temporary relief, but because they do not alter the potential for the occurrence of conflicts or tension, they are viewed as palliatives at best.

Classifications of Psychotherapy

One way in which psychotherapy has been classified is according to the objectives of therapy. There are two broad categories: *supportive therapy* and *insight therapy*. Insight therapy is further subdivided into insight therapy with reeducative goals and insight therapy with reconstructive goals.

Supportive therapy involves an attempt to restore a person's adaptive capacities by teaching him new ways to maintain control and by strengthening his existing defenses against anxiety. Supportive therapy may involve persuasion, pressure and coercion, reassurance, environmental manipulation, prestige suggestion, suggestive hypnoses, muscular relaxation, and the use of drugs or electric shock. When people undergo spontaneous cures without receiving any treatment it is usually due to the fact that the anxiety-producing stimuli have ceased to exist or the person's defenses against anxiety have been strengthened as a result of changing conditions in his life. Supportive techniques attempt to accomplish the same thing only in a systematic way, typically by the strengthening of the person's defensive resources, because the elimination of the anxiety-producing stimuli is often impractical.

The insight therapies differ from supportive therapies in that they try to release what has been variously called the self-actualizing tendency (see p. 377), the individual's creative potential, or simply the growth tendencies of the individual. Some theories postulate positive growth motives which are released by the therapeutic process. Others, like the psychoanalytic theory, postulate simply a change in the availability of adaptive energies. The same observations can be recast into the terminology of learning theory: When conditioned emotional responses are extinguished, behavior with an avoidance history decreases and positively reinforced behavior tends to increase.

Thus the insight therapies, instead of trying to remove anxiety-producing sources of stimulation, or bolster behavior that allows the individual to cope with anxiety, attempt to extinguish anxiety and to alter the individual's perception. When the goals of insight therapy are *reeducative*, therapeutic techniques are directed toward producing insight into the person's more conscious conflicts. The goal is to allow insight into the faulty mechanisms and goals of his present adjustment and a subsequent strengthening of attempts at new modes of behavior and evidences of his making use of his creative potentialities. Carl Roger's client-centered therapy is a prime example of insight therapy with reeducative goals. Although space does not permit a detailed discussion of them, a number of other therapeutic approaches have similar goals, e.g., therapeutic counseling, casework therapy, and reconditioning.

Insight therapy with *reconstructive* goals has as its objective insight into unconscious conflicts, the extensive alteration of the individual's character structure, and the release of energies for the development of new adaptive capabilities. Therapeutic approaches with reconstructive goals include Freudian analysis and psychoanalysis and psychoanalytic-oriented approaches to psychotherapy. Several of Freud's followers eventually broke from him and developed their own theories of personality. They are called *psychoanalytic revisionists*, or *neo-Freudians*. Karen Horney's theory of neurotic character development, which we discussed in the previous chapter, is one such theory. Again, space does not permit an outline of the major neoanalytic theories of personality. Nevertheless, the reader can grasp the great diversity among psychoanalytic theories by comparing the classical Freudian theory (outlined on pp. 429–432) with Horney's theory (summarized on pp. 404–417), or with the brief account of Adler's theory given on pp. 403–404.

Although the differences in thinking are at times striking, the actual technique of therapy has remained much the same. Thus, through free association, the analysis of dreams, the analysis of behavioral episodes in the patients life, and the like, resistances are overcome. The individual is enabled to have insight into his unconscious motives; previously unconscious motives become conscious. Therefore, energy previously devoted to repressive defenses is released for more constructive, adaptive responses.

General psychotherapy often combines the techniques of reconstructive and reeducative approaches. Many therapists are *eclectic*. They vary their technique with the personality of the patient, the particular problem which must be dealt with, practical considerations, and the aims and objectives they consider feasible and desirable under the circumstances. Thus, they may talk face to face with the patient, and clarify and interpret by way of reflecting his feelings and attitudes back to him. At other times they may have the same patient free-associate or report his dreams in order to get at unconscious processes. At times with the same patient they may try to strengthen defenses and play a supportive role.

Unfortunately, the practice of psychotherapy remains more an art than a science. Although it has been commonly recognized that the basic principles of the psychology of learning apply not only to the acquisition of maladjustive behavior, but also to the treatment of it, the actual practice of psychotherapy has not, until recently, involved the direct application of these principles. Also, in general the practitioners of psychotherapy have not been research-oriented. As a result, the process of psychotherapy has not been open to objective analysis and the results of it have been difficult to appraise.

Research in Psychotherapy: An Evaluation

Neither the psychoanalysts nor the general practitioners have been noted for their interest in research. Rogers and his group of co-workers have been an exception. Since the beginning they have attempted to make their therapeutic techniques and the behavior of the therapist and of the client during therapy a matter of public record. Snyder (1945), for example, made tape recordings of complete sessions to obtain a record. He analyzed thousands of client's statements in six recorded cases by placing them into a number of relevant categories. He found that statements classifiable under the heading "discussion of problems" declined steadily as therapy progressed, while statements indicative of "insight" increased irregularly during the therapeutic interviews and showed an upward spurt at the end. There was also a sharp increase in the discussion of plans for the future and in decision-making in the later sessions.

Seeman (1949) sampled interview sessions of ten clients and attempted to derive a quantitative measure of the changes that took place. He counted the frequency of statements belonging to the following categories: (1) expression of problems or symptoms; (2) acceptance of the therapist's responses; (3) understanding or insight; (4) discussion of plans for the future. Seeman's results are shown in Fig. 12-1. There the percentage of the client's responses are plotted against fifths of the counseling process. It can be seen in Fig. 12-1 that the statement of problems or symptoms declined as therapy progressed. At the same time understanding or insight increased. Planning increased only slightly over the sessions. Acceptance of the therapist's responses at first increased and then decreased.

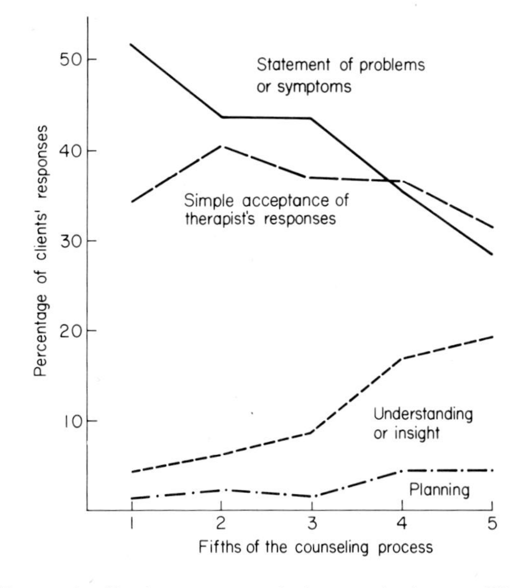

FIG. 12-1. Changes in client's statements during psychotherapy. These data, based on 10 clients, indicate that statements of problems and symptoms tend to decline, and statements of insight and plans tend to emerge as therapy proceeds. (After Seeman, 1949.)

A number of investigators of the client-centered approach pooled their findings and reported them in a volume called *Psychotherapy and personality change* edited by Carl Rogers and Rosaland Dymond (1954). The result was one of the first relatively thorough and systematic scientific appraisals of the therapeutic process available in psychological literature. The project, to which each investigator contributed a part, involved the measurement of various characteristics of a group of clients before therapy, after the completion of therapy, and a follow-up 6 months to 1 year later. A therapy group of about 30 clients was designated as the experimental group. Sixteen therapists were chosen on the basis of their being average (representative) counselors, with 1 to 6 year's experience. The clients ranged from 21 to 40 years of age. They presented initial problems ranging from borderline psychosis and homosexuality; through marital, school, or job failures; to general problems such as indecisiveness and social inadequacy.

Part of the experimental group was given a battery of psychological tests and asked to wait for a 60-day period. They were then given the battery a second time before the beginning of counseling. Thus, the investigators were able to test whether changes occurred over the 60-day waiting period simply because these individuals were motivated for help.

Still a third group of individuals, not to be given therapy and not desirous of it, were matched with the therapy group with respect to age, economic status, and other variables thought to be relevant. This control group was given all of the research instruments at the same times that they were given to the therapy group in order to detect differences between the two groups or any changes that might occur over time. They included a variety of psychological tests and rating scales. The persons doing the ratings were not informed whether the material they were rating was produced before or after therapy or came from a client or a control individual.

Attempts to measure the client's perception of himself at various stages indicated an initial discrepancy between the self-concept and the self-ideal for individuals in the therapy group. The discrepancy tended to disappear as therapy proceeded. It was especially marked for clients rated as showing considerable improvement. Data by one investigator (Rudikoff) indicated that the change in the SC/SI discrepancy during the course of therapy resulted from the SC changing in the direction of a more positive self-evaluation.

Butler and Haigh similarly found that whereas the control group had a small discrepancy between the self-concept and the self-ideal to begin with and there was no significant change in it over time, the client group had a large initial discrepancy, and a statistically significant change toward a lessening of the discrepancy after therapy. The 60-day waiting group began with an initial discrepancy and showed no change after the 60-day waiting period.

Other techniques to decide whether the therapy group had improved included having a trained psychologist rate the client's responses to the Thematic Apperception Test blind, without knowing which group the subject came from or what time the test had been given. There was clear evidence of improvement during therapy which was maintained through the follow-up (reported by Dymond).

Another interesting aspect of this research was its attempt to quantify changes in behavior resulting from therapy. The client himself and two of his friends were asked to indicate specific behaviors that were characteristic of the client from a list of behaviors. The list had been evaluated by 100 experienced clinicians in terms of the degree of maturity of behavior each represented. No significant changes were found over a period of time for either the control or for the therapy group during the period of no therapy. However, in instances where therapy was rated by the counselor as having been at least moderately successful, there was a significant increase in the maturity of the rated daily behaviors whether the behavior was judged by the client himself or on the basis of ratings made by his friends. On the other hand, for those persons for whom therapy was rated as unsuccessful a deterioration in the maturity of behavior was indicated by the ratings of the observer-friends. In these cases personality defense was also apparent in that the unsuccessful clients rated themselves as much *more* mature. This study by Rogers was one of the first to suggest that sometimes insight therapy makes people worse instead of better.

In terms of factors thought to have prognostic value in psychotherapy these studies reported some unexpected findings. For example, at least for the 21 to 40 age range, age was found to be unrelated to therapeutic movement. Equally surprising was Seeman and Dymond's report that the initial adjustment or integration of the client had no relationship to gains made in therapy.

The 60-day wait group did not improve during the waiting period. However, Seeman and Grummon found that the better adjusted among them tended to drop out of therapy after just a few interviews, a fact that might indicate that spontaneous recovery occurred for this group of clients. This group of clients was also reported to like the counselor less when they began their interviews, to show less benefit from therapy, and to be less likely to become involved in it. They were also reported to be more extreme in their social attitudes. It is tempting to hypothesize that when there is an immediate warm interest taken in clients the therapeutic process is already beginning to work, whereas any indication of a lack of interest, such as having them wait, automatically has negative therapeutic value.

An earlier study by Haimowitz and Haimowitz (1952) indicated that specific personality characteristics were likely to make for constructive change in therapy as opposed to certain others. Most importantly, the intrapunitive–extrapunitive dimension of personality appears to differentiate the two. Those who tend to project aggression outward and to have a blaming, punitive attitude appear to be more likely to drop out of therapy or less likely to profit from it. On the other hand, self-effacing, intrapunitive individuals were found to be more likely to show therapeutic gains. Data from the Rogers and Dymond (1954) series of studies further suggest that constructive change is more likely for the poorly adjusted who are aware of considerable inner tension and dissatisfaction; those who are aware of less internal tension appear to profit less from therapy. This suggests that sensitizers are better therapeutic risks than repressors.

The Rogers and Dymond series also attempted to differentiate successful from unsuccessful clients. One of the outcomes of therapy is a decreased dependence upon authority, which is coincidental with increased feelings of self-acceptance and self-respect. Therefore it was argued that authoritarian traits involving overconventionalism, overdependency upon experts, intolerance for ambiguity, etc., would indicate a poor therapeutic risk.

Gordon and Cartwright (Rogers & Dymond, 1954) used a single inventory, called the self-other attitude scale (the SO scale). The scale appeared to measure the degree to which the individual accepted others. They found that high scorers and low scorers both tended to be failures or only moderate successes, and that they tended to intensify their attitudes as a result of therapy. That is, if they scored high their scores tended to increase and if they scored low their scores tended to decrease.

Those who scored in the medium ranges tended to be therapeutic successes. After therapy they tended to change in the direction of deemphasizing their attitudes. The scores of subjects within this middle range who initially scored high tended to decrease with therapy and those within this middle range who initially scored low tended to increase with therapy.

Another study in this series, by Tougas, used the ethnocentrism scale devised by Adorno, Frenkel-Brunswik, Levison, and Sanford (1950) in their investigation of the *authoritarian personality*. Ethnocentric individuals are characterized by rigid ingroup–outgroup attitudes involving a positive evaluation and submissive attitudes toward ingroup members (members of their own group) and stereotyped negative evaluations and hostile attitudes toward members of outgroups. They are prejudiced personalities who tend to see the groups with whom they identify as "good" and "strong" and outsiders as subordinate and weak.

Figure 12-2 shows two groups of data collected by these investigators. One group was treated by their own client-centered approach, labeled the Chicago Group, and a second was treated using another therapeutic orientation (Sullivanian, labeled the California group). The ethnocentric scale (E-scale) scores are plotted on the abscissa in both cases. Frequency is plotted on the ordinate. The symbol X stands for an improved client. The symbol 0 stands for an unimproved client.

Note that each of these groups tended to score toward the lower end of the E scale. It appears that self-referred patients as a group tend to be less ethnocentric than normal individuals. The interesting comparison, however, is between the improved and the unimproved cases. There is a statistically reliable tendency for those clients rated as improved to have low E-scale scores. For both of these groups combined the probability of improvement decreases as the individual's E-scale score rises above the group mean of 45. The authors say:

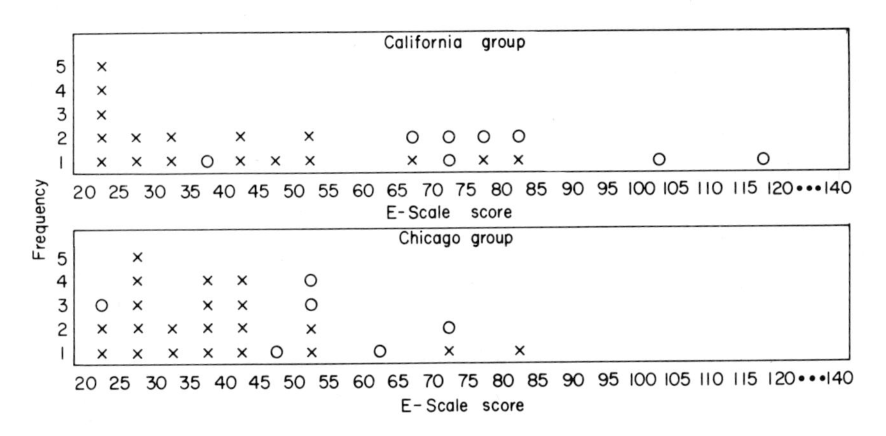

FIG. 12-2. Frequency of improved and unimproved clients in relation to E-scale scores. The California group (N = 25) were rated by other than their own psychiatrists; the Chicago group were rated by their own psychiatrists. X, improved; 0, unimproved. (Adapted from Rogers & Dymond, 1954.)

> The implied limitation this imposes on individual therapy generally is an important problem, especially since ethnocentric attitudes have also been shown to be widespread in the population and their influence on interpersonal and group relations anything but salutary. [Tougas, in Rogers & Dymond, 1954, p. 208]

A more fine-grain analysis indicated that the lowest scorers were actually rated as only marginally successful, those at the group mean as successful, and those beyond the group mean as unsuccessful. Tougas argued that the extremely low E-scale scorer was likely to have a highly idiosyncratic view of the social world and of himself because in order to score so low he would have to reject practically all cultural stereotypes. The extremely high scorer on the other hand would show very rigid conventionalized stereotypy. Whereas the low scorer tended to be introspective, withdrawn, socially isolated, and depressive, extreme high scorers were characterized by projection, expressive hostility, and "bound anxiety." Thus, both the defensively isolated individuals and the rigidly conventional individuals (who seem to actually be segregated from social contacts) tended to lack the capacity for change. It was further found that success in therapy was related to the capacity of the clients to form a meaningful interpersonal relationship. Both extremes of the E-scale scorers appeared to lack this capacity.

Other Data on Therapeutic Outcomes

Ratings of therapeutic success or of improvement vary with both the raters and the criteria used to define improvement. The greatest amount of variability occurs in the criteria. For example, Kelman and Parloff (1957) found 20 of 21 intercorrelations of measurements of improvement to be statistically insignificant. The criteria upon which these measurements were based involved all of the commonly accepted indexes of improvement, including increased awareness, relief from felt dissatisfaction, and an increased personal effectiveness.

Given the fact that various criteria of improvement are essentially unrelated, at least it would seem that when individuals requesting psychotherapy but not receiving it are compared, at the end of therapy, with similar individuals who had received therapy, the therapy group would show improvement. As a matter of fact, a number of studies employing such a comparison have seriously questioned claims of improvement as a result of psychotherapy. As an example, Levitt (1957) reported that 67% of over 4000 children improved as a result of psychotherapy. When improvement was measured some time after the termination of therapy the percentage of improvement rose to 78%, indicating that there is an increase in improvement when therapy is over. The rate of improvement of untreated controls, however, was 72%.

Denker (1961) studied a group of individuals, diagnosed as neurotics, who had disabilities severe enough to warrant insurance payments for nonemployment. No professional psychotherapeutic treatment was administered to these individuals other than normal reassurance, suggestion, and sedatives administered by their personal physicians. Furthermore, since they were receiving insurance payments while they were disabled these patients had some incentive to remain "sick."

Figure 12-3 shows a plot of the improvement rate of these (500) severe neurotics not receiving psychotherapy. This analysis of Denker's data was provided by Eysenck (1961). It can be seen from Fig. 12-3 that after 2 years 72% were classified as improved; after three years, 82%; and so forth. In order for psychotherapy to be judged effective it would seem that improvement rates ought to be at least significantly better than the improvement rate shown by these untreated individuals.

Eysenck (1961) used Denker's figures as a base rate in order to compare the recovery rate of patients treated psychoanalytically and another group of patients treated by ecletic methods. The respective recovery rates were 44% and 64%, thus providing little evidence for the effectiveness of therapy.

Successful Behavior Change in Therapy

The evidence is that the therapeutic procedures which we have been discussing are not effective enough to warrant their use for the general population of neurotic patients. However, they do produce fundamental

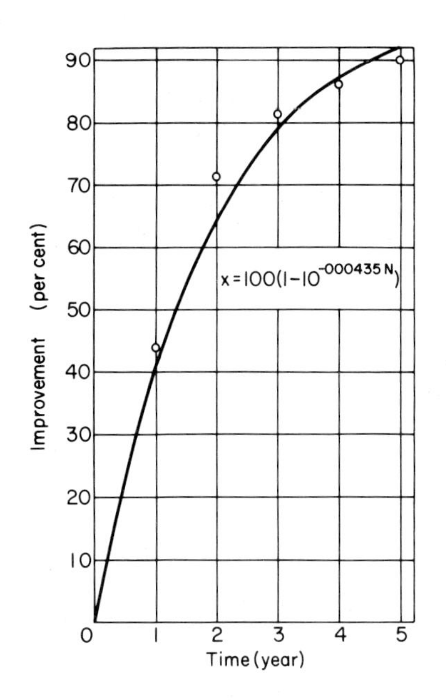

FIG. 12-3. Improvement shown by 500 severe neurotics not receiving psychotherapy. The formula describes the function which best fit the data points. In the formula x denotes the proportional improvement while N denotes the number of weeks elapsing from the beginning of the experiment. Notice that 90% improved over the 5-year period. (Eysenck, 1961.)

changes in personality organization and functioning for *some* individuals. It will be instructive to look at the process of therapeutic change for these people.

When therapy does produce significant improvement, therapists of almost all orientations seem to agree that the vehicle through which it occurs is the *therapeutic relationship*. The establishing of a working relationship appears to be at the very core of constructive therapeutic change with insight therapies. Before change can occur the patient must experience a willingness to experiment with a new way of life. Neither time nor effort nor insight in themselves accomplish the desire for change. Wolberg (1954) says:

> One great disappointment to the beginning psychotherapist is that painstaking and elaborate exploration of a patient's problem, and a most thorough investigation of the underlying dynamics, may sometimes fail to influence the emotional ailment in the least....When we investigate the causes of such treatment failures, we often find that the patient has been unable to absorb, to integrate or to utilize insight due to anxiety. This has prevented the

ego from evolving more effective ways of adjusting the person to his inner conflicts and to external demands. Investigation usually shows that what is basically lacking in the therapeutic situation, and what probably has been missing from the inception of therapy is the proper kind of working relationship between the patient and the therapist. This relationship is a unique interpersonal experience in which the patient feels a quality of warmth, trust, acceptance and understanding such as he has never before encountered with any human being. [Wolberg, 1954, p. 317]

According to Wolberg the most expedient way of creating incentive for change is through rapport with the therapist. Exactly what is involved in rapport that makes it a necessary ingredient for therapeutic change is difficult to specify. Some attempts that have been made to do so will be discussed below. In any event, even given optimal conditions for rapport, the patient always tries to satisfy a variety of needs in therapy, some of which are inimical to it. Most of these needs arise from character problems which are a part of the patient's history of disturbed relationships with people. Common among these, and not the least of them in terms of the difficulties required to overcome it, is a clinging-dependent attitude toward the therapist. The role of the therapist as an expert with the knowledge and skills to cure the patient helps to mobilize this kind of attitude. It is even more basically rooted in the real or imagined helplessness of the patient. Should the therapist try to bring the patient to the conviction that he has abilities and strengths of his own too quickly in the face of these strong unsatisfied dependency strivings he may produce only resentment and a resistance to a working relationship.

A similar need on the part of many patients is the need for unqualified understanding and approval. The anticipated rejection and condemnation of most neurotics makes these needs especially urgent. A lack of sensitivity to them on the part of the therapist, or even worse, the slightest indication of disapproval or rejection may seriously interfere with the working relationship.

Many neurotics expect and anticipate condemnation. They may show an automatic distrust of the therapist. When the therapist does not behave punitively toward them they may condemn him for not being strong and try to seek another therapist.

Still another need of the patient which interferes with rapport is the need for detachment and to avoid feelings of being trapped and hurt. Such patients may experience tension shortly after signs of good rapport occur and attempt to leave the therapeutic relationship, thus typically duplicating their life pattern.

We noted previously that Rogers places particular emphasis on specifying the necessary and sufficient conditions for therapeutic change.

There is some recent research supporting the importance of his factors of empathic understanding, unconditional positive regard, counselor genuineness, and the like. Halkides (1958) studied these variables, along with the extent to which the counselor matched the client's intensity of affective expression. Ratings of these variables were obtained by judges who did not know the degree of success or from which of two sessions, early vs. late, the statements to be rated were obtained. Surprisingly, a high degree of reliability among the judges was demonstrated in their ratings of these somewhat loosely defined constructs. The three variables of empathic understanding, unconditional positive regard, and ratings of the counselor's genuineness were highly associated with the more successful cases. The matching of the intensity of affective expression by the counselor was the only variable for which the results were ambiguous.

Rogers has extensively quoted a number of related studies to support his viewpoints. Although good systematic research in this area is not as plentiful as would be desirable, taken together these studies do present a consistent picture. Examples include a study by Heine (see Rogers, 1961) of individuals who had received psychoanalytic, client-centered, and Adlerian therapies. Regardless of the orientation of the therapists the patients reported similar changes in themselves. Although they accounted for the changes which occurred by giving explanations in keeping with the orientations of their therapists, there was agreement that their improvement stemmed from the fact that they trusted the therapist, felt understood by him, and felt that they had been independent in making choices and decisions. They also reported that they had found it particularly helpful that the therapist had clarified and interpreted feelings of which the client had only been partially aware. Fiedler (1953) found that expert therapists with different orientations formed similar relationships with their clients. He also found that the difference between them and less expert therapists involved the ability to understand the client's meanings and feelings, sensitivity to his attitudes, and a warm interest without overemotional involvement.

It is clear that the therapist's job in establishing a working relationship is a complex one. It is also evident why the technique of psychotherapy is frequently thought of as being more of an art than a science. There is every reason to expect that this will change with continued research, although it is equally clear that insight therapy is not effective for everyone. The standard practice of psychotherapy will undoubtedly have to be reevaluated.

The direct application of learning theory offers considerable promise. The complexity involved in its application, however, must be more thoroughly recognized than it is by many therapists who are strongly oriented toward the application of learning techniques. This is illustrated in an experiment by Ends and Page (1957), who made a comparison of three different methods of psychotherapy—one based on learning theory, one based on a client-centered approach, and a third based on a psychoanalytically oriented approach. Their patients were chronic hospitalized alcoholics in a state hospital setting. They found that therapy based upon the learning approach gave poorer results than the other two approaches. In fact, compared to a control group there appeared to be a negative effect from it. Rogers has commented on this study (1961), indicating that he was puzzled that the approach to which the authors were committed proved least effective. He pointed out that the aim of the learning in the therapy based on the learning approach was to be impersonal and that the therapist tried to avoid impressing the patient with his own personality characteristics. It may well be that treating the individual as an impersonal object, at least with certain types of people under certain conditions, destroys whatever value the actual technique used might otherwise have.

The quality of the working relationship may be central to therapeutic change, at least given the present techniques of the insight therapies. Therapists of widely differing orientation appear to be in agreement on this aspect of therapy. Unfortunately few people other than Rogers have given this theoretical attention. He notes that when in his therapeutic experience he has acted consistently acceptant, when in fact he actually felt annoyed, skeptical, or had some other nonacceptant feeling, then in the long run it hindered rapport. He uses the term *congruent* to describe the matching of the feeling or attitude that he is actually experiencing with an awareness of that attitude. When congruence exists the therapist (or any other person) is integrated. It appears to create a reality which clients and others experience as dependable and trustworthy.

He notes further that in our culture people are often loathe to let themselves experience their positive feelings toward others for fear that they may be disappointed, that demands will be made upon them, or that they will become trapped in some way. He believes that the therapist is most helpful if he behaves in just the opposite way: to feel that it is safe to care and to relate his own positive feelings to the client and to others. Similarly he believes that the therapist ought to feel trust enough in his own person to be able to feel separate—not to be adversely affected by the client's depression, fear, morbid dependency, or whatever. A similar kind of security within the therapist ought to permit the client to be what *he* is, instead of attempting to mold him into conformity with the therapist's ideas of what he ought to or should be. In a similar vein, it is most desirable, according to Rogers, that the therapist be able to empathize with the feelings of the client sensitively and without advising, evaluating, or judging.

Finally, he definitely visualizes a relationship between learning principles and the therapeutic process, and he emphasizes that which attitudes the therapist chooses to reinforce is of vital importance. Rogers is worth quoting on this point. He says:

> If I see a relationship as only an opportunity to reinforce certain types of words or opinions in the other, then I tend to confirm him as an object—a basically mechanical manipulable object. And if I see this as his potentiality, he tends to act in ways which support this hypothesis. If, on the other hand, I see a relationship as an opportunity to "reinforce" *all* that he is, the person that he is with all his existent potentialities, then he tends to act in ways which support *this* hypothesis. [Rogers, 1961, pp. 55–56]

The Mechanics of Therapeutic Change

Before turning to the behavioral science approaches to therapy, the manner and kinds of changes that occur under the optimal therapeutic conditions that have been discussed above deserve mention. Here also there appears to be considerable agreement among therapists of different orientations as to how the successful client or patient behaves. Wolberg and Rogers happen to be two experienced therapists who have ventured a detailed description of the process. Roger's description is within the framework of client-centered therapy. Basically Rogers sees the therapeutic process as involving a shift from incongruence to congruence. It also involves certain other changes. One of these is a shift from the construing of experience in relatively rigid ways toward an increasing openness to the reality and novelty of each new experience. There is also a change in the individual's relationship to his problems. Whereas at first they seem largely unrecognized and there is little desire for change, as the process continues there is a "subjective living" of his problems and a feeling of responsibility for the contribution he makes to their development and maintenance. A change also occurs in the way the individual relates to people, from his initial avoidance of close relations, which may be perceived as dangerous, to spontaneousness and "immediateness of experiencing" in his relationships with others.

Rogers has attempted to divide the therapeutic process into stages, which really represent points along a continuum. Changes in the experience of the individual's relation to himself, his problems, and others are seen by Rogers as evolving in a lawful process. Central to this is the loosening of feelings. According to Rogers:

> At the lower end of the continuum they are described as remote, unowned, and not now present objects with some sense of ownership by the individual. Next they are expressed as owned feelings closer to their immediate experiencing. Still further up the scale they are experienced and expressed in the immediate present with a decreasing fear of this process.

Also, at this point, those feelings which have been previously denied to awareness bubble through to awareness, are experienced, and increasingly owned. At the upper end of the continuum living in the process of experiencing a continuously changing flow of feelings becomes characteristic of the individual. [Rogers, 1961, p. 156]

Wolberg also believes the process is a systematic, orderly one. He has outlined 15 component parts or sequences that are usually encountered. These are presented below. For him the patient's verbalization of his feelings and of his problems produces a basis for his gaining insight into them, and with the help of the working relationship the patient is able to resolve resistances that allow him to abandon paralyzing fears and inhibitions. The *mechanics of therapeutic change* according to Wolberg are as follows:

1. The patient, concerned with his symptoms and complaints, elaborates on these.
2. He discusses feelings which disturb him and which are associated with his symptoms.
3. He realizes that his feelings are related to certain dissatisfactions with his environment, and that they are somehow conditioned by a mysterious turmoil that rages within him.
4. Along with his feelings, he recognizes patterns of behavior which frustrate him and which are repetitive, compulsive, and even automatic. Soon he starts to appreciate that some of these patterns are responsible for his disturbed feelings. This causes him to doubt their value.
5. As he becomes aware of how dissatisfied he is with his impulses and behavior, he begins to try to stop their operation; yet he finds that they persist in spite of himself.
6. He slowly realizes, then, that his behavior serves a function of some sort, and that he cannot give it up easily. Indeed, his patterns repeat themselves in various settings, perhaps even with the therapist.
7. If he has the incentive to explore his patterns, he finds that they have a long history, issuing from attitudes originating in his early relationships with people, particularly his parents.
8. Gradually he recognizes that he is governed by impulses and feelings such as were present in him as a child. He sees that by carrying over certain attitudes into his present life, he is reacting to people as if they were facsimiles of his parents, siblings, and other significant persons.
9. With great trepidation, he begins to challenge his early attitudes; progressively, he inhibits automatic and repetitive behavior patterns, slowly mastering his anxieties as he realizes that fantasied dangers and expectations of injury do not come to pass. In the therapeutic relationship, particularly, he begins to change, especially in his attitudes toward the therapist.
10. He begins to entertain hopes that he is not the weak and contemptible person who has constituted his inner self-image, that he actually has value and integrity, that he need not be frustrated in the expression of his needs, and that he can relate himself happily to people.
11. This causes him to resent all the more the patterns of living he customarily employs, which are products of devaluated feelings toward himself and devastating fears of his environment.
12. Slowly he begins to experiment with new modes of behavior which are motivated by a different conception of himself as a person.

13. Finding gratifications in these new patterns, he becomes more and more capable of liberating himself from his old goals and styles of action.

14. Growing strengths within himself contribute to a sense of mastery and produce healthy changes in his feelings of security, his self-esteem, and his attitudes toward others.

15. He liberates himself more and more from attitudes and anxieties related to past experiences and misconceptions. He approaches life as a biologic being, capable of gaining satisfactions for his personal impulses and demands, and as a social being who participates in community living and contributes to the group welfare. [Wolberg, 1954, pp. 155–156]

THE APPLICATION OF LEARNING PRINCIPLES TO PSYCHOTHERAPY

It cannot be stressed too much that whatever occurs as a result of verbal therapy must result from the behavior of the persons involved in the therapeutic relationship towards each other. Maher (1966) has argued that the interpersonal relationship in psychotherapy might be defined by the kind of dominant pattern of responses that emerges from both the therapist and the patient. In this connection he says:

> With long experience the persons may begin to respond to stimuli provided by each other, such as facial expressions, posture, manner of lighting a cigarette, etc. When this stage is reached it is easy to assume that some unusual kind of communication has been established, and that this kind of communication occurs mainly in something defined as psychotherapy. From such assumptions it is natural to conclude that a psychotherapeutic relationship is different *in kind* from the interaction that takes place between two travelers conversing during a chance meeting on a train. There is no scientific basis for this assumption. We suggest strongly at this point that the variables operating in a psychotherapeutic relationship are those that operate in any social interaction, and that a special set of concepts is not necessary to describe or understand them. (Maher, 1966, p. 454)

Accordingly, learning theorists view the changes which take place in psychotherapy in terms of the acquisition and the extinction of certain kinds of behavior. The therapist, through the working relationship, acts as a nonpunishing audience. Anxiety produced by thoughts and feelings connected with the discussion of the patient's thoughts and problems tends to become extinguished.

We have seen in Chapter 8 that warm and acceptant parental figures tend to be imitated under the normal conditions of social learning. It would not be surprising to find that many of the therapist's attitudes and behaviors toward the patient's problems, and toward the patient himself, tend to be incorporated into the patients behavioral repertoire. Consciously or unconsciously, the therapist may actively reinforce certain attitudes and behaviors on the part of the patient. It was noted above that Rogers, a therapist with a distinctively nonlearning orientation, is readily able to view the changes that take place during therapy as a result of the direct reinforcement of the therapist.

The traditional therapeutic exchange between a patient and a psychotherapist who may be a psychiatrist, clinical psychologist, or counselor. (Photograph by Tor Eigland.)

The Dollard and Miller Model

As we have seen a number of times in this text, the learning explanation of neurotic behavior assumes that symptoms develop from a conflict between approach and avoidance tendencies. Dollard and Miller were among the first to reinterpret the psychotherapeutic process (psychoanalytic psychotherapy) in learning terms (Dollard & Miller, 1950). They proposed that the conflict model (see pp. 163–170) be applied to human neurotic disturbances. They conceived of the avoidance gradient as largely learned fear. Typical human neurotic conflicts were seen as involving sexual excitement, anger, resentment, hostility, and dependency. It was hypothesized that many conflicts, especially those involving parent-child interactions such as feeding, weaning, and early cleanliness training, can be learned by the young child unconsciously. This is because there are no verbal labels available to describe his experiences, and what is not verbalized at the time cannot be well reported later. Nevertheless the emotional conflicts survive and dictate the individual's behavior. Early sexual conflicts involving these dynamics have already been discussed (pp. 279–281).

When conflicts are cued off in the adult the fact that they were never labeled and/or repression of the thoughts associated with them make the neurotic "stupid." He is incapable of representing his own feelings on a verbal or a mental level. The emotional turmoil, stimulated by cues of which the individual is unaware, produces a state of high drive with its accompanying subjectively felt misery. At the same time any of a great variety of symptoms may be learned on the basis of anxiety-reduction.

These basic factors involved in neurotic conflict interact with one another. Thus, the misery must be reduced by further symptoms. Since the neurotic cannot represent certain aspects of his conflict in consciousness, he is less able to be on guard against certain aspects of his experience and he is likely to have less adequate solutions to his problems.

His "stupidity" makes him unable to discriminate circumstances under which being sexy, angry, or whatever, will not lead to punishment from those under which such behavior will. The probability of his fears becoming extinguished, therefore, is extremely low.

The misery, symptoms, and stupidity may increase his fears, which, in turn, increase his need for repression, which in a vicious circle contributes to his stupidity and misery.

Dollard and Miller outline in detail the analytically oriented therapist's role in alleviating these disturbances. The therapist must establish himself as a strong reinforcer. His reinforcing potency comes from such factors as his prestige, his sympathetic understanding, the fact that he holds out hope by having faith in an eventual cure, the reassuring effect of his composure, and his permissiveness. The patient is eventually able to talk about frightening topics because the therapist encourages him to talk freely and he consistently fails to punish him. Talking about frightening topics under such conditions leads to an extinction of his fears and to a generalization of extinction which weakens his motivation to repress other related topics.

The therapist further aids the process by being skilled in decoding conflict and in dosing anxiety. Dollard and Miller stress the fact that the therapist presents the patient with a graded series of learning situations. They also see one of his major functions to be that of helping the patient to discriminate. In particular, he contrasts the disproportionate fears and anxieties with the realities of the patient's actual living situation. He also provides verbal contrasts between the helpless past conditions of childhood and present social conditions. When the process is successful the gradual extinction of fears and the generalization of extinction is accompanied by a gradual increase in the patient's attempts to try better solutions. At first there is a restoration of his ability to think and plan. Then there are attempts to try out different solutions in real life. When these responses produce satisfaction an incremental reduction occurs for other fears of the patient. The process is continued until the patient is able to function normally or to make a superior adjustment.

Dollard and Miller have not added anything new. They have simply pointed out that traditional psychotherapy operates according to established principles of learning. It ought also to be pointed out that traditional therapy, because of the very way that it is structured, must remain far from the most ideal situation in which to modify the patient's behavior. There are a number of places where it is particularly weak. Because the therapist has little or no control over the conditions of the patient's life outside of therapy to enable the manipulation of rewards and punishments directly, whatever effectiveness he has as a therapist must rest largely on his personal skill in handling the therapeutic relationship. Similarly, if he fails to establish rapport with the patient, or if for any reason the patient becomes frightened or negative, the therapist is helpless to prevent him from leaving therapy.

Supposing success in reducing the patient's fears and anxieties in therapy, it is *assumed* that the effects of the extinction of fear will generalize to situations in the patient's life and that his behavior will change accordingly. The generalization from therapy to the patient's life situation may occur in successful cases, but classical therapeutic procedures are not noteworthy for their degree of success. Learning psychologists know very well that much of our behavior is situation-specific. To expect a patient to react differently toward, let us say, his business associates, on the basis of his attitude toward conflicts that they may engender within him having changed significantly during the therapeutic hours may be somewhat overoptimistic. Even if it is argued that therapy reduces his tension and defensiveness, in order for the patient to change in the direction of behaving in a constructive way in particular settings, the cues for initiating new responses must be present at the appropriate time and in the appropriate situation. It will be seen below that many psychologists are beginning to suggest the direct manipulation of symbolic and other cues in order to make behavior change more situation-specific.

Rotter's Social-Learning Approach

An attempt is made to correct some of these deficiencies of the traditional therapeutic approach in Rotter's social-learning theory (Rotter, 1954). He argues that the patient has learned certain expectancies, largely connected with neurotic avoidances. He has typically learned, for example, to have a high expectancy for punishment or failure should he do anything but avoid certain situations. Thus, he expects that his behavior (which observers view as maladjusted) will lead to less punishment or to gratification. Therapy consists in changing expectancies by manipulating rewards and punishments within a social context. Rotter stresses the direct application of learning principles in order to change the patient's expectancies. He places heavy emphasis upon reinforcing the response of looking for alternative solutions to the patient's problems.

Traditional therapies appear to accomplish their results through the extinction of fear. Rotter feels that this is inadequate because, first, symptomatic behavior and the fear underlying it are very difficult to extinguish. Secondly, in traditional therapies the patient has nothing better to substitute for his inadequate behavior so that he tends to "hold on to what he has."

Rotter favors the face-to-face discussion of the patient's problems without stressing such techniques as free association or the interpretation of dreams. He feels that the patient could only accept interpretations based upon such techniques slowly and inefficiently. It is far better to deal with the individual's real-life experiences and problems that are present and meaningful to him. He particularly stresses the interpretation of the patients behavior in relation to the people around him, with whom he has daily contact.

The therapist assumes an active role. He does not always wait for the patient to discover important relationships all by himself. Rotter sees the therapist's role as one of providing new ideas and new experiences for the patient. Acceptance, understanding, and reassurance are emphasized. Particularly important kinds of reassurance are, first, the reassurance that the patient's problems are genuine and that his attempt to do something about it in seeking psychological therapy is justified, and second, reassurance in the form of building up in the patient expectations that there are alternative ways of behaving, that he is able to do something about his problems, and that therapy will be successful.

Structuring. In this latter connection the social-learning approach places value on *structuring* therapy, where by structuring is meant

> the discussion about the therapy that takes place between the therapist and the patient. It is a discussion that is concerned with the purposes and goals of therapy both specific and general, the plans of the therapist, the respective roles and responsibilities of the patient and the therapist, and the attitudes both have toward the therapy at any time. [Rotter, 1954, pp. 351–352]

Rotter says further:

> As in any conflict-learning situation, successful solution is considerably enhanced by the proper set or by attention to the proper cues. As we have already discussed, one of the most efficient and easiest ways to attain this set is through verbal communication. [pp. 352–353]

As to the content of structuring he has the following to say:

> From a social-learning point of view structuring places a major emphasis on behavior. The purpose of therapy is to help the patient change his way of reacting to outside pressures, his way of behaving, his attitudes (or his feelings and emotional responses if this is the language the patient is likely to use). The goal is the ability on the patient's part to deal more effectively with his problems or his long-term life problems, while at the same time he experiences greater subjective satisfaction or happiness. The therapist may help him do this by helping him to discover how his present attitudes and reactions developed, which ones are appropriate and which ones are inappropriate for his present-day life, and also what alternative ways there are of dealing with his problems. It should be made clear at the outset that such understanding in and of itself does not lead to change or "cure," but that the patient himself must take the responsibility of frankness, must have the motivation to change, and above all, must be willing to try out new behaviors . . . . It is the therapist's intent only to provide the patient with information or to help him to see things in a different light so that he might make better decisions than he has made in the past. [pp. 354–355]

Structuring is successive. In the beginning the patient is given realistic expectations of what he can expect from therapy, along with appropriate reassurance. Problems are dealt with first which are perhaps less basic, partly because they are of immediate importance to the patient and evidence of success is particularly gratifying. In this connection the therapist may allow himself to become a strong source of reinforcement for the patient in the middle of therapy; however, toward the end he deliberately reinforces independent behavior, indicating to the patient that it is important in order for him to obtain satisfactions.

Changing Expectancies. A number of techniques are used in the social-learning approach to change expectancies. If possible direct reinforcement for the patient's behavior is arranged either by the therapist himself or by having the individual's wife, husband, boss, or others related to him behave toward him in a therapeutic manner. Another technique is to place the individual in situations where he can observe the relationship between the way others behave and the consequences of their behavior. The person with poor social skills, for example, might be encouraged to enter into group situations whenever possible. Rotter says:

> When alternative behaviors do not occur frequently enough to be reinforced directly in the patient's own behavior repertoire or in the people around him, and when the behaviors are not discovered by the patient himself in his discussing possible solutions, then the therapist may suggest them directly. He does this by discussing with the patient, if necessary, both how the specific behavior may be carried out and the potential consequences of the behavior. [Rotter, 1954, pp. 341]

The therapist may use many other techniques to change the patient's behavior. He may point out to the patient the differences between conditions in the past, when behavior was punished, and presently existing conditions. He may nonreinforce expectancies by failing to react with sympathy, concern, attention, etc. He may attempt to change the values of the patient's goals, set more realistic standards, or change the way the patient goes about satisfying particular needs.

Rotter points out that it is often easier and desirable to get the patient to change the way he goes about reaching certain goals or satisfying certain needs than it is to get him to change those goals or needs. Thus, becoming president of the local Ladies Aid Society leads to satisfactions and to fewer problems than attempts to dominate one's husband. By structuring, by changing the patient's expectancies, taking the fullest possible advantage of the social sources of reinforcement available, this approach attempts to provide alternative solutions to maladjustment and to increase the patient's ability to solve his own problems. In this type of therapy there is much more communication from the therapist to the patient.

Structuring allows for certain verbal connections to be established by the therapist which otherwise might be only very slowly learned by the patient or never learned at all.

Philip's Assertion Therapy

Another approach which emphasizes structuring and a more active role on the part of the therapist in communicating *his* insights to the patient has been formulated by Lakin Philips. He calls it *assertion therapy.*

Philip's theoretical position (1956) is that the crucial point in most maladjustment involves certain beliefs or assertions the person makes and which the environment disconfirms, or which are contradictory to other beliefs of the person. This state of affairs can be described by the conflict paradigm. Therapy consists in reducing the avoidance gradient by *hypothesis structuring.* That is to say, feedback from the therapist enables the patient to structure his own decisions. The very fact that he comes to perceive his decisions as based upon assumptions serves to reduce the noxious and avoidance features of the person's life situation, and allows him to see his interpersonal relations with some perspective. It also sets going processes that signal an increase in his resourcefulness in dealing with his problems.

Philips argues that the height of the avoidance gradient (e.g., *X* amount of anxiety) prevents the person from hitting upon alternative solutions to his problems in living. This is because the tension produced by the conflict limits his flexibility and tends to increase his rigidity. When the person is made behaviorally narrow it tends to set up a steady state of maladjustment. That is, his assertion–disconfirmation–tension becomes a self-sustaining process which gets activated anew each time he meets the relevant situation.

Such tension-producing circular behavior does not necessarily involve the total personality. Problem areas may be highly focal and they may center around certain aspects of the patient's living or around the behavior of others.

Space permits us to select but one of any number of examples of assertions in interpersonal relations. Thus, Philips describes portions of an interview with the mother of an overprotected child. Among the mother's assumptions was that she should supervise and control the child's every act. Philips summarizes the interaction as follows:

> The conflict for the mother is between what she *wants* the child to do and be, and what he is doing (in his play). The conflict for the child is between his "natural" desires in play (according to age, sex, social status, experience) and the mother's goal-setting activities which are mostly distasteful to him. The mother's behavior and the child's anticipation of her attitude set the boy against her and her demands: He is henceforth judged as "stubborn" or "dull" by the mother who then redoubles her efforts to get him to do things her way (so he won't appear "dumb" or "stubborn"). The mother's assumptions are narrow; so are the child's, at least where the mother–child interactions take place. The child is reluctant to play with other children, shys away from adult company, is not spontaneous in play with other children unless he knows them well. The assumptions that he has led to

develop about play are such that he is under so much scrutiny that he cannot "be himself" in play. Thus a vicious circle is instigated. [Phillips, 1956, p. 236]

Philips points out that in this case the mother's conflictful involvement can be reduced if the potency of her assumption (that she must see to it that her son does everything as she wishes) can be reduced. When this is accomplished both the mother and the child can live more easily with each other.

The therapist's role in dealing with such problems is described by Phillips as follows:

> The therapist's interaction with the client is directed to the end that the client's assumptions are to be structured; and all the therapist's responses are efforts to encourage hypothesis-structuring. Then, in this way, the therapist may react to the parent with a question such as "What would happen if you did not try to get Charles to ride the tricycle when he didn't want to?" Or with a reflection: "You feel that he just won't get going in his efforts to ride unless you urge him on?" Or, "He's probably reacting not to your wish that he ride the tricycle well, but to your insistence that he ride it right now." [Philips, 1956, p. 237]

Hence, the major role of the therapist in assertion therapy is to focus the attention of the patient onto what he is doing and on the inadequacy of his current solutions. A reorganization of the patient's attitudes and attempts at trying new solutions follows, as the reality of his assertions and the results of them become clear.

Cybernetic Concepts Applied to Psychotherapy and Behavior Change

Phillips has also been instrumental in employing certain aspects of cybernetics as a model for neuroses and behavior change (see Phillips & Wiener, 1966). *Cybernetics* comes from the Greek word meaning "steersman." It is generally considered to be the study of control and communication systems. It has been applied in areas such as human engineering and human factor analysis, and man–machine systems. It has only been recently used as a model for psychotherapy.

Only a few of the possible cybernetic concepts which can be applied to behavioral processes will be noted here. The concept of *feedback* is one of them. Feedback means that the device or system in question has an effect that acts back on one of its causes. This usually involves a sequence of causes and effects. Such a sequential feedback chain is called a *loop,* or a *feedback loop.*

Positive elements of a loop are those which directly facilitate another element. This facilitation is sometimes called *deviation-amplifying.* Fig. 12-4, adapted from Phillips and Wiener (1966), illustrates the operation of two feedback loops, an efficient one and an inefficient one. Both of these loops begin and end in the same place at point A, with a student's decision to take more courses. However, loop ABCA begins positively and ends positively, whereas loop ADEFA begins positively but ends negatively. The positive loop is said to have an *amplifying* effect, because the increased confidence enables the student to further increase the number of courses taken. On the other hand, the negative loop ends in a *counteracting* influence, with poorer grades having a counteracting influence on increasing the number of courses.

Loops containing only amplifying factors, like the loop to the right in Fig. 12-4 are positive loops because the elements feed back and amplify the system. However, the introduction of one or more negative elements into the system increases the likelihood that the loop will be counteractive. In this example the counteractive effect is negative in character.

Phillips and Wiener point out that feedback loops representing mental health problems often develop as a result of the *initial kick*. The individual gets off to a bad start, consequently he receives negative feedback of some variety from his environment. At this point he may stop trying and act too soon as if he were defeated. Hence he develops attitudes and behavior that result in much more serious maladjustment.

Whereas in conventional psychotherapy the original stimulus situation has been studied as if it were entirely responsible for the deviant behav-

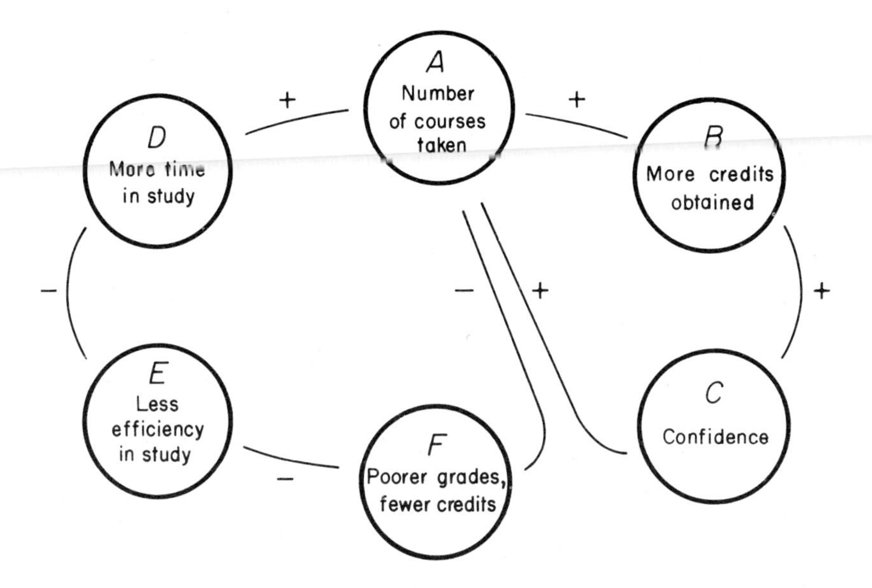

FIG. 12-4. A comparison of efficient with inefficient feedback loops in studying. (Phillips & Wiener, 1966.)

ior, the cybernetic model stresses the amplifying factor involved in the feedback loop. As an example, Phillips and Wiener point out that psychotherapy with an unreliable, antisocial young man would be predicted to have little effect, because outside of therapy so much of his antisocial behavior, especially his seeking of immediate gratification, is strongly reinforced. They say:

> The therapist who fails to exercise some control over the continuing choices and actions of such a patient is unlikely to create any change in his behavior. If these important deviation-amplifying processes are over-looked, or if they are treated as symptomatic of a more basic problem, the therapist wastes his basic tools and resources. [Phillips & Wiener, 1966, p. 102]

They consequently view part of the therapist's task as that of identifying the deviation-amplifying processes in the patients life and helping him to bring them under control. In doing this the therapist introduces a *change plan* which involves a deliberate effort to interfere with the deviation-amplifying processes by creating deviation-counteracting processes. For example, in a case similar to the one mentioned above involving an individual about to be expelled from school as a result of irresponsible and wild behavior, a deviation-counteracting element was introduced in the form of a detailed daily schedule which included studying and recreational activities and the avoidance of self-defeating friendships and temptation-producing stimuli. Thus, in changing deviation-amplifying loops the therapist tries to build up desirable alternatives to the unwanted behavior, while at the same time reducing the likelihood of occurrence of the undesired behavior. Many of the techniques that are broadly classified as behavior therapies are used. These are given some attention below.

Behavior Therapy

Recently there has been a proliferation of techniques and demonstrations of the direct counterconditioning of specific behaviors or the extinction of them, with the latter usually based on negative practice. As early as 1924, Jones demonstrated that the counterconditioning of human behavior was entirely feasible by treating a 3-year-old boy (Peter) for his learned fear of animals and other furry objects. The counterconditioning technique involved feeding Peter in the presence of these anxiety-provoking stimuli. Thus, a rabbit was placed in a cage at some distance from the boy, and each day it was brought nearer to him as he was eating. Finally, the rabbit could be placed in his lap without his showing any signs of anxiety. Furthermore, as a result of this procedure his fear of furry objects and other generalized stimuli also disappeared.

Wolpe (1954, 1958, 1962) has developed a very similar technique that he calls *systematic desensitization.* The problem is to arrange for the reciprocal inhibition of anxiety. Wolpe employs three types of responses which are incompatible with anxiety: approach responses, sexual responses, and relaxation responses. Under the appropriate conditions these responses are capable of inhibiting anxiety. Wolpe has the patient rank stimuli that seem to produce anxiety for him in a hierarchical order from the most to the least anxiety-arousing. Usually the items in the hierarchy vary along some dimension as, for example, "Fear of rejection by authority figures." Once the anxiety-arousing episodes are categorized from highest to lowest the patient is put into a deep state of relaxation by the therapist and the least anxiety-arousing item on the list is introduced by having the patient think about it. At any indication of anxiety the item is withdrawn. When relaxation as the dominant response has become associated with this least-disturbing item the next highest item in the list is introduced, and so on, until, after a number of sessions relaxation responses become strongly attached to the highest ranking items. This is what is meant by systematic desensitization.

The technique does not always rely upon generalization from ideationally presented anxiety-producing stimuli to real-life situations. A hierarchy of actual behavioral, real-life situations may also be established and the patient directed to make assertive responses that are incompatible with anxiety in the least threatening of them first. Thus, he may begin by asserting his emotional rights with the neighborhood grocer and at the end of therapy be capable of doing so in the presence of his boss.

Desensitization using the sexual response has been successfully employed in the case of impotency. The cooperation of the sexual partner is sought and the patient is instructed simply to remain in the partner's presence under sexually arousing conditions (e.g., both nude), until the patient begins to feel sexually aroused. At that point the session is broken off so that cues of fear which might follow if the patient responded normally never become attached to the cues of sexual arousal. In later sessions the patient is able to come closer and closer to consummating the sexual act before the fear gradient intervenes, until finally the fear is completely counterconditioned.

Aversive stimuli are also commonly used to countercondition learned responses. Thus, a relatively common treatment of alcoholism involves giving the patient a glass of his favorite alcoholic beverage in which emetine, a nauseous drug, has been dissolved. According to Bandura (1961) eight or ten pairings of the sight, smell, and taste of alcohol and the extremely unpleasant feelings of nausea that immediately follow are enough to produce total abstinence. Follow-up studies have indicated that 60% of the patients continue to abstain after treatment.

Painful shock is also commonly used as an aversive stimulus. For example, Blakemore, Thorpe, Baker, Conway, and Lavin (1963) reported the case of a patient who compulsively engaged in transvestite behavior (wearing clothes of the opposite sex). These investigators described the case and its subsequent treatment as follows:

The patient was a male, aged thirty-three, married, who had engaged in transvestite behavior from approximately the age of twelve years. He had dressed in his mother's and sister's clothes before that age, but the frequency of the compulsion to do so did not become clear until this point. His transvestism was interrupted by military service, but was replaced during that time by transvestite fantasies. During this period he developed a duodenal ulcer.

He subsequently married, but found that it was necessary for him to cross-dress in order to engage in normal sexual relations. He frequently went about in public at night dressed in female attire and wearing a female wig. His motivations for treatment appeared to be strong: they included a fear of legal difficulties were he discovered cross-dressed in public, consideration for his wife, and concern about his young son should the latter discover his problem. The patient had had six years of psychotherapy and had been using sodium Amytal to reduce tension. He had become addicted to the drug, and the symptoms remained unchanged.

The treatment consisted essentially of repeated trials wherein the patient received painful shock at a point where he had partially completed the sequence of getting dressed in female clothes. During treatment the patient was standing on a grid through which shock could be delivered. [The details of the method are given by the authors as follows]:

"At the beginning of each session the patient, dressed only in a dressing gown, was brought into a specially prepared room and was told to stand on the grid behind the screen. On the chair beside him was his 'favorite outfit' of female clothing, which he had been told to bring with him at his first attendance. These clothes had not been tampered with for the most part, with the exception that slits had been cut in the feet of nylon stockings and a metal plate fitted into the soles of the black court shoes to act as a conductor. Each trial commenced with the instruction to start dressing, at which he removed the dressing gown and began to put on female clothing. At some point during the course of dressing, he received a signal to start undressing irrespective of the number of garments he was wearing at that time. This signal was either a shock from the electric grid or the sound of a buzzer, and these were randomly ordered on a fifty per cent basis over the 400 trials. The shock or buzzer recurred at intervals until he had completed his undressing. He was allowed a one minute rest period between each of the five trials which made up a treatment session.

"In order that the number of garments put on should not be constant from trial to trial, the time allowed from the start of dressing to the onset of the recurrent shock or buzzer was randomly varied between one and three minutes among the trials. It was possible, also, that the patient's undressing behavior would become stereotyped if the interval between the successive shocks or soundings of the buzzer remained constant from trial to trial, therefore, while these intervals were kept constant within a trial, they were randomly varied between 5, 10 and 15 seconds among trials. The explanation of the procedure given to the patient before treatment commenced did not contain any reference to the randomization of these variables, and did not include any instructions pertaining to the speed of undressing. At the start of each new trial the patient did not know, therefore, how long it would be before he received the signal to undress, whether this would be a shock or the buzzer, or the frequency with which these would recur during his undressing." [Blakemore *et al.*, 1963, pp. 31–32]

A follow-up on this patient indicated that there had been no recurrence of transvestism 6 months after the end of treatment. This case also reflected a typical and important finding in the use of aversive counterconditioning techniques. With the removal of the patient's symptoms he reported himself generally less anxious than he had been for several years. Thus, there is strong indication that the symptoms themselves generate anxiety. It is one of the deviation-amplifying aspects of the steady state of maladjustment. This is quite the reverse of traditional thinking where symptoms are supposed to be a consequence of anxiety insofar as they are thought to have an anxiety-reduction function. If this theory were true, removal of symptoms should increase anxiety or lead to the development of other kinds of symptomatic behavior.

Still another variety of behavior therapy that is increasingly used is that of operant conditioning. The Allyon and Michael's technique (reported on p. 17 and in Fig. 1-7) is typical of these procedures.

The latest suggestion in the application of operant techniques to behavior change is the programming of psychotherapy (McKee, 1963). This can be done by aural, visual, or audiovisual methods. As an example, Phillips and Wiener (1966) have suggested that a few frames of a program for teaching good living habits might read as follows:

An emotional problem or a problem in handling a situation arises because the person lacks the desired behavior at a particular time. To handle a situation better, one must determine what is needed and act in the d________ w______ in that situation.

desired way

Emotionally disturbed people do not appear to regulate small daily matters very efficiently, or as they would like. For example, they may not do what they want to do about jobs, daily responsibilities, or meeting deadlines or appointments. This failure to r________ one's daily life leads to confusion, piling up of one's commitments, and personal dissatisfaction and emotional problems.

regulate

A very simple way of regulating your daily life is to make an hour-by-hour schedule of specific activities that are most important and can be done. This kind of r________ makes you decide what is most important to get done and makes you see how much you can reasonably expect of yourself.

regulation

You should also keep a record of how well you have followed your daily schedule. This schedule may have to be adjusted to make it more realistic, particularly if you tend to schedule too much or too little or if some important things are left out. This plan allows you to regulate your life better and to get the most important things done most efficiently. The use of a d______ s________ serves not only as a planning guide but also as a way of determining your progress and success.

daily schedule

After a few days' or a few weeks' experience with a d______ s________, you can usually judge with considerable accuracy your strengths and weaknesses and decide where more p________ and r________ are needed. [Phillips & Wiener, 1966, Pp. 121–122.]

daily schedule
planning
regulation

THE LIMIT OF VERBAL THERAPIES: PSYCHOTIC BEHAVIOR

The seriously disordered personality is least likely to profit from verbal psychotherapy. This is primarily because the severely disordered person is often not in contact with reality, at least reality as the more normal person experiences it. Present-day psychiatric classification lists the psychoses under the heading: "Disorders of psychogenic origin or without clearly defined physical cause or structural change in the brain." These include involutional psychotic reactions, affective reactions, schizophrenic reactions, and paranoid reactions.

The involutional psychotic reaction has been directly associated with menopause in women. It affects them beginning in their forties. Men also occasionally experience it about 10 years later. The patient is depressed, irritable, and anxious. He also tends to experience symptoms such as loss of appetite, weeping, and delusions.

The affective reactions are composed largely of manic states, and depressive reactions. *Manic* reactions typically appear in situations of unrelieved tension and anxiety. The manic state itself is characterized by excitement. The person is extremely restless, appears to have boundless energy, and his verbal output is rapid and intense. Sometimes delusions are present, particularly delusions of grandeur.

Psychotic *depressive* episodes result in the opposite symptoms for the most part. In general, there is a retardation of behavior, sadness, and despondency. The person appears overwhelmingly unhappy, to the point where he is unable to work or even eat. In a severe depressive stupor the person will not even talk. In some cases, called *cyclic* or *manic-depressive,* periods of mania alternate with periods of depression. Mental hospital statistics have shown a decline in the frequency of manic-depressive disorders over the years. By 1947 it was reduced to 3.8% of first admissions (Bellak, 1952). More recent statistics are not available.

The overwhelming majority of hospitalized mental patients are diagnosed as schizophrenics. Perhaps the most common symptom leading to this diagnosis is desocialization. The individual fails to respond to people around him, neglects social responsibilities, his personal appearance, or even his biological needs. He appears to withdraw from external reality and to be preoccupied with his own thoughts. He may be notably inactive, sitting by himself for hours on end. Typically he also shows inappropriate affect. Although there is a "flattening" of his emotional responses and he may appear to be emotionally unresponsive in general, at other times his responses are inappropriate. He may, for example, laugh at bad news or smile mysteriously as if he were responding to his own thoughts. Anger, rage, and violent reactions may inexplicably occur. He may show odd mannerisms, postures, grimaces, stereotyped and repetitious motions.

Similar oddities occur in his speech, which may be incomprehensible because of the private meanings that words appear to have for him. His speech may also be characterized by associational intrusions, by an abrupt shifting of topics, and by a blocking of thoughts.

Still another characteristic of the disorder is the appearance of hallucinations. The patient may hear voices giving him commands or directions, experience vague visual images, taste poison in his food, etc. Delusions may also be present. He may begin to suspect people of trying to poison him. He may believe that he is some important personage or an agent of God.

Various combinations of symptom patterns may predominate, giving rise to a number of subclassifications. For example, among the official psychiatric diagnostic listings are included the simple type, hebephrenic type, catatonic type, paranoid type, acute and undifferentiated type, chronic undifferentiated type, schizoaffective type, and residual type. Little will be gained by describing these subtypes in detail here.

The Mental Hospital

Because of his inability to cope with the demands of normal living, and because he may be dangerous to himself and to others, care and keeping in a mental hospital is usually recommended for the psychotic patient. There, although he may at first reject his status as a mental patient, eventually he learns that normal hospital privileges can be obtained only through obedience and acceptance of it. This adjustment to the hospital routine has been termed *institutionalism.* Aspects of it have been commented upon by many observers, but there have been few empirical studies of it. One investigation by Wing (1962) is an exception. He found that there was a shift as the time of hospitalization increased in the proportion of the patients with a definite wish to leave the hospital. The majority of patients eventually become indifferent or express a clear wish to stay. A similar trend occurs in the percentage of patients who are visited by someone from outside and in the frequency of visits. Both of these decline with time in the hospital. Wing characterizes the patient's adjustment as involving resignation, depersonalization, apathy, and reliance on fantasy.

Some psychotics are responsive to psychotherapy; there are also some spontaneous recoveries. For the most part, however, establishing a working relationship so that therapeutic movement can occur is either impossible, or requires so many hundreds of hours of the therapist's time on an individual basis that it is impractical. However, there is a continuing attempt to reach as many individuals as possible as a part of the standard operating program in most mental hospitals.

The causes of the psychoses, and of schizophrenia in particular, are only poorly understood at the present time. They most certainly result from a very complex interaction of factors, both biological and social. For an exhaustive and excellent account of the known facts and theories of schizophrenia the reader is referred to Professor Maher's volume entitled *Principles of psychopathology* (Maher, 1966). Despite the extensive and detailed research and a bewildering array of theories that are reported by Maher, he is compelled to say:

> Human effort in research and treatment of the schizophrenic patient probably exceeds that devoted to any other behavior deviation. As a condition or group of conditions, as a process or several processes, the puzzles presented by the behavior of schizophrenic patients remains to a significant extent unsolved. [Maher, 1966, p. 299]

Somatic Methods of Treatment

The somatic treatment of deviant behavior may include the use of tranquilizing drugs, shock therapy, or psychosurgery. Tranquilizing drugs and their effects have been discussed (p. 95). Wortis (1959) has presented data that reflect the popularity of the therapeutic use of drugs as compared with electroconvulsive therapies. In recent years drugs have become of major therapeutic interest, whereas interest in the shock therapies, especially insulin therapy, has declined. Electroconvulsive shock therapy, however, remains of interest. It is still widely used, in the treatment of psychosis.

Insulin Shock Therapy. Insulin shock therapy was first introduced by Sakel in 1938. It was the therapy most commonly used in the years immediately following. Although the initial enthusiasm for its use sharply declined thereafter, it is still used, particularly with cases of schizophrenia that can be treated within a reasonably short time after the onset of the more severe symptoms.

The procedure involves the injection of increasingly larger doses of insulin into the patient over a period of several weeks (1 or 2 months) until the dosage results in the patient remaining in a coma state for about 1 hour, 2 or 3 hours after an injection. Investigators have reported that they have observed "improvement" after this procedure. However, even Sakel insisted that the technique must be combined with psychotherapy to have lasting effects. Malzberg (1940) reported that 13 percent of his sample of 1000 patients treated with insulin fully recovered, and 54 percent improved. A number of others have reported about the same.

Electroconvulsive Shock Treatment. The technique of electroshock (ECT) was first introduced by Cerletti and Rini in 1938. The common procedure is to attach electrodes to both temples of the patient's head and to apply a current of from 70 to 130 volts through the brain for a period of up to .5 seconds. This is followed by a convulsive seizure of short duration, and then by a stuporous state which in turn is followed by temporary confusion and amnesia. A series of treatments may be given as often as three or more times a week.

There is a vast literature on the therapeutic effects of ECT, as well as a diversity of theories to explain them. A critical evaluation of them has led many authorities to the conclusion that the technique is not defensible. Although there are few claims of any benefit from it in the case of schizophrenia, many practitioners apparently believe that it is of value in cases of depression. There are many psychological theories of a literary character to the effect that treatment satisfies unconscious motives for these patients (e.g., constitutes punishment for a sense of guilt). We can do little better that then to quote Maher's final evaluation of this treatment:

> Electroconvulsive therapy is a purely empirical technique. At this time there is no acceptable theoretical rationale for its use, little knowledge of its physical effects, considerable ambiguity in our understanding of its behavioral effects, and inadequate substantiation of its therapeutic efficiency. Nevertheless, the present and probabable future use of ECT is well indicated by Kalinowsky's conclusion that "the place of ECT in the therapeutic armamentarium of the psychiatrist is definitely established in all mental institutions." [Maher, 1966, p. 500]

Psychosurgery

The logic behind psychosurgery is the modification of behavior by direct surgical interference with certain pathways of the brain. One of the most commonly used techniques is the *prefrontal leucotomy.* This operation involves drilling small holes in either side of the head and by means of a sharp cutting instrument, called a leucotome, severing the major neural pathways between the frontal area of the brain and the lower brain centers. This technique was developed by a Portuguese neurologist named Egas Moniz (1937). A second technique, which is a modification of Moniz's procedure, was developed in the United States by Freeman and Watts (1948). This technique is called a *prefrontal lobotomy.* It involves severing connections between the nuclei of the thalamus and the frontal areas of the brain. There are other variations of psychosurgery. All of them are designed to alter the patient's emotional, and hence, pathological behavior by direct interference with emotional circuits and pathways. These include transorbital lobotomy, cerebral topectomy, and thermocoagulation.

Freeman and Watts (1948) have described the personality characteristics that follow a prefrontal lobotomy as *affective bleaching,* by which they mean that the patient's fundamental distress disappears, although the cognitions which apparently tended to produce it remain. Thus, the individual may still have systematized delusions that there is a plot against him, but he seems uninterested in it any more. Freeman claims actual recovery for about one-third of the patients and improvement for about another third of those undergoing psychosurgery. Other investigators have reported similar improvement rates (e.g., Jenkins, Holsopple, & Lorr, 1954; Kalinowsky & Hoch, 1961). This is not surprising to many observers, since only the most hopeless, deteriorated schizophrenics are selected for psychosurgery to begin with, improvement may appear dramatic. The patient may be released from the hospital only to be kept under constant watch at home and an eventual routine of care and concern by close relatives. On the other hand, the unimproved group includes many who become worse; completely irresponsible psychopathy, as well as a helpless vegetative state have both been reported. Since the advent of tranquilizing drugs, psychosurgery has become less common. Today, it is rarely recommended unless all forms of treatment have failed, including tranquilizing drugs and shock therapy (Noyes & Kolb, 1963).

SUMMARY

In this chapter certain theories, goals, and practices related to the treatment of emotional disturbances and the modification of behavior have been discussed. An attempt was made to evaluate the theory and practice of psychotherapy, to indicate how the general principles of learning are related to it, and to illustrate its evolution.

Although the objective judgement as to whether or not an individual is maladjusted is invariably based on the observation of his behavior, traditional conceptualizations of maladjustment have not emphasized this. Instead, certain conceptual models have been used which interpret the maladjustive responses of the individual in terms of analogies from other hypothesized or observed processes. Most notable among these are the medical model and the dynamic model. The former treats maladjustment as a disease process. The person is described as mentally ill, his deviant behavior as symptoms, and so on. The latter draws analogies from homeostatic processes in biology or physics in which there is a dynamic balance of forces. The individual is viewed as striving to maintain certain steady states, as, for example, the steady state of self-esteem.

Psychoanalysis postulates that the individual has a constant amount of energy—psychic energy—available to him, that the energy can be discharged by investment in certain objects, and that the objects in which the energy is invested can change. Enough of the details of the psychoanalytic theory have been outlined so that the reader could (hopefully) grasp the essentials of the psychoanalytic interpretation of maladjustment. Psychoanalytic therapy consists basically of the techniques of free association and of dream analysis. The aim of psychoanalytic therapy is to make the unconscious conscious.

A rival psychotherapeutic orientation is Carl Roger's *client-centered* approach. In this approach neither free association nor the analysis of dreams is employed. Instead, a change in the client's awareness of himself, or his thoughts and feelings, occurs in the direction of making his perception more congruent with his actual experiences. This is accomplished by the continual reflecting of the client's thoughts and feelings back to him in an atmosphere of warmth, permissiveness, and positive regard for him on the part of the therapist.

General psychotherapy is largely eclectic. A modified form of the free-association technique is commonly used. However, the technique employed and the therapeutic goals are modified to fit the demands of the circumstances and the problems and personality characteristics of the patient. Accordingly, the goals may vary from *supportive,* where an attempt is made to strengthen the patient's defenses against anxiety and to remove external sources of anxiety insofar as it is possible to do so; through *reeducative,* in which an attempt is made to give the patient insight into his more conscious conflicts and consequently to alter the resolution of them; to *reconstructive,* where the goal is the analysis of unconscious motivation and the reconstruction of personality. Psychoanalytic psychotherapy is largely based upon reconstructive goals.

Research using objective criteria of improvement and the appropriate controls has consistently found little evidence that psychotherapy as it is traditionally practiced is any more effective than no therapy. Clinical theorists and practitioners in the past have not been research-oriented. Rogers and his co-workers are an exception to this. Much of their research was reviewed in order to give a more meaningful research account of the psychotherapeutic process. From this research a fairly consistent picture emerges, indicating the following: (1) Whether or not this variety of psychotherapy leads to improvement is dependent upon certain personality characteristics of the client. Thus, those high in ethnocentrism, those who are primarily extrapunitive, and those with repressive-type defenses appear either to benefit little or to get worse. (2) When improvement does occur the nature of the therapeutic relationship appears to be very important in sponsoring it. Therapist attitudes described by such phrases as unconditional positive regard, empathic understanding, and counselor genuineness appear to be highly relevant to it. Given the fact that under some conditions some types of personalities show improvement, details of the process of therapeutic change were discussed.

The essentially literary theorizing of the psychoanalysts has been translated into learning terms by Dollard and Miller. They, and learning theorists generally, conceive of conflict and avoidance learning as at the core of most neurotic behavior. Newer approaches to psychotherapy and to behavior change tend to actively apply learning principles to modify the patient's behavior. One of these is Rotter's social-learning approach, which approach emphasizes the importance of the deliberate reinforcement of certain kinds of behavior in therapy. The most important point where reinforcement is applied is to strengthen the patient's attempts to find and try alternative solutions to his problems. Emphasis is also placed upon changing the patient's learned expectancies that create problems for him and on a continual structuring of the therapeutic program and of the therapeutic relationship.

Phillip's approach, called *assertion therapy*, applies the conflict paradigm to the beliefs, or assertions, that patients make. Tension is produced when the environment disconfirms the persons assertions. Therapy consists of hypothesis structuring whereby the patient comes to see that his behavior and his decisions are based upon certain assumptions. Once his assertions are structured the patient can deal with them, as he can with any other clear reality, and take a problem-solving approach to them.

Phillips and Wiener, as well as other psychologists, have employed a *cybernetic model* to interpret maladjustive behavior. Certain relevant cybernetic concepts were discussed. One of the major advantages of this conceptual model that it emphasizes amplifying factors involved in both maladjustment and in programs designed to counteract maladjusted behavior. Therapy based upon the model also heavily emphasized direct intervention and the introduction of a change plan, with amplifying effects of its own.

Recently techniques have been introduced to directly countercondition behavior based upon fear and neurotic avoidance. Desensitization and counterconditioning by the use of aversive stimuli were discussed. There is some suggestion that the programming of psychotherapy might be possible in the future. A sample program was provided.

In psychotic behavior the use of verbal therapies is seriously limited. Our scientific understanding of the problems of the psychosis despite extensive research and clinical observation, remains poor indeed. The use of shock therapies and psychosurgery were discussed, along with institutionalism.

REFERENCES

Adorno, T. W., Frenkel-Brunswik, Else, Levison, D. J., & Sanford, R. N. *The authoritarian personality.* New York: Harper, 1950.

Bellak, L. *Manic-depressive psychosis and allied conditions.* New York: Grune & Stratton, 1952.

Blakemore, C. B., Thorpe, J. F., Baker, J. C., Conway, C. F., & Lavin, N. T. The application of paradic aversion conditioning in a case of transvestism. *Behav. Res. Ther.*, 1963, **1**, 63–68.

Denker, W. In H. J. Eysenck (Ed.), *Handbook of abnormal psychology.* New York: Basic Books, 1961. Pp. 710–712.

Dollard, J., & Miller, N. E. *Personality and psychotherapy: An analysis of learning, thinking, and culture.* New York: McGraw-Hill, 1950.

Ends, E. J., & Page, C. W. A study of three types of group psychotherapy with hospitalized male inebriates. *Quart. J. Stud. Alc.*, 1957, **18**, 263–279.

Eysenck, H. J. Classification and the problem of diagnosis. In H. J. Eysenck (Ed.), *Handbook of abnormal psychology.* New York: Basic Books, 1961. Pp. 1–31.

Fiedler, F. E. Quantitative studies on the role of therapists feelings toward their patients. In O. H. Mowrer (Ed.), *Psychotherapy: Theory and research.* New York: Ronald Press, 1953. Pp. 296–315.

Freeman, W., & Watts, J. W. Pain mechanisms and frontal lobes: A study of prefrontal lobotomy for intractable pain. *Ann. intern. Med.*, 1948, **28**, 747–754.

Haimowitz, M. L., & Haimowitz, Natalie R. Personality change in client-centered therapy. In W. Wolff (Ed.), *Success in psychotherapy.* New York: Grune & Stratton, 1952. Pp. 63–93.

Halkides, G. An experimental study of four conditions necessary for therapeutic change. Unpublished doctoral dissertation, Univer. of Chicago, 1958. (Cited in Rogers, 1961)

Jenkins, R. L., Holsopple, J. Q., & Lorr, M. Effects of prefrontal lobotomy on patients with severe chronic schizophrenia. *Amer. J. Psychiat.*, 1954, **111**, 84–90.

Jones, M. C. The elimination of children's fear. *J. exp. Psychol.*, 1924, **7**, 382–390.

Kalinowsky, L. B., & Hoch, P. H. *Somatic treatments in psychiatry; Pharmacotherapy, convulsive, insulin, surgical and other methods.* New York: Grune & Stratton, 1961.

Kelman, H. C., & Parloff, M. B. Interrelations among three criteria of improvement in group therapy: Comfort, effectiveness, and self awareness. *J. abnorm. soc. Psychol.*, 1957, **54**, 281–288.

Levitt, E. E. The results of psychotherapy with children: an evaluation. *J. consult. Psychol.*, 1957, **21**, 189–196.

Maher, B. A. *Principles of psychopathology.* New York: McGraw-Hill, 1966.

Malzberg, B. *Social and biological aspects of mental disease.* Utica, N. Y.: State Hospital Press, 1940.

McKee, J. Programed instruction as a therapeutic tool. *Amer. Psychologist,* 1963, **18**, 385.

Mehlman, B. The reliability of psychiatric diagnosis. *J. abnorm. soc. Psychol.*, 1952, **47**, 577–578.

Moniz, E. Psycho-chirurgie. *Nervenarzt,* 1937, **10**, 113–118.

Noyes, A. P., & Kolb, L. C. *Modern clinical psychiatry.* (6th ed.) Philadelphia: Saunders, 1963.

Phillips, E. L. *Psychotherapy: A modern theory and practice.* Englewood Cliffs, N. J.: Prentice-Hall, 1956.

Phillips, L. E., & Wiener, D. N. *Short-term psychotherapy and structured behavior change.* New York: McGraw-Hill, 1966.

Raimy, V. (Ed.), *Training in clinical psychology.* Englewood Cliffs, N. J.: Prentice-Hall, 1950.

Raines, G. N., & Rohrer, J. H. The operational matrix of psychiatric practice. I. Consistency and variability in interview impressions of different psychiatrists. *Amer. J. Psychiat.,* 1955, **110**, 721–733.

Rogers, C. R. *On becoming a person: A therapist's view of psychotherapy.* Boston: Houghton, 1961.

Rogers, C. R., & Dymond, R. F. (Eds.), *Psychotherapy and personality change.* Chicago: Univer. of Chicago Press, 1954.

Rotter, J. B. *Social learning and clinical psychology.* Englewood Cliffs, N. J.: Prentice-Hall, 1954.

Sakel, M. *The pharmacological shock treatment of schizophrenia.* New York: Nervous and Mental Disease Publ. Co., 1938.

Schmidt, H. O., & Fonda, C. P. The reliability of psychiatric diagnosis: A new look. *J. abnorm. soc. Psychol.,* 1956, **52**, 262–267.

Seeman, J. A study of the process of nondirective therapy. *J. consult. Psychol.,* 1949, **13**, 157–168.

Snyder, W. U. An investigation of the nature of nondirective therapy. *J. gen. Psychol.,* 1945, **33**, 193–223.

Wing, J. K. Institutionalism in mental hospitals. *Brit. J. soc. clin. Psychol.,* 1962, **1**, 38–51.

Wolberg, L. R. *The technique of psychotherapy.* New York: Grune & Stratton, 1954.

Wolpe, J. Reciprocal inhibition as the main basis of psychotherapeutic effects. *A.M.A. Arch. Neurol. Psychiat.,* 1954, **72**, 205–226.

Wolpe, J. *Psychotherapy by reciprocal inhibition.* Stanford, Calif.: Stanford Univer. Press, 1958.

Wolpe, J. The experimental foundations of some new therapeutic methods. In A. J. Bachrach (Ed.), *Experimental foundations of clinical psychology.* New York: Basic Books, 1962. Pp. 554–575.

Wortis, J. *History of insulin shock.* New York: Philosophical Library, 1959.

Author Index

Numbers in italics refer to pages on which the references are listed.

H

J

K

L

M

S

T

U

V

Subject Index